PROCESS MODELING AND SIMULATION FOR CHEMICAL ENGINEERS

PROCESS MODELING AND SIMULATION FOR CHEMICAL ENGINEERS

THEORY AND PRACTICE

Simant Ranjan Upreti

Department of Chemical Engineering,
Ryerson University,
Toronto, Canada

This edition first published 2017
© 2017 John Wiley & Sons Ltd

The right of Simant Ranjan Upreti to be identified as the author of this work has been asserted in accordance with law.

Registered Offices
John Wiley & Sons, Inc., 111 River Street, Hoboken, NJ 07030, USA
John Wiley & Sons Ltd, The Atrium, Southern Gate, Chichester, West Sussex, PO19 8SQ, UK

Editorial Office
The Atrium, Southern Gate, Chichester, West Sussex, PO19 8SQ, UK

For details of our global editorial offices, customer services, and more information about Wiley products visit us at www.wiley.com.

Wiley also publishes its books in a variety of electronic formats and by print-on-demand. Some content that appears in standard print versions of this book may not be available in other formats.

Library of Congress Cataloging-in-Publication Data

Names: Upreti, Simant Ranjan, author.
Title: Process modeling and simulation for chemical engineers : theory and
  practice / Simant Ranjan Upreti.
Description: Chichester, UK ; Hoboken, NJ : John Wiley & Sons, 2017. |
  Includes bibliographical references and index.
Identifiers: LCCN 2016053339| ISBN 9781118914687 (cloth) | ISBN 9781118914663
  (epub)
Subjects: LCSH: Chemical processes–Mathematical models. | Chemical
  processes–Data processing.
Classification: LCC TP155.7 .U67 2017 | DDC 660/.284401–dc23
LC record available at https://lccn.loc.gov/2016053339

Cover image: shulz/Gettyimages
Cover design by Wiley

Set in 10/12pt, NimbusRomNo by SPi Global, Chennai, India.
Printed and bound in Malaysia by Vivar Printing Sdn Bhd

10 9 8 7 6 5 4 3 2 1

to my wife and kids

CONTENTS

PREFACE

I am delighted to present this book on process modeling and simulation for chemical engineers. It is a humble attempt to assimilate the amazing contributions of researchers and academicians in this area.

The goal of this book is to provide a rigorous treatment of fundamental concepts and techniques of this subject. To that end, the book includes all requisite mathematical analyses and derivations, which could be sometimes hard to find. Target readers are those at the graduate level. This book endeavors to equip them to model sophisticated processes, develop requisite computational algorithms and programs, improvise existing software, and solve research problems with confidence.

Chapter 1 provides the groundwork by introducing the terminology of process modeling and simulation. Chapter 2 presents the fundamental relations for this subject. Chapter 3 incorporates important constitutive relations for common systems. Chapter 4 presents model formulation with the help of several examples. Transformation techniques are introduced in Chapter 5. Model simplification and approximation methods are discussed in Chapter 6. The numerical solution of process models is the theme of Chapter 7. Review of important mathematical concepts is provided in Chapter 8.

This book can be used as a primary text for a one-semester course. Alternatively, it could serve as a supplementary text in graduate courses related to modeling and simulation. Readers could also study the book on their own. During an initial reading, one could very well skim quickly through a derivation, accept the result for the time being, and learn more from applications. Computer programs for the solutions of book examples can be obtained from the publisher's website, www.wiley.com/go/upreti/pms_for_chemical_engineers.

I am grateful to the editorial team at John Wiley & Sons for providing excellent support from start to finish.

Finally, I am deeply indebted to my wife Deepa, and children Jahnavi and Pranav. I could not have completed this book without their unsparing support and understanding.

Ryerson University, Toronto   *Simant R. Upreti*

Notation

Symbol	Description	Units
a	surface area per unit volume	$\mathrm{m^{-1}}$
A_{m}	area of moving surfaces	$\mathrm{m^2}$
A_{p}	area of a port of flow	$\mathrm{m^2}$
A	area	$\mathrm{m^2}$
$\mathbf{A}$	area vector	$\mathrm{m^2}$
c	average concentration of a mixture	$\mathrm{kmol\,m^{-3}}$
c_i	concentration of the ith species	$\mathrm{kmol\,m^{-3}}$
$\hat{C}_{\mathrm{P}}$	specific heat capacity of mixture	$\mathrm{J\,kg^{-1}K^{-1}}$
$\hat{C}_{\mathrm{P}i}$	$\hat{C}_{\mathrm{P}}$ of the ith species in pure form	$\mathrm{J\,kg^{-1}K^{-1}}$
$\underline{C}_{\mathrm{P},i}$	molar specific heat capacity of the ith species in a mixture	$\mathrm{J\,kmol^{-1}K^{-1}}$
$\tilde{C}_{\mathrm{P},i}$	partial specific heat capacity of the ith species in a mixture	$\mathrm{J\,kmol^{-1}K^{-1}}$
$\mathrm{d}x$	differential change in x	of x
D	diffusivity of species	$\mathrm{m^2\,s^{-1}}$
D_{AB}	binary diffusivity of A in a mixture of A and B	$\mathrm{m^2\,s^{-1}}$
$\mathbf{D}$	matrix of multicomponent diffusivities	$\mathrm{m^2\,s^{-1}}$
e_i	the ith component of energy flux	$\mathrm{J\,m^{-2}\,s^{-1}}$
E	sum of the squared errors in y in a population	of y^2
E	total energy of a system	J
E	activation energy of reaction	$\mathrm{J\,kmol^{-1}}$
$\hat{E}$	energy per unit mass	$\mathrm{J\,kg^{-1}}$
f_i	fugacity of the ith species	Pa
F	volumetric flow rate	$\mathrm{m^3\,s^{-1}}$
$\mathbf{f}_i$	mass flux of the ith species	$\mathrm{kg\,m^{-2}\,s^{-1}}$
$\mathbf{f}$	overall mass flux of a mixture	$\mathrm{kg\,m^{-2}\,s^{-1}}$

Symbol	Description	Units
$\mathbf{F}_i$	molar flux of the ith species	$\text{kmol}\,\text{m}^{-2}\,\text{s}^{-1}$
$\mathbf{F}$	overall molar flux of a mixture	$\text{kmol}\,\text{m}^{-2}\,\text{s}^{-1}$
$\mathbf{F}$	force vector	N
G	Gibbs free energy	J
$\hat{G}_i$	Gibbs free energy per unit mass of the ith species	$\text{J}\,\text{kg}^{-1}$
$\mathbf{g}$	gravity, 9.806 65	$\text{m}\,\text{s}^{-2}$
h	heat transfer coefficient	$\text{W}\,\text{m}^{-2}\,\text{K}^{-1}$
H	enthalpy	J
$\hat{H}$	enthalpy per unit mass	$\text{J}\,\text{kg}^{-1}$
H_i	Henry's law constant for the ith species	Pa
$\hat{H}_i$	enthalpy per unit mass of the ith species	$\text{J}\,\text{kg}^{-1}$
$\tilde{H}_i$	partial specific enthalpy of the ith species in a mixture	$\text{J}\,\text{kg}^{-1}$
$\bar{H}_i$	partial molar enthalpy of the ith species in a mixture	$\text{J}\,\text{kmol}^{-1}$
$\underline{H}_i$	molar enthalpy of the ith species in pure form	$\text{J}\,\text{kmol}^{-1}$
$\underline{H}_i^{\circ}$	standard $\underline{H}_i$	$\text{J}\,\text{kmol}^{-1}$
$\mathbf{H}$	Hessian matrix	
$\mathbf{I}$	identity matrix	
$\mathbf{j}_i$	diffusive mass flux of the ith species	$\text{kg}\,\text{m}^{-2}\,\text{s}^{-1}$
$\bar{\mathbf{j}}$	vector of $\mathbf{j}_i$	$\text{kg}\,\text{m}^{-2}\,\text{s}^{-1}$
$\mathbf{J}$	Jacobian matrix	
$\mathbf{J}_i$	diffusive molar flux of the ith species	$\text{kmol}\,\text{m}^{-2}\,\text{s}^{-1}$
$\bar{\mathbf{J}}$	vector of $\mathbf{J}_i$	$\text{kmol}\,\text{m}^{-2}\,\text{s}^{-1}$
k	reaction rate coefficient	as per reaction
k	thermal conductivity	$\text{W}\,\text{m}^{-1}\,\text{K}^{-1}$
k_0	frequency factor	as per reaction
k_c	mass transfer coefficient	$\text{m}\,\text{s}^{-1}$
K	equilibrium constant of a chemical reaction	
$\mathbf{L}$	lower triangular matrix	
m	mass	kg
m_i	mass of the ith species	kg
M_i	molecular weight of the ith species	$\text{kg}\,\text{kmol}^{-1}$
N_c	number of components or species	

Symbol	Description	Units
N_i	number of moles of the ith species	
N_{r}	number of chemical reactions	
$\hat{\mathbf{n}}, \tilde{\mathbf{n}}, \underline{\mathbf{n}}$	unit vectors	
p_i	partial pressure of the ith species	Pa
P	pressure	Pa
P_{c}	critical pressure	Pa
$\mathbf{p}$	momentum	kg m s^{-1}
q	rate of heat transfer	J s^{-1}
q_i	the ith component of conductive heat flux	$\text{J m}^{-2} \text{s}^{-1}$
Q	heat, i.e., energy in transit	J
$\mathbf{q}$	conductive heat flux	$\text{J m}^{-2} \text{s}^{-1}$
r	correlation coefficient	
r	rate of reaction	$\text{kg(kmol) m}^{-3} \text{s}^{-1}$
r	radial direction in cylindrical and spherical coordinates	m
r^2	coefficient of determination	
R	universal gas constant, $8.314\,46 \times 10^3$	$\text{J kmol}^{-1} \text{K}^{-1}$
$r_{\text{gen},i}$	mass rate of ith species generated per unit volume	$\text{kg m}^{-3} \text{s}^{-1}$
$R_{\text{gen},i}$	molar rate of ith species generated per unit volume	$\text{kmol m}^{-3} \text{s}^{-1}$
s_y	standard deviation in values of y	of y
S	entropy	J K^{-1}
$\hat{S}_i$	entropy per unit mass of the ith species	$\text{J K}^{-1} \text{kg}^{-1}$
S	sum of the squared errors from the average of y in a population	of y^2
t	time	s
T	temperature	K, °C
T_{c}	critical temperature	°C
U	internal energy	J
$\hat{U}$	internal energy per unit mass	J kg^{-1}
$\hat{U}_i$	internal energy per unit mass of the ith species	J kg^{-1}
$\mathbf{U}$	upper triangular matrix	
v	magnitude of velocity	m s^{-1}
v_i	the ith component of $\mathbf{v}$, or average velocity along the x_i-direction	m s^{-1}

Symbol	Description	Units
V	volume	m^3
$\underline{V}$	molar volume	$m^3\,kmol^{-1}$
$\hat{V}$	specific volume	$m^3\,kg^{-1}$
$\hat{V}_i$	specific volume of the ith species	$m^3\,kg^{-1}$
$\mathbf{v}$	mass average velocity	$m\,s^{-1}$
$\tilde{\mathbf{v}}$	molar average velocity	$m\,s^{-1}$
W_s	shaft work	J
$\dot{x}$	change of x per unit time	x-units s^{-1}
x_{ext}	amount of x from external source	of x
x_{gen}	generated amount of x	of x
X	extent of chemical reaction	
$\tilde{X}$	extent of chemical reaction per unit volume	m^{-3}
$\mathbf{x}^{\top}$	transpose of $\mathbf{x}$ (vector or matrix)	
$\hat{\mathbf{x}}_i$	unit vector along the x_i-direction	
$\|\mathbf{x}\|$	norm or magnitude of $\mathbf{x}$ (vector or matrix)	
z	axial direction in cylindrical coordinates	m

Greek Symbols

Symbol	Description	Units
$\dot{\boldsymbol{\gamma}}$	rate-of-strain tensor	s^{-1}
δ_{ij}	Kronecker delta	
δQ	net Q involved in differential changes, dU and dS	J
$\boldsymbol{\delta}$	unit dyadic (unit tensor)	
ΔH_i°	standard heat of formation of the ith species	$J\,kmol^{-1}$
ΔH_r	heat of reaction	$J\,kmol^{-1}$
ΔH_r°	standard heat of reaction	$J\,kmol^{-1}$
Δx	small amount of, or change in x	of x
$\Delta[x]$	loss of x from a system through its ports	of x
∇	gradient operator (a vector)	
η	Non-Newtonian viscosity	$Pa\,s$
θ	θ-direction in cylindrical and spherical coordinates	$rad, ^{\circ}$
κ	dilatational viscosity	$Pa\,s$
μ	viscosity	$Pa\,s$

Symbol	Description	Units
ν_i	stoichiometric coefficient of ith species in a chemical reaction	
$\boldsymbol{\pi}$	molecular stress tensor	Pa
Π	dimensionless number	
ρ	density, mass concentration	kg m^{-3}
ρ_i	density, mass concentration of the ith species	kg m^{-3}
$\tau_{ij}, \pi_{ij}, \phi_{ij}$	the jth component of stress acting on the ith area component	Pa
τ	time period	s
$\boldsymbol{\tau}$	viscous stress tensor	Pa
ϕ	ϕ-direction in spherical coordinates	rad, °
$\boldsymbol{\phi}$	overall stress tensor	Pa
$\hat{\Phi}$	potential energy per unit mass	J kg^{-1}
ω	mass fraction	
ω	acentric factor	
$\boldsymbol{\omega}$	vorticity tensor	s^{-1}

1

Introduction

Process modeling and simulation is our intellectual endeavor to explain real-world processes, foresee their effects, and improve them to our satisfaction. Using foundational rules and the language of mathematics, we describe a process, i.e., develop its model. Depending on what needs to be known, we pose the model as a problem. Its solution provides the needed information, thereby simulating the process as it would unfold in the real world.

This chapter lays the groundwork for process modeling and simulation. We explain the basic concepts, and introduce the involved terminology in a methodical manner. Our starting point is the definition of a system.

1.1 System

A **system** is defined as a set of one or more units relevant to the knowledge that is sought. Eventually, that knowledge is obtained as system characteristics, and their behavior in time and space.

We specify a system based on what we want to know about it. Consider for example a well-mixed reactor shown in Figure 1.1 below. The reactor is fed certain amounts of non-volatile species A and B in a liquid phase. Inside the reactor, the species react to form a non-volatile liquid product C. Given that we wish to know the concentration of C in the liquid phase,

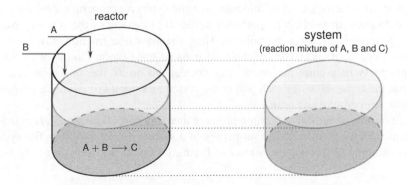

Figure 1.1 A system of reaction mixture in a reactor

Process Modeling and Simulation for Chemical Engineers: Theory and Practice, First Edition. Simant Ranjan Upreti.
© 2017 John Wiley & Sons Ltd. Published 2017 by John Wiley & Sons Ltd.
Companion website: www.wiley.com/go/upreti/pms_for_chemical_engineers

the system is precisely the reaction mixture as shown in the figure. Anything not relevant – such as the reactor wall, and the vapor phase over the mixture – is not included in the system.

Everything external to a system constitute its **surroundings**. A region of zero thickness in the system separating it from the surroundings is called the **boundary**. Any interaction between a system and its surroundings requiring physical contact takes place across the boundary. For instance, this interaction could be transfer of mass.

For the above system of reaction mixture, the surroundings comprise the reactor wall, and the vapor phase over the reaction mixture. The system boundary is made of the surface of mixture in contact with (i) the reactor wall, and (ii) air. An example of interaction between this system and its surroundings is the evaporation of the species from the mixture through its top surface (i.e., across the boundary) to air.

1.1.1 Uniform System

A system is said to be **uniform** or homogenous if it stays the same, regardless of any recombination of its parts. As an illustration, consider a system in the shape of a cube. We split it into a set of arbitrary number of small cubes of identical size. Next, we recombine them in all possible ways to form the initial cube. The system would be uniform if each recombination (for each set of small cubes) resulted in the original system. If even one recombination produced a different system, the system would be non-uniform or heterogenous.

1.1.2 Properties of System

We associate a system with the properties it possesses. By **property** we mean any measurable characteristic that is related to matter, energy, space, or time. Some common examples of property are mass, concentration, temperature, enthalpy, pressure, volume, diffusivity, etc. With the help of the properties of a system, we can keep track of it, and compare it to other systems of interest.

System properties can be classified into intensive and extensive properties. Given a uniform system, an **extensive property** is proportional to the size or extent of the system. Examples of extensive properties are mass and volume. Thus, the mass of a fraction (say, $1/10^{th}$) of a system is the same fraction ($1/10^{th}$) of the total mass of the system. On the other hand, an **intensive property** of a uniform system does not depend on its size or extent, and is the same, i.e., has the same value, for each part of the system. Examples of intensive property are concentration, temperature and pressure.

Thus, if a uniform system is at a certain pressure then any part of the system is at the same pressure. Equivalently, if all intensive properties of a system do not vary then the system is uniform. An example is the reaction mixture of Figure 1.1 on the previous page. The mixture has

1. the same value of concentration of the species A throughout the system (or uniform concentration of A),

2. uniform concentration of each of the remaining species B and C, and

3. a similar uniformity of any other intensive property, e.g., temperature.

For a non-uniform system, one or more intensive properties vary within the system. More precisely, the properties vary with space inside the system. For example, if the reaction mixture of Figure 1.1 on p. 1 were not well-mixed then the species concentrations, and temperature would not be the same throughout the mixture. In that situation, the mixture would be a non-uniform system.

1.1.3 Classification of System

Based on whether or not the intensive properties of a system vary with space, we can call a system uniform, or non-uniform. A non-uniform system is also known as a **distributed-parameter** system. If the intensive properties have variations that are small enough to be insignificant then the system may be considered as a uniform system by taking into account the space-averaged values of intensive properties. This system is then called a **lumped-parameter** system. Thus, the reaction mixture of Figure 1.1 would be a lumped-parameter system if the temperature varied slightly within the mixture but the latter was considered to be at some average temperature throughout.

Depending on the degree of separation from the surroundings, systems are also classified into open, closed and isolated systems. An **open system** allows exchanges of mass and energy with the surroundings. On the other hand, a **closed system** allows only the exchange of energy. An **isolated system** does not allow any exchange of mass, or energy.

Thus, the reaction mixture shown in Figure 1.1 would be an open system if it were heated, or cooled, and the species volatilized to the air. The mixture would be a closed system if it were heated, or cooled, but the top surface were covered to prevent any escape of the species. With the top surface covered and perfectly insulated along with the reactor vessel, the mixture would become an isolated system.

1.1.4 Model

The reason we conceive a system is that we want to learn about it. This learning is synonymous with figuring out relations between system properties. These relations give rise to a model. In mathematical terms, a **model** is a set of equations that involve system properties. A simple example of a model is the ideal gas law,

$$PV = RT$$

where P, V and T are, respectively, the pressure, molar volume, and temperature (properties) of a system at sufficiently low pressure, and R is the universal gas constant. The properties in the model do not depend on time, and the associated system is unchanging or at equilibrium to be exact.

When a system undergoes a change, it appears as an effect on one or more of the system properties. This is where the notion of process emerges.

1.2 Process

A **process** is defined as a set of activities taking place in a system, and resulting in certain effects on its properties. A process is either natural, or man-made. Natural processes – such as blood circulation in a human body, photosynthesis in plants, or planetary motion in the solar system – happen without human volition, and are responsible for certain effects on the associated systems. For instance, the process of blood circulation in a human body primarily involves pulmonary circulation between the heart and lungs, and systemic circulation between the heart and the rest of the body excluding lungs. This process results in, among other things, specific levels of oxygen and carbon dioxide concentrations in different parts of the body, i.e., the system.

Man-made processes on the other hand are contrived by human beings to produce results of utility. Common examples include processes to produce various synthetic chemicals and materials, to extract and refine natural resources, to treat gaseous emissions and wastewaters, and to control climate in living spaces. The reactor shown in Figure 1.1 on p. 1 enables a man-made process. It involves a chemical reaction between the reactants A and B, which results in the product C.

1.2.1   Classification of Processes

Based on how system properties change with time, processes are classified into unsteady state and steady state processes. A process that changes any property of a system with time is called an **unsteady state process**, and the system is called unsteady. Thus, the process of chemical reaction in the reactor of Figure 1.1 is an unsteady state process. It causes the reactant and product concentrations to, respectively, decrease and increase with time.

In contrast, a **steady state process** does not result in any change in system properties with time. The reason is that any decrease in a property gets instantaneously offset by an equal increase in the same.

A simple example of a steady state process is the filling of a tank with a non-volatile liquid, and its simultaneous drainage from the tank at the same flow rate as that of filling. Here, the system and the property of interest are, respectively, the liquid and its volume inside the tank. Due to equal inflow and outflow rates, the volume does not change with time.

Another example of a steady state process is a chemical reaction that continues after a transient period in a constant volume stirred tank reactor with constant flow rates of incoming reactants, and the outgoing mixture of residual reactants and products. In this process, the species concentrations, and the volume of the reaction mixture inside the tank do not change with time.

Steady State versus Equilibrium

It may be noted that the time derivatives of all properties are zero in a system under the influence of a steady state process. That is also true for a system at equilibrium. But there is a subtle difference. A system at equilibrium does not sustain any process since the gradients (i.e., spatial derivatives) of all potentials in the system have decayed to zero. For this reason, the properties in such a system do not have any propensity to change. However, the properties in a system under a steady state process undergo simultaneous increments and decrements in such a way that the net change in each property is zero. In other words, the properties end up being time-invariant. If a steady state process is stopped then the properties of the system would begin to change with time until the system eventually arrived at equilibrium.

1.2.2 Process Model

A **process model** is a set of equations or relations involving properties of the system under the influence of a process. The properties represent observable occurrences or phenomena classifiable into the categories of (i) initiating events, (ii) specifications, and (iii) effects. Thus, a process model is a scheme according to which a process with given specifications and initiating events would generate effects in the system.

Figure 1.2 below shows the concept of the model as a triangle, each side of which represents the relations between the two ends or vertices denoting the categories. If we know these relations, i.e., the process model, we can use it to unravel one or more unknown phenomena when the remaining ones are known. And we can do this without having to execute the process in the real world. Of course, the better the process model the better is our ability to explain the involved phenomena.

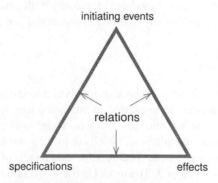

Figure 1.2   Concept of a process model

In particular, we can predict the effects of a process based on its model, and the knowledge of initiating events, and specifications. As a matter of fact, we can predict the effects for different initiating events, and specifications. Doing that enables us to isolate desirable effects, and the related initiating events, and specifications. We can then apply the latter two in the real world to achieve the desirable effects from the process. This exercise basically is process optimization and control.

Referring to Figure 1.1 on p. 1, if we know the model of the process taking place in the reaction mixture (system) then we can predict the species concentrations (effects) corresponding to different initial concentrations of A and B (initiating events), and reaction temperatures (specifications). From the predictions over a given time of reactor operation, we can pick out a desirable effect, say, the maximum concentration of C, and the related (optimal) set of initial concentrations, and reaction temperature. Based on this exercise, we can then expect to achieve the maximum concentration of C in the real world by feeding the reactor with A and B such that the mixture has the optimal initial concentrations, and is maintained at the optimal temperature for the given operation time.

In general, using a process model we can predict initiating events, specifications, or effects in the system. The prediction requires the following courses of action:

1. Process modeling – the development of a process model, and

2. Process simulation – the solution or simulation of the model, which mimics the process as it would unfold in the real world.

Types of Process Models

Process models can be categorized based on the process being represented. Thus, a **steady state model** represents a steady state process, and has no time derivatives. An **unsteady state model** represents a unsteady process, and involves one or more time derivatives.

Based on the nature of system properties, a process model may be a **lumped-parameter model**, or **distributed-parameter model**. The former involves uniform properties while the latter has at least one property that varies with space.

Based on the type of equations involved, process models are also classified into algebraic, differential, or differential-algebraic models. Moreover, if all involved equations are linear then the process model is called **linear**. Otherwise, the model is called **non-linear**.

1.3   Process Modeling

Process modeling is essentially an exercise that involves relating together the properties of a system influenced by a process. Represented as mathematical symbols, the properties are associated with each other using relevant relations under one or more assumptions. The outcome is a set of mathematical equations, which is a process model. The system properties – and through them, the initiating events, specifications and effects of the process – are expected to abide by the model thus developed. The model is therefore said to represent or describe the process.

As an example, consider the reaction process in the reactor shown in Figure 1.1 on p. 1. The involved properties are the concentrations of A, B and C, which can be represented as c_A, c_B and c_C, respectively, with initial values $\bar{c}_A$, $\bar{c}_B$ and $\bar{c}_C = 0$. If we assume that during the process the reaction mixture

1. is well-mixed,

2. has constant volume, and

3. is at constant temperature

then based on certain relations, the concentrations can be associated with each other at any time t through the following equations:

$$\frac{dc_i}{dt} = -r, \qquad c_i(0) = \bar{c}_i ; \qquad i = A, B \tag{1.1}$$

$$\frac{dc_C}{dt} = r, \qquad c_C(0) = 0 \tag{1.2}$$

$$\text{where} \qquad r = k_0 \exp\left(-\frac{E}{RT}\right) c_A^a c_B^b \tag{1.3}$$

where (i) r, k_0, E and T are, respectively, the rate, frequency factor, activation energy, and temperature of the reaction, (ii) a and b are reaction-specific parameters, and (iii) R is the universal gas constant.

The above set of equations is the model of the reaction process taking place in the reactor. This model relates the initiating events, specifications and effects of the process with each other [see Figure 1.2, p. 5]. The initiating events are represented by the initial concentrations, $\bar{c}_A$, $\bar{c}_B$ and $\bar{c}_C$. The specifications are the values of a, b, k_0, E, R and T. Finally, the process effects at any time t, are represented by the concentrations $c_A(t)$, $c_B(t)$ and $c_C(t)$. Changes in these properties, or, equivalently, the phenomena they represent, are expected to be in accordance with the model. Thus, for a given set of initiating events, and specifications, we expect to find the process effects from the model.

1.3.1 Relations

Relations are the ground rules that are used in process modeling to interlink the properties of a system under the influence of a process. These rules comprise **fundamental relations** based on scientific laws, and **constitutive relations**. The rules manifest as equations that constitute process models.

A scientific law is a statement that is generally accepted to be true about one or more phenomena on the basis of repeated experiments and observations. A common example is the law of conservation of mass. This law states that if a system does not exchange mass, or energy (a form of mass) with the surroundings then the mass of the system stays the same, or is conserved. By applying this law to an individual species, and accounting for its generation, or consumption in a system, we derive a fundamental relation called the mass balance of species.

A **constitutive relation** on the other hand is a rule that is true for systems with a specific makeup or constitution. An example is Newton's law of viscosity, which relates shear stress to strain for a system that is a Newtonian fluid.

In the model given by Equations (1.1)–(1.3) on the previous page, the last equation is a constitutive relation. It is valid for the specific reaction mixture of species A, B and C. The two differential equations are the mass balances of the species.

1.3.2 Assumptions

Assumptions of a process model are the necessary conditions that must be satisfied during the execution of the process in order for it to be represented by the model. In equivalent terms, if any assumption of a model is not satisfied during the execution of the process then the latter is not represented by the model.

For example, the assumptions of the model given by Equations (1.1)–(1.3) are perfect mixing, and constant volume as well as temperature. If any of these assumptions is not satisfied during the reaction process then it would become different from the process 'assumed' by the model, and would not be represented by the model. For instance, if mixing is not sufficiently close to perfect then there would be spatial changes in the concentrations of species, and temperature. Consequently, the reaction process would get dominated by diffusion, convection of species, and heat transfer, which are not accounted for by the model.

The above example shows that the assumption of perfect mixing excludes from the model the sub-processes of diffusion, convection and heat transfer in the reaction mixture. In general, an assumption implies the exclusion of one or more sub-processes from the model. Hence, an assumption restricts the model to a specific type of process, and in doing so simplifies the model. Note that the assumption of perfect mixing of the reaction mixture markedly simplifies the model by obviating the need for the diffusive fluxes of the species, and partial differential equations of momentum and heat transfer. Because of this assumption, the concentrations of species, and temperature can be considered uniform throughout the reaction mixture.

Making an assumption is justifiable under one or more of the following circumstances:

1. It is realistic to satisfy the assumption during the execution of the process.

2. Any sub-process excluded by the assumption has negligible influence on the overall process.

3. Without the assumption, any sub-process that should be included in the model can increase its complexity unnecessarily.

Thus, the assumption of perfect mixing can be justified for the reaction process utilizing a good agitator that sufficiently homogenizes the reaction mixture so that the sub-processes of diffusion, convection and heat transfer can be excluded from the model. Dropping the assumption would make the model very complex. Instead it is far more practical to retain the assumption, and satisfy it by having a good agitator for the reaction process.

Remarks

An assumption limits the scope of a model by making it ignore certain sub-processes or details. Therefore, it follows that the fewer the assumptions the more detailed the model. A model with no assumptions would be the most comprehensive, or perfect model with no limitations whatsoever. This of course is not possible.

As a consequence, no model can be derived without an assumption. It may be noted that some assumptions are implied, and not mentioned explicitly. For example, the model for the reaction process given by Equations (1.1)–(1.3) on p. 6 is based on the assumption that there is no intermediate reaction, or product.

1.3.3   Variables and Parameters

A **variable** is any property that could vary with time during the execution of a process in a system. On the other hand, a **parameter** is a property that is fixed or specified. For the reaction process modeled by Equations (1.1)–(1.3), variables are species concentrations, and parameters are the initial concentrations, and process specifications. From the standpoint of a process model, any symbol whose value is not fixed or specified denotes a variable. The remaining symbols represent parameters.

Note that for a system that is not isolated, a variable, or parameter may be a property of mass and (or) energy exchanged with the surroundings.

1.4   Process Simulation

As the name suggests, process simulation is the imitation of a real-world process. The imitation is carried out by actors, which are symbols that represent system properties. They play their roles, and express the effects according to a script, i.e., the process model. In direct terms, process simulation means solving the equations of a process model to find the values of the system properties that are unknown. These values tell us about the effects of a process.

Consider the process model given by Equations (1.1)–(1.3) on p. 6. Table 1.1 below lists the initiating events, and specifications as parameters, which are used for simulating the model for the given operation time t_f.

Table 1.1   Simulation parameters for the process model given by Equations (1.1)– (1.3)

parameter	value	parameter	value
a	$1/2$	b	$1/3$
$\bar{c}_A, \bar{c}_B$	$5\ \mathrm{kg\,m^{-3}}$	E/R	$500\ \mathrm{K}$
$\bar{c}_C$	$0\ \mathrm{kg\,m^{-3}}$	T	$303.15\ \mathrm{K}$
k_0	$4\ \mathrm{kg^{1/6}\,m^{-1/2}\,min^{-1}}$	t_f	$10\ \mathrm{min}$

The simulation results, i.e., the process effects are illustrated in Figure 1.3 on the next page with the help of a graph – a common visualization tool for such a purpose. Here, the graph shows how the concentration of species would change with time if the process were executed in reality with the specified parameters.

Depending on a process, simulations results can be shown using bar charts, graphs, images and animations with increasing level of visualization as needed.

1.4.1   Utility

The utility of process simulation lies in its ability to predict process effects. During the initial stage of model development, process simulation helps in improving the process model. The basis for improvement is the criterion that the better the model the closer the simulation (or predicted effects) to the actual process carried out in the real world (or experimental effects). Thus, if the predicted and experimental effects do not agree to a desired extent then we revise the model to achieve a better agreement.

Once the model of a process is sufficiently improved, process simulation is utilized to predict process effects under different circumstances, and develop the process as desired. For instance, if we want a certain effect from a process then we could narrow down predictions to obtain an appropriate (or optimal) set of initiating events, and specifications from process simulation. This of course requires process simulation to be done for different initiating events, and specifications of the process. Note that the process does not need to be executed in the real world except in the end to verify whether the optimal set leads to the desired effect. Without process simulation, we would need to carry out the process repeatedly in the real world right from the beginning, and spend time as well as resources to determine the optimal

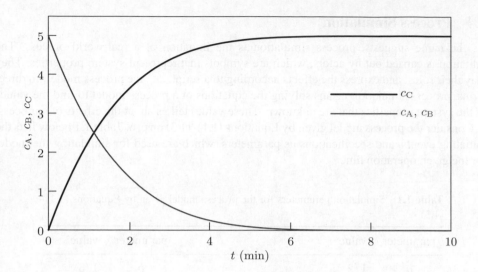

Figure 1.3 Simulation of the process model given by Equations (1.1)–(1.3), for the parameters of Table 1.1 on the previous page

set. Thus, process simulation helps obviate, or reduce initial experimentation. This is very advantageous, especially for those processes that are expensive, time consuming, or difficult to implement.

1.4.2 Simulation Methods

Simulation methods determine the process unknowns by solving mathematical equations, which constitute process models. The methods are of two types, analytical and numerical. **Analytical methods** determine the unknowns explicitly as analytical functions [see Section 8.3, p. 295] in terms of process parameters and variables. These methods involve straightforward manipulation of algebraic equations, analytical differentiation, and analytical integration to extricate the unknowns as analytical functions.

However, only simple process models are amenable to analytical methods. Most process models are sophisticated, and can only be solved by **numerical methods**. These methods determine the process unknowns directly in terms of numerical values. To that end, these methods approximate the mathematical expressions in process models as finite Taylor expansions [see Section 8.8, p. 322], and require iterative calculations. A desired accuracy of the numerical results is realized by selecting sufficiently small step sizes, and adequate number of terms in the expansions. In general, we can solve any process model using numerical methods with certain accuracy of results.

Consider for example, the process model given by Equations (1.1)–(1.3) on p. 6. The unknowns of the model are the concentrations of A, B and C at each time instant during the process, and are given by

$$c_i(t), \quad 0 < t \le t_f, \quad i = A, B, C$$

An analytical method would solve the equations, and provide analytical expressions or functions for the concentrations in terms of process parameters, and time. From the functions, we could then obtain the numerical values of species concentrations at any time. On the other hand, a numerical method would solve the equations, and directly deliver the numerical values of the concentrations at desired time instants.

1.5   Development of Process Model

The benchmark for the development of a process model is the agreement between process simulation, and the process in the real world. At the beginning, a process model is simulated to obtain the values of process variables that were previously unknown. These variables relate to the process effects. If they are dependent upon time and space then the variable values at different time instants, and spatial coordinates are obtained. Next, an experiment is performed by carrying out the process in the real world with the same parameters as used in the simulation. The unknown variables are determined from this experiment, and compared to those obtained earlier from the simulation. If the discrepancy between the experimental and simulated values of unknown variables is not as small as desired then it means that additional details or sub-processes need to be incorporated in the process model.

Since assumptions imply a lack of sub-processes, the aforementioned discrepancy specifically means that some assumptions of the process do not hold when it is carried out in reality. Such an assumption can sometimes be identified directly from the experiment. Otherwise, we need to identify an assumption that could be preventing an important sub-process from being included in the process model. This assumption is therefore made less restrictive, or simply eliminated from the process model. The corresponding sub-processes that were ignored earlier are incorporated in the process model. This is achieved by modifying the mathematical equations, and (or) including additional ones in the process model. Incorporation of new sub-processes necessitates the inclusion of associated, new assumptions. The resultant model is more rigorous and complex than the previous one.

Next, the process model thus enhanced is simulated to obtain the values of the unknown variables. The values should be closer to their experimental counterparts than they were before. In particular, we expect the discrepancy between the simulation and the experiment to reduce to a desired level. If that does not happen then another assumption is either moderated, or eliminated to further enhance the process model. The simulation of this further enhanced model is expected to bring down the discrepancy further. Elimination of another assumption, resultant model enhancement, and comparison between the simulation and experiment may be required until the discrepancy is as small as desired. This cycle of model development is shown in Figure 1.4 on the next page.

As an example, consider the model given by Equations (1.1)–(1.3) on p. 6 describing the reaction process shown in Figure 1.1 on p. 1. Suppose that there is some discrepancy between the simulated and experimental species concentrations. This discrepancy could be due to the fact that the constant volume assumption does not hold during the experiment, i.e., when the process is carried out in the real world. Note that this assumption implies that there is neither any volume change of mixing of the species nor any loss of any species to the vapor phase due to entrainment, or volatilization. In the first model development cycle, this assumption may be moderated by limiting it to no volume change of mixing only. Next, the model may be revised by including the sub-process of the loss of species to the vapor phase. This step

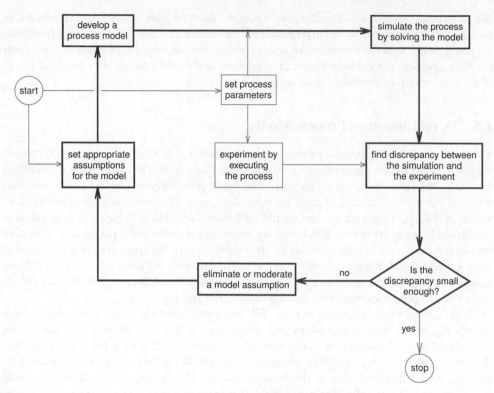

Figure 1.4   The cycle of model development

modifies Equations (1.1) and (1.2) to

$$\frac{dc_i}{dt} = -r \underbrace{- k_i a(c_i - \tilde{c}_i)}_{\text{loss to vapor phase}}, \quad c_i(0) = c_{i,0}; \qquad i = \text{A, B}$$

$$\frac{dc_C}{dt} = r \underbrace{- k_C a(c_C - \tilde{c}_C)}_{\text{loss to vapor phase}}, \quad c_C(0) = 0$$

by including the loss of species to the vapor phase. In the modified equations, a is the surface area of the reaction mixture exposed to the vapor phase per unit mixture volume, k_i is the mass transfer coefficient of the ith species (A, B, or C) for its transfer from the reaction mixture to the vapor phase, and $\tilde{c}_i$ is the vapor phase concentration of the species. This concentration is assumed to be constant for each species, implying thereby a steady flow of air across the top of the reaction mixture. Observe that this is a new assumption for the process model being enhanced in this model development cycle. The simulation of the revised model is expected to yield species concentrations that are closer to the experimental counterparts than they were before.

Remarks

It may be noted that when eliminating, or moderating an existing assumption, one or more new assumptions arise for each sub-process that gets included to enhance the process model.

Therefore, the set of model assumptions can never be null, no matter how many times the cycle of model development is continued, i.e., how much the model is enhanced.

Since an assumption implies one or more ignored sub-processes, there would always be some discrepancy, however small, between the simulation of a process model, and the experiment. It is up to us to decide a tolerable level of that discrepancy, and develop a process model accordingly.

1.6 Learning about Process

Besides predicting the effects of a process by simulating its model, we can learn about the process itself by utilizing the model. Recall from Figure 1.2 on p. 5 that a process model relates the phenomena grouped into certain initiating events, specifications and effects. Suppose that a process model is adequate to simulate the effects from the knowledge of the initiating effects, and specifications. We can then determine any unknown specifications on the basis of certain initiating events, and process effects, which are known from the real-world execution of the process, i.e., the experiment.

As an example, consider the model given by Equations (1.1)–(1.3) on p. 6 for the reaction process of Figure 1.1 on p. 1. Given the initiating events, and specifications, the model can simulate the process effects, e.g., the change in species concentrations with time. Now suppose that we do not know the frequency factor k_0 for the reaction process. Based on the model of the process, and the experiment, we can determine k_0 by

1. obtaining the species concentrations at different instants from the experiment, and

2. simulating the model with different trial values of k_0 until the predicted species concentrations are sufficiently close to the experimental concentration values.

The value of k_0 is determined when the last step is completed. This step is usually carried out using sophisticated optimization methods.

Note that unlike simulation, the above procedure solves an inverse problem of finding the unknowns of the given process model from experimental effects, or data. We can use this procedure to enhance our knowledge of the process. This is done by proposing different trial mathematical expressions that relate to the process, and finding the unknown specifications with the help of experimental data. A trial expression would try to capture more insights and details about the process. If a trial expression results in a better agreement between the predicted and experimental effects then it reflects an enhanced understanding of the process.

For the aforementioned example, if the agreement is better with the trial reaction rate expression

$$r = \frac{k_0 \exp\left(\dfrac{E}{RT}\right) c_A^a c_B^b}{(k_1 + c_A^c)}$$

where k_1 and c are two new unknowns that are determined then we could infer that a different reaction mechanism, and kinetics are in operation. Thus, this result would enhance our understanding of the reaction process.

Having covered the foundational aspects of process modeling and simulation, we now elucidate how a system is specified for process modeling.

1.7   System Specification

Specification of the right system is very important when modeling a process. From this standpoint, a system is the house of elements whose properties change with time under the action of a process. When interrelating the properties in a process model, it is a prerequisite that each intensive property be unique, i.e., uniform throughout the system at any time. Otherwise, say, if the temperature (intensive property) of a reaction mixture (system) in Figure 1.1 on p. 1 varies from 90°C at the center to 70°C at the periphery then there is no unique temperature as an intensive system property to begin with. In such a situation, the system must be specified in such a way that it is uniform at all times.

The above requirement is inherently satisfied for uniform and lumped-parameter systems, which involve uniform intensive properties per se. However, this is not the case for distributed-parameter systems. To model a process in such a system, the approach is to

1. split the system into sub-systems that are so small that all intensive properties within each sub-system are unique, i.e., uniform at any time, and

2. model the process for each sub-system thus determined.

A sub-system determined as above is called a differential system, or, simply, a **differential element**. Its thickness along each spatial direction of property variation is sufficiently small to house uniform properties. The thickness tends to zero (but is not zero), and as it does, the number of differential elements tends to infinity. With contiguous differential elements of identical shapes, we model the process on a single representative differential element, and extend the model to cover the entire system. In what follows, we explain the determination of differential elements by using examples based on the reaction mixture system shown in Figure 1.1.

As the first example, consider the reaction mixture system under the action of a process such that the species concentrations vary only along the vertical z-direction. For this system, Figure 1.5 below shows (rather exaggerated) the differential element, which is a circular disk of thickness Δz conceived to be so small that the species concentrations in it are uniform along the z-direction. Put differently, there is no concentration variation in the disk so that

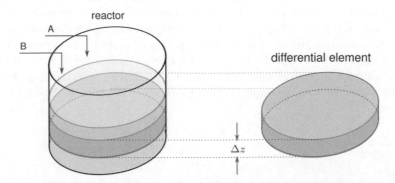

Figure 1.5   The differential element when a property changes vertically

there is only one representative concentration for each species therein. An infinite number of these disks fit along the z-direction starting from the bottom of the reaction mixture to its top surface. However, for process modeling, we only need to consider the representative differential element as shown in the figure, and two similar differential elements touching, respectively, the bottom and the top of the reaction mixture.

For the second example, consider the reaction mixture system influenced by a process, which causes the species concentrations to vary only along the radial r-direction. For this system, Figure 1.6 below shows the differential element, which is an annulus of thickness Δr conceived to be so small that the species concentrations in it are uniform along the r-direction. Thus, there are no concentration variations in the annulus so that there only one

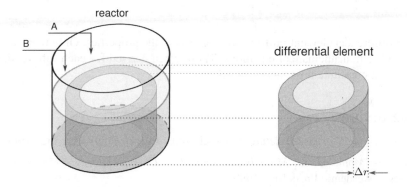

Figure 1.6   The differential element when a property changes radially

representative concentration for each species therein. An infinite number of these annuli fit along the r-direction starting from the center of the reaction mixture to its periphery. But we only need to model the process for the representative differential element as shown in the figure, and two similar elements touching, respectively, the center and periphery of the reaction mixture.

For the third example, consider the reaction mixture system under the influence of a process, which causes the species concentrations to vary along the vertical as well as radial direction. For this system, Figure 1.7a on the next page shows the differential element, which is a thin annulus of thicknesses Δz and Δr along the two directions, respectively. These thicknesses are conceived to be so small that the species concentrations in the differential element are uniform along each direction. To model the process, we need to consider this representative differential element, i.e., a thin annulus, and similar elements touching the system boundaries along the z- and r-directions.

In the final example, consider the reaction mixture system under the influence of a process, which causes the species concentrations to vary along the vertical, radial and θ-directions. For this system, Figure 1.7b on the next page shows the differential element, which resembles a horseshoe of thicknesses Δz, Δr and $\Delta \theta$ along the three directions, respectively. These thicknesses are conceived to be so small that the species concentrations in the differential element are uniform along each direction. To model the process, we need to consider this representative differential element, and similar elements that touch the system boundaries along each direction except the circular θ-direction.

To recapitulate, before modeling a process involving a distributed-parameter system, which has spatial changes in intensive properties, we need to determine a relevant sub-system. Its thickness along each direction of property variation should be small enough to render the

(a)   (b)

Figure 1.7 The differential element when a property changes (a) along the depth z, and the radius r, and (b) along z, r and the angle θ

resultant differential element uniform with respect to the properties. Only then can we relate them together in mathematical equations. These when derived for each differential element of the system constitute the process model.

Bibliography

[1]   M.M. Denn. *Process Modeling*. Massachusetts: Pitman Publishing Inc., 1986.

[2]   R. Aris. *Mathematical Modeling, Volume 1: A Chemical Engineer's Perspective*. San Diego: Academic Press Inc., 1999.

[3]   R.G. Rice and D.D. Do. *Applied Mathematics and Modeling for Chemical Engineers*. 2nd edition. New Jersey: John Wiley & Sons, 2012.

[4]   W.L. Luyben. *Process Modeling, Simulation and Control for Chemical Engineers*. 2nd edition. New York: McGraw-Hill, 1989.

[5]   A. Rasmuson et al. *Mathematical Modeling in Chemical Engineering*. Cambridge, U.K.: Cambridge University Press, 2014.

[6]   B.W. Bequette. *Process Dynamics Modeling, Analysis, and Simulation*. New Jersey: Prentice-Hall, Inc., 1998.

[7]   C. Boyadjiev. *Theoretical Chemical Engineering Modeling and Simulation*. Berlin: Springer, 2010.

Exercises

1.1 What combination of properties results in an intensive property?

1.2 In the model given by Equations (1.1)–(1.3) on p. 6, what sub-processes get excluded because of assumptions of constant volume and temperature of the reaction mixture?

1.3 What is the assumption that is common to any process model?

1.4 What would be a perfect model for a process? Justify whether or not such a model could be obtained?

1.5 Describe the differential elements needed to capture all possible spatial property variations in spherical coordinates.

2

Fundamental Relations

In this chapter, we present fundamental relations, which govern the transport of mass, momentum and energy. These relations are derived as differential equations, which describe changes in the involved properties with time and space. Process models rely profoundly on these relations for accurate simulations.

Fundamental relations make the continuum assumption, i.e., assume that systems are continuous. A continuous system is divisible into an infinite number of sub-systems such that the adjacent ones are almost the same while those separated by finite distances are different. In other words, there is a gradual transition from one point to another in the system without any abrupt change. Thus, system properties are assumed to be continuous functions of space.

Typically though, systems comprise matter, and are not continuous because of the presence of discrete molecules with vacant spaces between them. Nevertheless, fundamental relations are still applicable in many practical situations, and provide accurate results. The reason is that the volume of vacant spaces is significantly smaller than that of the involved system. Specifically, Knudsen number, i.e., the ratio of mean free path in the system to its characteristic length, is considerably smaller than unity.

2.1 Basic Form

A basic form of fundamental relations is a differential equation describing the change in an entity inside a system with time. This form is also known as a balance equation for an entity.

Consider an entity ϵ in the system shown in Figure 2.1 on the next page. This entity

1. enters and exits the system via its boundary surfaces as, respectively, ϵ_{in} and ϵ_{out},

2. enters otherwise as ϵ_{ext}, and

3. is generated within the system as ϵ_{gen}.

Then between any two time instants, t_1 and $t_2 > t_1$, the change in ϵ in the system is given by

$$\epsilon(t_2) - \epsilon(t_1) = \underbrace{[\epsilon_{\text{in}}(t_2) - \epsilon_{\text{in}}(t_1)]}_{\substack{\epsilon \text{ that entered via the boundary} \\ \text{surface of the system}}} - \underbrace{[\epsilon_{\text{out}}(t_2) - \epsilon_{\text{out}}(t_1)]}_{\substack{\epsilon \text{ that exited via the boundary} \\ \text{surface of the system}}}$$

$$+ \underbrace{[\epsilon_{\text{ext}}(t_2) - \epsilon_{\text{ext}}(t_1)]}_{\epsilon \text{ induced from external sources}} + \underbrace{[\epsilon_{\text{gen}}(t_2) - \epsilon_{\text{gen}}(t_1)]}_{\epsilon \text{ generated in the system}}$$

In the above equation, ϵ_{in} and ϵ_{out} represent the ϵ that must contact the boundary surface of the system in order to enter and exit, respectively. For example, ϵ_{in} could be (i) the energy

Process Modeling and Simulation for Chemical Engineers: Theory and Practice, First Edition. Simant Ranjan Upreti.
© 2017 John Wiley & Sons Ltd. Published 2017 by John Wiley & Sons Ltd.
Companion website: www.wiley.com/go/upreti/pms_for_chemical_engineers

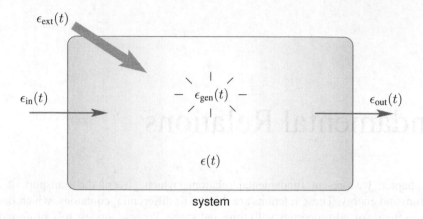

Figure 2.1 A system with different contributions to an entity ϵ at time t

of the mass that crosses the boundary surface to enter the system, or (ii) the momentum that enters the system due to shear stress acting on the boundary surface.

As opposed to ϵ_{in} and ϵ_{out}, the term ϵ_{ext} accounts for an external contribution to ϵ that does not involve any contact with the boundary surface. An example is the contribution from the gravitational field to the momentum of a system. Finally, the term ϵ_{gen} stands for the entity ϵ generated in the system, e.g., mass of a species produced in a chemical reaction. If ϵ is consumed in the system then ϵ_{gen} is negative. This generation term is zero for entities that are conserved such as the total mass, total energy, and momentum.

Let Δt be the time duration, $(t_2 - t_1)$. Then for sufficiently small Δt during which the rates of change of ϵ_{in}, ϵ_{out}, ϵ_{gen} and ϵ_{ext} with time t, – respectively, $\dot{\epsilon}_{in}$, $\dot{\epsilon}_{out}$, $\dot{\epsilon}_{gen}$ and $\dot{\epsilon}_{ext}$ – are constant, we can express the equation on the previous page as

$$\epsilon(t_1 + \Delta t) - \epsilon(t_1) = \dot{\epsilon}_{in}\Delta t - \dot{\epsilon}_{out}\Delta t + \dot{\epsilon}_{ext}\Delta t + \dot{\epsilon}_{gen}\Delta t$$

Applying the first order Taylor expansion of $\epsilon(t_1 + \Delta t)$ in the limit of Δt tending to zero [see Section 8.8.2, p. 323], we obtain after simplification

$$\left. \frac{d\epsilon}{dt} \right|_{t_1} = \dot{\epsilon}_{in} - \dot{\epsilon}_{out} + \dot{\epsilon}_{ext} + \dot{\epsilon}_{gen}$$

Since t_1 is arbitrary, we can write the above result as

$$\frac{d\epsilon}{dt} = \dot{\epsilon}_{in} - \dot{\epsilon}_{out} + \dot{\epsilon}_{ext} + \dot{\epsilon}_{gen} \qquad (2.1)$$

for any time instant. The above equation is a **general unsteady state balance**, which is applicable to any system entity. It could be scalar, or vector. The equation states that at any time, the rate of change of an entity in a system is the algebraic summation of different rates at which contributions to the entity are made.

2.1.1   Application

Equation (2.1) on the previous page is an ordinary differential equation. It is commonly used in conjunction with an initial condition, $\epsilon(t = 0) = \epsilon_0$, to determine the temporal change in entity ϵ of a system under the influence of a process. The application is straightforward if the intensive properties of the system either stay uniform, i.e., are space-independent, or are averaged over space. Figure 2.2a below shows such a system in Cartesian coordinates.

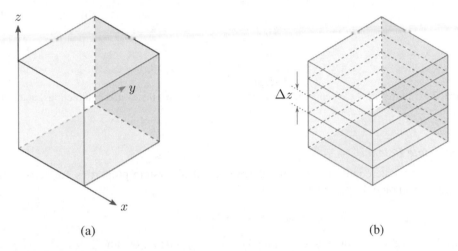

(a)	(b)

Figure 2.2   Schematic of (a) uniform system, and (b) non-uniform system (along the z-direction) in Cartesian coordinates

Distributed-Parameter Systems

In these systems, Equation (2.1) is applied to suitable differential elements [see Section 1.7, p. 14]. In Cartesian coordinates for instance, if intensive properties in such a system change along the z-direction then the system is divided into differential elements along that direction, as shown in Figure 2.2b above. The thickness of each element, Δz, is considered so small that the properties within an element are uniform.

If intensive properties in the system also change similarly along another direction, say, the x-direction, then each existing differential element is sub-divided along that direction, as shown in Figure 2.3a on the next page. The thickness Δx of each resultant differential element is considered small enough to ensure the uniformity of the properties within. Likewise, if intensive properties change also along the remaining y-direction then each existing differential element is further sub-divided along that direction, as shown in Figure 2.3b on the next page. The new thickness Δy of each resultant differential element is taken to be small enough to ensure the uniformity of the properties.

After sub-dividing the system in the above manner depending on the directions of property changes, Equation (2.1) is applied to a representative differential element. The equations of all differential elements can then be pieced together, or integrated over the entire system to simulate the process. For simplicity, we will consider rectangular or Cartesian coordinate system in this chapter.

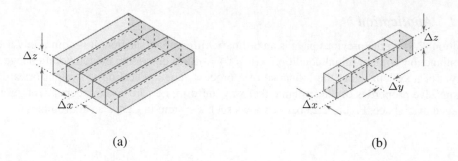

(a) (b)

Figure 2.3 Differential elements for property changes along (a) the z- and x-directions, and (b) all directions

Steady State Balance

Under steady state condition, the time derivatives of all system properties are zero so that Equation (2.1) on p. 18 becomes

$$\dot{\epsilon}_{in} - \dot{\epsilon}_{out} + \dot{\epsilon}_{ext} + \dot{\epsilon}_{gen} = 0$$

Multiplying the above equation by a time duration, say, Δt, we obtain

$$\epsilon_{in} - \epsilon_{out} + \epsilon_{ext} + \epsilon_{gen} = 0$$

which is the steady state balance in terms of various contributions made to an entity ϵ of a system.

Generalization

Equation (2.1) can be generalized for multiple ports of entry as well as exit of an entity, and multiple sources of external contribution, and internal generation. Let the rates of ϵ entering and exiting a system through the jth entry and exit ports be $\dot{\epsilon}_{in,j}$ and $\dot{\epsilon}_{out,j}$, respectively. Then the rate of loss of ϵ from the system through N_{in} entry and N_{out} exit ports is

$$\Delta[\dot{\epsilon}] \equiv \sum_{j=1}^{N_{out}} \dot{\epsilon}_{out,j} - \sum_{j=1}^{N_{in}} \dot{\epsilon}_{in,j} \tag{2.2}$$

Also, let the rates of external contribution, and internal generation of ϵ from their respective j^{th} sources (numbering N_{ext} and N_{gen}) be $\dot{\epsilon}_{ext,j}$ and $\dot{\epsilon}_{gen,j}$, respectively. The rates are positive or negative depending on whether they increase or decrease the ϵ in the system. With this arrangement, Equation (2.1) can be written in the following general form:

$$\frac{d\epsilon}{dt} = -\Delta[\dot{\epsilon}] + \sum_{j=1}^{N_{ext}} \dot{\epsilon}_{ext,j} + \sum_{j=1}^{N_{gen}} \dot{\epsilon}_{gen,j} \tag{2.3}$$

2.2   Mass Balance

Applying Equation (2.3) on the previous page to the entity that is the mass m_i of the ith species in a system, we obtain the individual mass balance,

$$\frac{dm_i}{dt} = -\Delta[\dot{m}_i] + \sum_{j=1}^{N_r} \underbrace{\dot{m}_{\text{gen},ij}}_{V r_{\text{gen},ij}} \tag{2.4}$$

where V is the system volume, and N_r is the number of chemical reactions with $r_{\text{gen},ij}$ as the rate of generation per unit volume of the mass of the ith species in the jth reaction. The rate of generation is positive (negative) if the species is produced (consumed) in the reaction. Note that there is no external contribution to m_i. Summing the above equation for all N_c chemical species yields the balance for the total mass m of a system, i.e.,

$$\frac{d}{dt} \underbrace{\sum_{i=1}^{N_c} m_i}_{m} = -\sum_{i=1}^{N_c} \Delta[\dot{m}_i] + \underbrace{V \sum_{i=1}^{N_c} \sum_{j=1}^{N_r} r_{\text{gen},ij}}_{=0} = -\Delta\left[\underbrace{\sum_{i=1}^{N_c} \dot{m}_i}_{\dot{m}}\right]$$

where $\dot{m}_i$ and $\dot{m}$ are the mass flow rates of, respectively, the ith species and all species through a port. The generation term in the above equation is zero since the total mass of a system is conserved. Thus, the **overall mass balance** of a system is

$$\frac{dm}{dt} = -\Delta[\dot{m}] \tag{2.5}$$

2.2.1   Microscopic Balances

These balances pertain to distributed-parameter systems. For such a system, let ρ_i and $\mathbf{v}_i$ be, respectively, the mass density and average velocity of the molecules of the ith species in the differential element, as shown in Figure 2.4 on the next page. This element is a cuboid of fixed volume with dimensions Δx_1, Δx_2 and Δx_1 along the x_1-, x_2- and x_3-directions, respectively. For the ith species in this element,

1. the mass is $\rho_i \Delta V$ where $\Delta V = \Delta x_1 \Delta x_2 \Delta x_3$,

2. the mass flux of along the jth direction is $\rho_i v_{ij}$,

3. the mass flow rates in and out of the system along the jth direction are, respectively, $[\rho_i v_{ij} \Delta A_j]_{x_j}$ and $[\rho_i v_{ij} \Delta A_j]_{x_j + \Delta x_j}$ where $\Delta A_j = \Delta x_k \Delta x_l$, $(j, k, l \in \{1, 2, 3\}$ and $j \neq k \neq l)$, and

4. the rate of generation per unit volume in a jth reaction (of total N_r reactions) is $r_{\text{gen},ij}$.

With these considerations, the application of Equation (2.1) on p. 18 for the mass of the ith species in the above element yields

$$\frac{d(\rho_i \Delta V)}{dt} = \Delta V \frac{\partial \rho_i}{\partial t} = \sum_{j=1}^{3} \left\{ \left[\rho_i v_{ij} \Delta A_j\right]_{x_j} - \left[\rho_i v_{ij} \Delta A_j\right]_{x_j + \Delta x_j} \right\} + \sum_{j=1}^{N_r} r_{\text{gen},ij} \Delta V \tag{2.6}$$

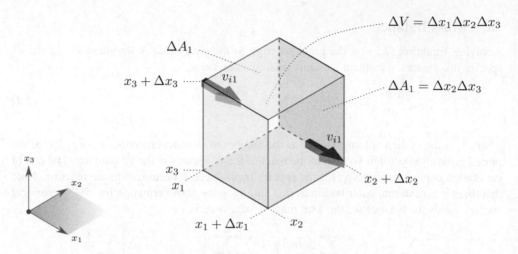

Figure 2.4   A differential element of fixed volume in Cartesian coordinates. v_{i1} is the component of $\mathbf{v}_i$ (average molecular velocity of the ith species) along the x_1-direction

In the limit of Δx_j tending to zero at any time t,

$$\left[\rho_i v_{ij} \Delta A_j \right]_{x_j + \Delta x_j} = \left[\rho_i v_{ij} \Delta A_j \right]_{x_j} + \frac{\partial (\rho_i v_{ij} \Delta A_j)}{\partial x_j} \Delta x_j$$

from the first order Taylor expansion. Since ΔA_j is independent of x_j, the above expansion leads to

$$\Delta A_j \left[\rho_i v_{ij} \right]_{x_j + \Delta x_j} = \Delta A_j \left[\rho_i v_{ij} \right]_{x_j} + \frac{\partial (\rho_i v_{ij})}{\partial x_j} \underbrace{\Delta A_j \Delta x_j}_{\Delta V}$$

where ΔV is the volume of the differential element. Substitution of the above result into Equation (2.6) on the previous page, and simplification yields the **individual microscopic mass balance**,

$$\frac{\partial \rho_i}{\partial t} = \underbrace{- \sum_{j=1}^{3} \frac{\partial (\rho_i v_{ij})}{\partial x_j}}_{\equiv \nabla \cdot \rho_i \mathbf{v}_i} + \sum_{j=1}^{N_r} r_{\text{gen},ij} = -\nabla \cdot \underbrace{\rho_i \mathbf{v}_i}_{\mathbf{f}_i} + \sum_{j=1}^{N_r} r_{\text{gen},ij} \quad (2.7)$$

where $\mathbf{f}_i$ is the mass flux of the ith species. The shorthand $\nabla \cdot \mathbf{z}$, which is called the divergence of vector $\mathbf{z}$, is the sum of partial derivatives of components of $\mathbf{z}$ with respect to collinear Cartesian coordinates.

Summing the above equation for all N_c species, we obtain

$$\frac{\partial}{\partial t} \Big(\underbrace{\sum_{i=1}^{N_c} \rho_i}_{\rho} \Big) = -\nabla \cdot \underbrace{\sum_{i=1}^{N_c} \rho_i \mathbf{v}_i}_{\equiv \mathbf{f}} + \underbrace{\sum_{i=1}^{N_c} \sum_{j=1}^{N_r} r_{\text{gen},ij}}_{=\,0}$$

In the above equation,

1. ρ is the mass density of the mixture of the N_c species,

2. $\mathbf{f}$ is the total mass flux defined as $\rho\mathbf{v}$ implying $\mathbf{v}$ to be the mass average velocity given by

$$\mathbf{v} = \frac{1}{\rho}\sum_{i=1}^{N_c}\rho_i\mathbf{v}_i = \sum_{i=1}^{N_c}\omega_i\mathbf{v}_i$$

   where $\omega_i = \rho_i/\rho$ is the mass fraction of the ith species, and

3. the last term is zero since the total mass of a system is conserved.

With these considerations, we obtain the **overall microscopic mass balance**

$$\frac{\partial\rho}{\partial t} = -\nabla\cdot\rho\mathbf{v} \qquad (2.8)$$

which is also known as the **equation of continuity**.

2.2.2 *Equation of Change for Mass Fraction*

To obtain this equation, we express the mass flux of the ith species as

$$\underbrace{\mathbf{f}_i}_{\rho_i\mathbf{v}_i} = \rho_i(\mathbf{v} + \underbrace{\mathbf{v}_i - \mathbf{v}}_{\mathbf{v}_{\text{rel},i}}) = \rho_i\mathbf{v} + \underbrace{\rho_i\mathbf{v}_{\text{rel},i}}_{\mathbf{j}_i} \qquad (2.9)$$

where $\mathbf{j}_i$ is the relative mass flux of the ith species arising from its velocity $\mathbf{v}_{\text{rel},i}$ (which is relative to $\mathbf{v}$), and $\rho_i\mathbf{v}$ is the bulk flux due to $\mathbf{v}$. Since $\rho_i = \omega_i\rho$, the above equation can also be expressed as

$$\mathbf{f}_i = \omega_i\underbrace{\rho\mathbf{v}}_{\mathbf{f}} + \mathbf{j}_i = \omega_i\mathbf{f} + \mathbf{j}_i \qquad (2.10)$$

where ω_i is the mass fraction of the ith species, ρ is the mass concentration of the fluid mixture, or its density, and $\mathbf{f}$ is the overall or bulk flux of the mixture. The above equation shows that the mass flux of a species is the sum of the contribution from bulk flux, and relative flux.

Substituting Equation (2.9) above in Equation (2.7) on the previous page, we obtain for the ith species,

$$\frac{\partial\rho_i}{\partial t} = -\nabla\cdot\rho_i\mathbf{v} - \nabla\cdot\mathbf{j}_i + \sum_{j=1}^{N_r}r_{\text{gen},ij}$$

Upon expressing ρ_i as $\omega_i\rho$, and expanding the time derivative, the above equation becomes

$$\rho\frac{\partial\omega_i}{\partial t} = -\omega_i\frac{\partial\rho}{\partial t} - \nabla\cdot\omega_i\rho\mathbf{v} - \nabla\cdot\mathbf{j}_i + \sum_{j=1}^{N_r}r_{\text{gen},ij} \qquad (2.11)$$

Utilizing Equation (2.8) above as well as the identity [see Equation (8.10), p. 317]

$$\nabla\cdot\omega_i\rho\mathbf{v} = \omega_i\nabla\cdot\rho\mathbf{v} + \rho\mathbf{v}\cdot\nabla\omega_i$$

where $\nabla \omega_i$ is the gradient, i.e., the vector given by $\left[\partial \omega_i / \partial x_1 \quad \partial \omega_i / \partial x_2 \quad \partial \omega_i / \partial x_3\right]^\top$, Equation (2.11) on the previous page can be written as

$$\frac{\partial \omega_i}{\partial t} = -\mathbf{v} \cdot \nabla \omega_i - \frac{1}{\rho}\left(\nabla \cdot \mathbf{j}_i - \sum_{j=1}^{N_r} r_{\text{gen},ij}\right)$$

which is the **equation of change for the mass fraction** of the ith species. The above equation can be written in terms of substantial derivative [see Section 8.7.4, p. 321] as

$$\frac{\mathrm{D}\omega_i}{\mathrm{D}t} = -\frac{1}{\rho}\left(\nabla \cdot \mathbf{j}_i - \sum_{j=1}^{N_r} r_{\text{gen},ij}\right) \tag{2.12}$$

where the left-hand side stands for $(\partial \omega_i / \partial t + \mathbf{v} \cdot \nabla \omega_i)$.

2.3   Mole Balance

It is convenient in reactive systems to have mass balances in terms of mole numbers of the involved species. To obtain this, we express m_i in Equation (2.4) on p. 21 as the molecular weight multiplied (M_i) by the mole numbers (N_i, $N_{\text{in},ij}$, $N_{\text{out},ij}$, or $N_{\text{gen},ij}$), and obtain

$$\frac{\mathrm{d}(M_i N_i)}{\mathrm{d}t} = -\Delta \left[\frac{\mathrm{d}(M_i N_i)}{\mathrm{d}t}\right] + \sum_{j=1}^{N_r} \frac{\mathrm{d}(M_i N_{\text{gen},ij})}{\mathrm{d}t}$$

Since M_i is constant, it drops out from the above equation, and we get the **individual mole balance**,

$$\frac{\mathrm{d}N_i}{\mathrm{d}t} = -\Delta\left[\dot{N}_i\right] + V \sum_{j=1}^{N_r} R_{\text{gen},ij} \tag{2.13}$$

where $R_{\text{gen},ij}$ is the molar rate of generation of the ith species per unit volume (V) of the system.

2.3.1   Microscopic Balances

The derivation here is analogous to that for the microscopic mass balances [see Section 2.2.1, p. 21]. For the differential element shown in Figure 2.4 on p. 22, let c_i be the molar concentration of the ith species with $R_{\text{gen},ij}$ as the molar rate of generation per unit volume. Then in that element, the number of moles of the species is $c_i \Delta V$. Along the x_j-direction, the molar flux is $c_i \mathbf{v}_{ij}$, and the molar flow rates in and out of the differential element are $[c_i \mathbf{v}_{ij} \Delta A_j]_{x_j}$ and $[c_i \mathbf{v}_{ij} \Delta A_j]_{x_j + \Delta x_j}$, respectively. The application of Equation (2.1) on p. 18 for the number of moles of the ith species, and subsequent simplification using the first order Taylor expansion leads to the **individual microscopic mole balance**

$$\frac{\partial c_i}{\partial t} = -\sum_{j=1}^{3} \frac{\partial(c_i v_{ij})}{\partial x_j} + \sum_{j=1}^{N_r} R_{\text{gen},ij} = -\nabla \cdot \underbrace{c_i \mathbf{v}_i}_{\mathbf{F}_i} + \sum_{j=1}^{N_r} R_{\text{gen},ij} \tag{2.14}$$

where $\mathbf{F}_i$ is the molar flux of the ith species. Summing the last equation for all N_c species yields

$$\frac{\partial}{\partial t}\underbrace{\left(\sum_{i=1}^{N_c} c_i\right)}_{c} = -\nabla\cdot\underbrace{\sum_{i=1}^{N_c} c_i\mathbf{v}_i}_{\equiv\mathbf{F}} + \sum_{i=1}^{N_c}\sum_{j=1}^{N_r} R_{\text{gen},ij}$$

In the above equation,

1. c is the molar concentration of the mixture of the N_c species,

2. $\mathbf{F}$ is the total molar flux defined as $c\tilde{\mathbf{v}}$, implying $\tilde{\mathbf{v}}$ to be the molar average velocity given by

$$\tilde{\mathbf{v}} = \frac{1}{c}\sum_{i=1}^{N_c} c_i\mathbf{v}_i = \sum_{i=1}^{N_c} y_i\mathbf{v}_i$$

where $y_i = c_i/c$ is the mole fraction of the ith species, and

3. the last term is not necessarily zero since the number of moles of species may change in chemical reactions.

With these considerations, we obtain the **overall microscopic mole balance**

$$\frac{\partial c}{\partial t} = -\nabla\cdot c\tilde{\mathbf{v}} + \sum_{i=1}^{N_c}\sum_{j=1}^{N_r} R_{\text{gen},ij} \tag{2.15}$$

2.3.2   *Equation of Change for Mole Fraction*

We proceed similar to the derivation for the equation for mass fraction, and express the molar flux of the ith species as

$$\underbrace{\mathbf{F}_i}_{c_i\mathbf{v}_i} = c_i(\tilde{\mathbf{v}} + \underbrace{\mathbf{v}_i - \tilde{\mathbf{v}}}_{\tilde{\mathbf{v}}_{\text{rel},i}}) = c_i\tilde{\mathbf{v}} + \underbrace{c_i\tilde{\mathbf{v}}_{\text{rel},i}}_{\mathbf{J}_i} \tag{2.16}$$

where $\mathbf{J}_i$ and $c_i\tilde{\mathbf{v}}$ are the relative and bulk molar fluxes, respectively. The above equation can also be expressed as

$$\mathbf{F}_i = y_i\underbrace{c\mathbf{v}}_{\mathbf{F}} + \mathbf{J}_i = y_i\mathbf{F} + \mathbf{J}_i$$

where y_i is the mole fraction of the ith species, c is the molar concentration of the fluid mixture, and $\mathbf{F}$ is the overall molar flux of the mixture, or the bulk molar flux. The molar flux of a species, like its mass flux, is the sum of the contribution from bulk flux, and relative flux.

Substituting Equation (2.16) above in Equation (2.14) on the previous page for the ith species, we obtain

$$\frac{\partial c_i}{\partial t} = -\nabla\cdot c_i\tilde{\mathbf{v}} - \nabla\cdot\mathbf{J}_i + \sum_{j=1}^{N_r} R_{\text{gen},ij}$$

In the last equation, we express c_i as $y_i c$ and utilize Equation (2.15) on the previous page as well as the identity, $\nabla \cdot y_i c\tilde{\mathbf{v}} = y_i \nabla \cdot c\tilde{\mathbf{v}} + c\tilde{\mathbf{v}} \cdot \nabla y_i$, [see Equation (8.10), p. 317] to obtain

$$\frac{\partial y_i}{\partial t} = -\tilde{\mathbf{v}} \cdot \nabla y_i - \frac{1}{c}\left(\nabla \cdot \mathbf{J}_i + y_i \sum_{k=1}^{N_c} \sum_{j=1}^{N_r} R_{\text{gen},kj} - \sum_{j=1}^{N_r} R_{\text{gen},ij} \right) \qquad (2.17)$$

which is the **equation of change for mole fraction** of the ith species.

2.4 Momentum Balance

This balance involves forces acting on boundary surfaces of a system. These entities are conceived as pressure and viscous stress, both of which are forces per unit area. Relative to the surface, while pressure is perpendicular, viscous stress is at an arbitrary angle. Figure 2.5 below shows these forces per unit area (P and $\boldsymbol{\tau}_2$, respectively) acting on the $x_1 x_3$-plane* of a system. While P has only one component (i.e., itself) along the x_2-direction, $\boldsymbol{\tau}_2$

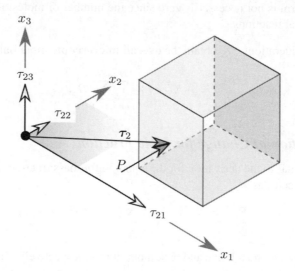

Figure 2.5 Pressure P and viscous stress $\boldsymbol{\tau}_2$ acting on the $x_1 x_3$-plane of a system in Cartesian coordinates

has components τ_{21}, τ_{22} and τ_{23} along the x_1-, x_2- and x_3-directions, respectively. For convenience, we organize these components into a row vector of molecular stress along the x_2-direction

$$\boldsymbol{\pi}_2 \equiv \left[\underbrace{\tau_{21}}_{\substack{x_1\text{-component} \\ (\pi_{21})}} \qquad \underbrace{P + \tau_{22}}_{\substack{x_2\text{-component} \\ (\pi_{22})}} \qquad \underbrace{\tau_{23}}_{\substack{x_3\text{-component} \\ (\pi_{23})}} \right]$$

---

*It is an area vector collinear with the x_2-direction. This direction is identified by the subscript '2' in $\boldsymbol{\tau}_2$.

Similarly, the molecular stress vectors along the x_1- and x_3-directions are, respectively,

$$\boldsymbol{\pi}_1 \equiv \left[\underbrace{P + \tau_{11}}_{\pi_{11}} \quad \underbrace{\tau_{12}}_{\pi_{12}} \quad \underbrace{\tau_{13}}_{\pi_{13}} \right] \quad \text{and} \quad \boldsymbol{\pi}_3 \equiv \left[\underbrace{\tau_{31}}_{\pi_{31}} \quad \underbrace{\tau_{32}}_{\pi_{32}} \quad \underbrace{P + \tau_{33}}_{\pi_{33}} \right]$$

Together, the three row vectors above form the **molecular stress tensor**

$$\boldsymbol{\pi} \equiv \begin{bmatrix} \boldsymbol{\pi}_1 \\ \boldsymbol{\pi}_2 \\ \boldsymbol{\pi}_3 \end{bmatrix} = \begin{bmatrix} \pi_{11} & \pi_{12} & \pi_{13} \\ \pi_{21} & \pi_{22} & \pi_{23} \\ \pi_{31} & \pi_{32} & \pi_{33} \end{bmatrix} = \begin{bmatrix} P + \tau_{11} & \tau_{12} & \tau_{13} \\ \tau_{21} & P + \tau_{22} & \tau_{23} \\ \tau_{31} & \tau_{32} & P + \tau_{33} \end{bmatrix}$$

In terms of the unit tensor $\boldsymbol{\delta}$ [see Equation (8.8), p. 312], the above equation can be written compactly as

$$\boldsymbol{\pi} = \boldsymbol{\delta} P + \boldsymbol{\tau} \tag{2.18}$$

Stress as Momentum Flux

From Newton's second law of motion, force is the rate of change of momentum. Thus each element of $\boldsymbol{\pi}$, which is stress, or force per unit area, is in fact the rate of change of momentum per unit area, or momentum flux due to molecular forces. To derive momentum balance, it is convenient to combine $\boldsymbol{\pi}$ with the momentum flux due to bulk motion of the material, i.e., the convective momentum flux.

2.4.1 Convective Momentum Flux

Consider a material crossing a plane of area $\mathbf{A}$ with velocity $\mathbf{v}$, as shown in Figure 2.6 on the next page. Along the x_1-direction, the components of $\mathbf{A}$ and $\mathbf{v}$ are A_1 and v_1, respectively. Let ρ be the density of the material. Then in time t, the mass of the material that crosses A_1 is $\rho v_1 A_1 t$. The convective momentum of this material, written as a row vector, is

$$\mathbf{p}_1 = (\rho v_1 t A_1) \mathbf{v}^\top$$

The corresponding convective momentum flux is then given by

$$\frac{\mathbf{p}_1}{t\, A_1} = \rho v_1 \mathbf{v}^\top = \left[\underbrace{\rho v_1 v_1}_{x_1\text{-component}} \quad \underbrace{\rho v_1 v_2}_{x_2\text{-component}} \quad \underbrace{\rho v_1 v_3}_{x_3\text{-component}} \right]$$

Likewise, the convective momentum fluxes of the material across A_2 and A_3 (the components of $\mathbf{A}$, respectively, along the x_2- and x_3-directions) are given by, respectively,

$$\rho v_2 \mathbf{v}^\top = \left[\rho v_2 v_1 \quad \rho v_2 v_2 \quad \rho v_2 v_3 \right] \quad \text{and} \quad \rho v_3 \mathbf{v}^\top = \left[\rho v_3 v_1 \quad \rho v_3 v_2 \quad \rho v_3 v_3 \right]$$

Together, the three row vectors above form the convective momentum flux tensor written as

$$\rho \mathbf{v v} = \begin{bmatrix} \rho v_1 \mathbf{v}^\top \\ \rho v_2 \mathbf{v}^\top \\ \rho v_3 \mathbf{v}^\top \end{bmatrix} = \rho \begin{bmatrix} v_1 v_1 & v_1 v_2 & v_1 v_3 \\ v_2 v_1 & v_2 v_2 & v_2 v_3 \\ v_3 v_1 & v_3 v_2 & v_3 v_3 \end{bmatrix} = \rho \mathbf{v v}^\top$$

Note that $\mathbf{vv}$ in tensor notation implies $\mathbf{vv}^\top$, i.e., the matrix multiplication of $\mathbf{v}$ and its transpose.

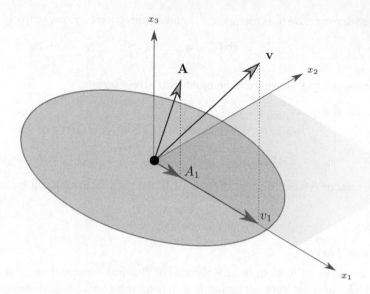

Figure 2.6   Velocity **v** of material across a plane of area **A**

2.4.2   *Total Momentum Flux*

The total momentum flux at the surface of a system is the sum of the molecular and convective momentum fluxes, and is given by

$$\boldsymbol{\phi} \;=\; \boldsymbol{\pi} + \rho\mathbf{v}\mathbf{v} \;=\; \boldsymbol{\delta}P + \boldsymbol{\tau} + \rho\mathbf{v}\mathbf{v}$$

$$= \begin{bmatrix} \underbrace{P + \tau_{11} + \rho v_1 v_1}_{\phi_{11}} & \underbrace{\tau_{12} + \rho v_1 v_2}_{\phi_{12}} & \underbrace{\tau_{13} + \rho v_1 v_3}_{\phi_{13}} \\[2ex] \underbrace{\tau_{21} + \rho v_2 v_1}_{\phi_{21}} & \underbrace{P + \tau_{22} + \rho v_2 v_2}_{\phi_{22}} & \underbrace{\tau_{23} + \rho v_2 v_3}_{\phi_{23}} \\[2ex] \underbrace{\tau_{31} + \rho v_3 v_1}_{\phi_{31}} & \underbrace{\tau_{32} + \rho v_3 v_2}_{\phi_{32}} & \underbrace{P + \tau_{33} + \rho v_3 v_3}_{\phi_{33}} \end{bmatrix}$$

$$= \begin{bmatrix} \underbrace{\begin{bmatrix} \phi_{11} & \phi_{12} & \phi_{13} \end{bmatrix}}_{\phi_1} \\[2ex] \underbrace{\begin{bmatrix} \phi_{21} & \phi_{22} & \phi_{23} \end{bmatrix}}_{\phi_2} \\[2ex] \underbrace{\begin{bmatrix} \phi_{31} & \phi_{32} & \phi_{33} \end{bmatrix}}_{\phi_3} \end{bmatrix} = \begin{bmatrix} \phi_1 \\[2ex] \phi_2 \\[2ex] \phi_3 \end{bmatrix} \qquad (2.19)$$

Thus, $\boldsymbol{\phi}$ is a tensor made of three row vectors, $\boldsymbol{\phi}_1$, $\boldsymbol{\phi}_2$ and $\boldsymbol{\phi}_3$. The vector $\boldsymbol{\phi}_1$, for example, is the total momentum flux acting on the x_2x_3-plane whose area is along the x_1-direction. The three components of ϕ_1 are: (i) ϕ_{11} along the x_1-direction, (ii) ϕ_{12} along the x_2-direction, and (iii) ϕ_{13} along the x_3-direction.

On the other hand, the ith column of ϕ is the set of total momentum fluxes acting on all three planes, and along the x_i-direction of motion. For example, the first column comprises ϕ_{11}, ϕ_{21} and ϕ_{31}, which

1. act, respectively, on the x_2x_3-, x_1x_3- and x_1x_2-planes, and

2. are along the x_1-direction.

2.4.3 *Macroscopic Balance*

This balance pertains to a macroscopic system, which is a uniform system with a number of ports for the entry and exit of mass, momentum, or energy. The system in this case is of mass m, and is subject to gravity **g** as well as a net external force **F** [see Figure 2.7 below].

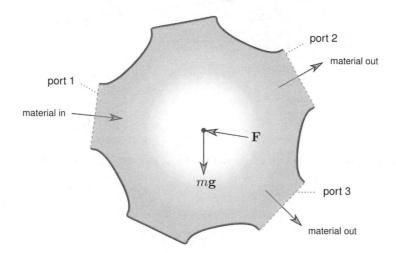

Figure 2.7 A macroscopic system with material flowing in and out through a number of ports, and subject to body forces

Applying Equation (2.3) on p. 20 to the momentum (**p**) of material in the system, we get

$$
\frac{d\mathbf{p}}{dt} = -\Delta[\dot{\mathbf{p}}] + \underbrace{\sum_{j=1}^{N_{\text{ext}}} \dot{\mathbf{p}}_{j,\text{ext}} + m\mathbf{g}}_{\mathbf{F}}
\tag{2.20}
$$

where $\dot{\mathbf{p}}_{j,\text{ext}}$ is the rate of momentum stemming from the jth external force (not due to gravity), **F** is the sum of such forces, and $m\mathbf{g}$ is the external force due to gravity.

Note that there is no generation term, and $\dot{\mathbf{p}}$ stands for the rate of momentum of material through a port of area **A**. A component of $\dot{\mathbf{p}}$, say, $\dot{p}_1$ along the x_1-direction, is obtained by summing the rates of momentum along the x_1-direction, and across all components (A_1, A_2

and A_3) of the port area $\mathbf{A}$. These momentum rates are contributed by the components in the first column of ϕ, which is the total momentum flux tensor defined by Equation (2.19) on p. 28. Figure 2.8 below shows these components (ϕ_{11}, ϕ_{21} and ϕ_{31}) for the port of area $\mathbf{A}$.

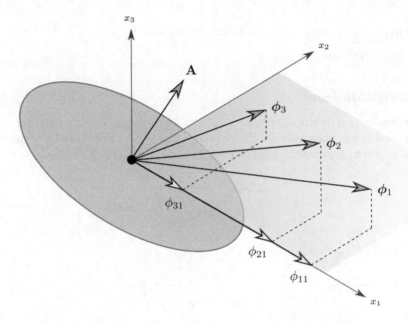

Figure 2.8 Components of ϕ in the x_1-direction – ϕ_{11}, ϕ_{21} and ϕ_{31} – for a port of area $\mathbf{A}$ in a macroscopic system

Thus, we can write

$$\dot{p}_1 \;=\; \sum_{j=1}^{3} A_j \phi_{j1} \;=\; A_1 P + \sum_{j=1}^{3} A_j (\rho v_j v_1 + \tau_{j1})$$

and similar equations for the other two components of $\dot{\mathbf{p}}$, namely, $\dot{p}_2$ and $\dot{p}_3$. The equations can be written together in the tensor notation as [refer Section 8.6.6, p. 310]

$$\dot{\mathbf{p}} \;=\; \mathbf{A}P + \mathbf{A} \cdot (\rho \mathbf{v}\mathbf{v} + \boldsymbol{\tau}) \;=\; \mathbf{A}P + (\mathbf{A} \cdot \rho \mathbf{v})\mathbf{v} + \mathbf{A} \cdot \boldsymbol{\tau}$$

The material is generally considered to flow perpendicularly across a port. Hence, $\mathbf{v} = v\hat{\mathbf{n}}$, where v is the average velocity, and $\hat{\mathbf{n}}$ is the unit vector normal to the port. Thus, $\mathbf{A} = A\hat{\mathbf{n}}$, where A is the magnitude of $\mathbf{A}$. Furthermore, shear stresses at a port are considered to be insignificant relative to pressure and convective flux. With these simplifications, we can write

$$\dot{\mathbf{p}} \;=\; A\hat{\mathbf{n}}P + (A\hat{\mathbf{n}} \cdot \rho v\hat{\mathbf{n}})v\hat{\mathbf{n}} \;=\; (P + \rho v^2)A\hat{\mathbf{n}}$$

Substituting the above equation in Equation (2.20) on the previous page, we finally obtain

$$\frac{d\mathbf{p}}{dt} = -\Delta\big[(P + \rho v^2)A\hat{\mathbf{n}}\big] + \mathbf{F} + m\mathbf{g} \tag{2.21}$$

which is the **macroscopic momentum balance**.

2.4.4 Microscopic Balance

Consider the momentum flux vectors – ϕ_1, ϕ_2 and ϕ_3 – acting on the six faces of differential element of fixed volume, as shown in Figure 2.9 below. The vector ϕ_2, which is along the x_2 direction, acts on the two parallel faces of area $\Delta A_2 = \Delta x_1 \Delta x_3$, and at locations x_2 and $x_2 + \Delta x_2$. The momentum rates into the system at x_2 due to ϕ_2 along the x_1-, x_2- and x_3-directions are, respectively, $\phi_{21}\Delta A_1$, $\phi_{22}\Delta A_1$ and $\phi_{23}\Delta A_1$. Note that there is no generation of momentum. Applying Equation (2.1) on p. 18 for the momentum $\Delta V \rho v_i$ of the differential

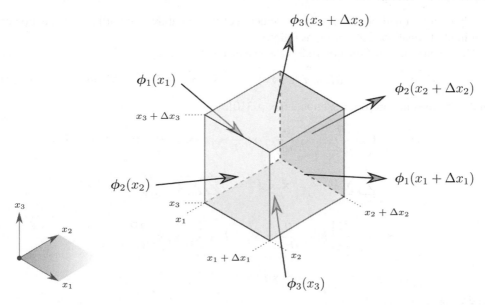

Figure 2.9 Molecular flux vectors – ϕ_1, ϕ_2 and ϕ_3 – acting on the six faces of the differential element of fixed volume in Cartesian coordinates

element along the x_i-direction, we obtain

$$\frac{d}{dt}(\Delta V \rho v_i) = \Delta V \frac{\partial}{\partial t}(\rho v_i) = \sum_{j=1}^{3}\Big[\phi_{ji}(x_j) - \phi_{ji}(x_j + \Delta x_j)\Big]\Delta A_j + \rho g_i \Delta V \tag{2.22}$$

where $\Delta V = \Delta x_1 \Delta x_2 \Delta x_3$ is the fixed volume of the element. In the limit of Δx_j tending to zero at any time t,

$$\phi_{ji}(x_j + \Delta x_j) = \phi_{ji}(x_j) + \frac{\partial \phi_{ji}}{\partial x_j}\Delta x_j$$

from the first order Taylor expansion. Using this result, Equation (2.22) on the previous page simplifies to

$$\frac{\partial(\rho v_i)}{\partial t} = -\sum_{j=1}^{3} \frac{\partial \phi_{ji}}{\partial x_j} + \rho g_i; \qquad i = 1, 2, 3$$

for the three directions. In the matrix notation the above set of equations can be written as

$$\frac{\partial(\rho \mathbf{v})}{\partial t} = -\nabla \cdot \phi + \rho \mathbf{g} \qquad (2.23)$$

The above equation is known **microscopic momentum balance**, or the **equation of motion**.

Equation of Change for Velocity

To obtain this equation, we expand the equation of motion above, and utilize the equation of continuity [Equation (2.8), p. 23] as follows.

The second term of Equation (2.23) above expands to

$$-\nabla \cdot \phi = -\nabla \cdot (\delta P + \tau + \rho \mathbf{vv}) = -\nabla \cdot \delta P - \nabla \cdot \tau - \nabla \cdot \rho \mathbf{vv} \qquad (2.24)$$

In the above equation [refer Section 8.6.6, p. 310],

$$\nabla \cdot \delta P = \left(\hat{\mathbf{x}}_1 \frac{\partial}{\partial x_1} + \hat{\mathbf{x}}_1 \frac{\partial}{\partial x_2} + \hat{\mathbf{x}}_3 \frac{\partial}{\partial x_3} \right) \cdot (\hat{\mathbf{x}}_1 \hat{\mathbf{x}}_1 P + \hat{\mathbf{x}}_2 \hat{\mathbf{x}}_2 P + \hat{\mathbf{x}}_3 \hat{\mathbf{x}}_3 P)$$

$$= \sum_{i=1}^{3} \sum_{j=1}^{3} \hat{\mathbf{x}}_i \frac{\partial}{\partial x_i} \cdot \hat{\mathbf{x}}_j \hat{\mathbf{x}}_j P = \sum_{i=1}^{3} \sum_{j=1}^{3} \hat{\mathbf{x}}_i \cdot \frac{\partial}{\partial x_i} (\hat{\mathbf{x}}_j \hat{\mathbf{x}}_j P)$$

$$= \sum_{i=1}^{3} \sum_{j=1}^{3} \left[\hat{\mathbf{x}}_i \cdot \underbrace{\frac{\partial}{\partial x_i} (\hat{\mathbf{x}}_j \hat{\mathbf{x}}_j)}_{\substack{=0 \\ \text{since } \partial \hat{\mathbf{x}}_j / \partial x_i = 0}} P + \underbrace{\hat{\mathbf{x}}_i \cdot \hat{\mathbf{x}}_j}_{= \delta_{ij}} \hat{\mathbf{x}}_j \frac{\partial P}{\partial x_i} \right]$$

where

$$\delta_{ij} = \begin{cases} 1, & \text{if } i = j \\ 0, & \text{if } i \neq j \end{cases}$$

Thus,

$$\nabla \cdot \delta P = \hat{\mathbf{x}}_1 \frac{\partial P}{\partial x_1} + \hat{\mathbf{x}}_2 \frac{\partial P}{\partial x_2} + \hat{\mathbf{x}}_3 \frac{\partial P}{\partial x_3} = \nabla P$$

To expand $\nabla \cdot \rho \mathbf{vv}$ in Equation (2.24) above, we use the following identity (see p. 315):

$$\nabla \cdot \mathbf{ab} = \mathbf{a} \cdot \nabla \mathbf{b} + \mathbf{b}(\nabla \cdot \mathbf{a}) \qquad (8.9)$$

With $\mathbf{a}$ as $\rho \mathbf{v}$, and $\mathbf{b}$ as $\mathbf{v}$

$$\nabla \cdot \rho \mathbf{vv} = \rho \mathbf{v} \cdot \nabla \mathbf{v} + \mathbf{v}(\nabla \cdot \rho \mathbf{v})$$

Thus, Equation (2.24) on the previous page becomes

$$-\nabla \cdot \phi \;=\; -\nabla P - \nabla \cdot \boldsymbol{\tau} - \rho \mathbf{v} \cdot \nabla \mathbf{v} - \mathbf{v}(\nabla \cdot \rho \mathbf{v})$$

Finally, applying the product rule of differentiation to the left-hand side of Equation (2.23) on the previous page, we get

$$\frac{\partial(\rho \mathbf{v})}{\partial t} \;=\; \underbrace{\frac{\partial \rho}{\partial t}}_{-\nabla \cdot \rho \mathbf{v}} \mathbf{v} + \rho \frac{\partial \mathbf{v}}{\partial t} \;=\; (-\nabla \cdot \rho \mathbf{v})\mathbf{v} + \rho \frac{\partial \mathbf{v}}{\partial t}$$

[from Equation (2.8), p. 23]

With the help of the last two equations, Equation (2.23) on the previous page simplifies to

$$\frac{\partial \mathbf{v}}{\partial t} \;=\; -\mathbf{v} \cdot \nabla \mathbf{v} - \frac{1}{\rho}(\nabla \cdot \boldsymbol{\tau} + \nabla P) + \mathbf{g} \qquad (2.25)$$

which is the **equation of change for velocity**.

2.5   Energy Balance

In this section, we will first derive the microscopic balance. We will then integrate it over the volume of the system to obtain the macroscopic balance. This approach helps in appreciating certain terms that arise at the molecular level, and that are better conceived in the microscopic balance.

2.5.1   Microscopic Balance

Consider the differential element of fixed volume, as shown in Figure 2.10 on the next page where e_i denotes the total energy flux across the differential area element ΔA_i perpendicular to the x_i-direction. Note that $\Delta A_i = \Delta x_j \Delta x_k$ where i, j and k are different from each other, and take values from the set $\{1, 2, 3\}$. Then the energy flux is given by

$$e_i \;=\; \underbrace{\rho\!\left(\frac{v^2}{2} + \hat{\Phi} + \hat{U}\right)\! v_i}_{\substack{\text{energy flux due to} \\ \text{moving mass}}} \;+\; \underbrace{\boldsymbol{\pi}_i \cdot \mathbf{v}}_{\substack{\text{energy flux due to} \\ \text{molecular forces}}} \;+\; \underbrace{q_i}_{\substack{\text{energy flux due to} \\ \text{heat conduction}}} \qquad (2.26)$$

where the energy flux due to moving mass stems from the kinetic, potential and internal energy of the mass crossing ΔA_i, where $\hat{\Phi}$ and $\hat{U}$ are the potential and internal energy per unit mass, respectively. The energy flux in the x_i-direction due to molecular forces is the dot product of the molecular stress $\boldsymbol{\pi}_i$ (i.e., shear stress acting on ΔA_i), and velocity.

Applying Equation (2.3) on p. 20 to the total energy of the differential element, we get

$$\frac{d}{dt}\underbrace{\left[\Delta V \rho\!\left(\frac{v^2}{2} + \hat{\Phi} + \hat{U}\right)\right]}_{\text{total energy of the element}} \;=\; \sum_{i=1}^{3}[e_i(x_i) - e_i(x_i + \Delta x_i)]\Delta A_i$$

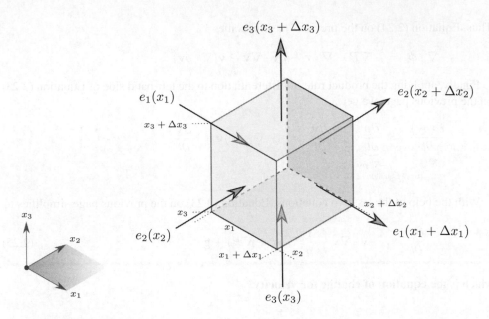

Figure 2.10 Components of the total energy flux vector – e_1, e_2 and e_3 – acting on the six faces of the differential element of fixed volume in Cartesian coordinates

Since $\Delta V = \Delta x_1 \Delta x_2 \Delta x_3$ is fixed, the above differential equation can be written as

$$\Delta V \frac{\mathrm{d}}{\mathrm{d}t}\left[\rho\left(\frac{v^2}{2} + \hat{\Phi} + \hat{U}\right)\right] = \sum_{i=1}^{3}[e_i(x_i) - e_i(x_i + \Delta x_i)]\Delta A_i$$

Using the first order Taylor expansion of $e_i(x_i + \Delta x_i)$ in the limit of Δx_i tending to zero at any time t, the above equation simplifies to

$$\frac{\partial}{\partial t}\left[\rho\left(\frac{v^2}{2} + \hat{\Phi} + \hat{U}\right)\right] = -\sum_{i=1}^{3}\frac{\partial e_i}{\partial x_i} \tag{2.27}$$

We express the right-hand side of the above equation with the help of Equation (2.26) on the previous page as

$$-\sum_{i=1}^{3}\frac{\partial e_i}{\partial x_i} = -\sum_{i=1}^{3}\frac{\partial}{\partial x_i}\left[\rho\left(\frac{v^2}{2} + \hat{\Phi} + \hat{U}\right)v_i + \boldsymbol{\pi}_i \cdot \mathbf{v} + q_i\right]$$

$$= \underbrace{-\sum_{i=1}^{3}\frac{\partial}{\partial x_i}\rho\left(\frac{v^2}{2} + \hat{\Phi} + \hat{U}\right)v_i}_{\nabla\cdot\rho\left(\frac{v^2}{2}+\hat{\Phi}+\hat{U}\right)\mathbf{v}} \underbrace{-\sum_{i=1}^{3}\frac{\partial}{\partial x_i}\sum_{j=1}^{3}\pi_{ij}v_j}_{\nabla\cdot P\mathbf{v}+\nabla\cdot\boldsymbol{\tau}\cdot\mathbf{v}} \underbrace{-\sum_{i=1}^{3}\frac{\partial q_i}{\partial x_i}}_{\nabla\cdot\mathbf{q}}$$

where, using the definition of $\boldsymbol{\pi}$ [see Equation (2.18), p. 27],

$$\sum_{i=1}^{3}\frac{\partial}{\partial x_i}\sum_{j=1}^{3}\pi_{ij}v_j = \sum_{i=1}^{3}\frac{\partial}{\partial x_i}\sum_{j=1}^{3}(\delta_{ij}P+\tau_{ij})v_j = \sum_{i=1}^{3}\frac{\partial}{\partial x_i}(Pv_i)+\sum_{i=1}^{3}\frac{\partial}{\partial x_i}\underbrace{\sum_{j=1}^{3}\tau_{ij}v_j}_{\tau_i\cdot\mathbf{v}}$$

$$= \nabla\cdot P\mathbf{v}+\nabla\cdot\boldsymbol{\tau}\cdot\mathbf{v}$$

Utilizing the above results, Equation (2.27) on the previous page can be finally written as

$$\frac{\partial}{\partial t}\left[\rho\left(\frac{v^2}{2}+\hat{\Phi}+\hat{U}\right)\right] = -\nabla\cdot\left[\rho\left(\frac{v^2}{2}+\hat{\Phi}+\hat{U}\right)\mathbf{v}\right]-\nabla\cdot P\mathbf{v}-\nabla\cdot\boldsymbol{\tau}\cdot\mathbf{v}-\nabla\cdot\mathbf{q}$$

$$(2.28)$$

The above equation is the **microscopic energy balance**.

2.5.2 *Macroscopic Balance*

This energy balance can be obtained by integrating Equation (2.28) above over the entire macroscopic system of which the differential element is a part. Figure 2.11 below shows a macroscopic system, which has a number of ports for the entry and exit of mass and energy. At any port, the velocity and heat flux are perpendicular to the cross-section area of the port. As shown in the figure, the boundary surface of the system is made of the fixed surface of wall, its conductive part or port, the ports of material transfer, and one or more moving surfaces.

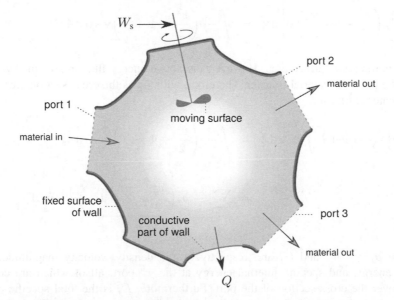

Figure 2.11 A macroscopic system with material flowing in and out through a number of ports and subject to the transfer of heat (Q) and shaft work (W_s)

While the ports are open to the exchange of mass, momentum and energy, moving surfaces such as the surface of the impeller of the mixer only allow momentum to be transferred. This momentum contributes shaft work to the system. Of course, nothing crosses the fixed surface.

Integrating Equation (2.28) on the previous page over the macroscopic system yields

$$\underbrace{\int_V \frac{\partial}{\partial t}\left[\rho\left(\frac{v^2}{2}+\hat{\Phi}+\hat{U}\right)\right]dV}_{\text{first term}} = \underbrace{\int_V -\nabla\cdot\left[\rho\left(\frac{v^2}{2}+\hat{\Phi}+\hat{U}\right)\mathbf{v}\right]dV}_{\text{second term}} + \underbrace{\int_V -\nabla\cdot P\mathbf{v}\,dV}_{\text{third term}}$$

$$+ \underbrace{\int_V -\nabla\cdot\tau\cdot\mathbf{v}\,dV}_{\text{fourth term}} + \underbrace{\int_V -\nabla\cdot\mathbf{q}\,dV}_{\text{fifth term}} \qquad (2.29)$$

Applying the general transport theorem for fixed volume [see Equation (2.65), p. 53] to the first term in the above equation, we obtain

$$\int_V \frac{\partial}{\partial t}\left[\rho\left(\frac{v^2}{2}+\hat{\Phi}+\hat{U}\right)\right]dV = \frac{d}{dt}\overbrace{\int_V\left[\rho\left(\frac{v^2}{2}+\hat{\Phi}+\hat{U}\right)\right]dV}^{E} \equiv \frac{dE}{dt}$$

where E is the total energy of the macroscopic system.

Applying the divergence theorem [see Equation (2.55), p. 48] to the second term of Equation (2.29) above, we obtain

$$\int_V -\nabla\cdot\left[\rho\left(\frac{v^2}{2}+\hat{\Phi}+\hat{U}\right)\mathbf{v}\right]dV = \int_A -\rho\left(\frac{v^2}{2}+\hat{\Phi}+\hat{U}\right)\mathbf{v}\cdot\hat{\mathbf{n}}\,dA$$

where A is the total surface area. However, the above energy flux crosses the system only through the ports of material transfer. Therefore, with A_p as the cross-section area of all N_p ports of material transfer,

$$\int_A -\rho\left(\frac{v^2}{2}+\hat{\Phi}+\hat{U}\right)\mathbf{v}\cdot\hat{\mathbf{n}}\,dA = \int_{A_p} -\rho\left(\frac{v^2}{2}+\hat{\Phi}+\hat{U}\right)\mathbf{v}\cdot\hat{\mathbf{n}}\,dA_p$$

$$= \sum_{j=1}^{N_p}\left[-\rho_j\underbrace{\left(\frac{v_j^2}{2}+\hat{\Phi}_j+\hat{U}_j\right)}_{\hat{E}_j}\mathbf{v}_j\cdot\underbrace{\hat{\mathbf{n}}_j A_j}_{\mathbf{A}_j}\right]$$

where the ρ_j, v_j, $\hat{\Phi}_j$ and $\hat{U}_j$ are, respectively, the density, velocity magnitude, specific potential energy, and specific internal energy at the jth port; all of which are considered uniform over the cross-section of the port. Furthermore, $\hat{E}_j$ is the total specific energy of material, which moves with velocity $\mathbf{v}_j$ through the jth port. As shown in Figure 2.12 on the next page, the area of the port is $\mathbf{A}_j = A_j\hat{\mathbf{n}}_j$ where A_j is the magnitude, and $\hat{\mathbf{n}}_j$ is unit vector normal to the port. Note that $\mathbf{v}_j$ and $\mathbf{A}_j$ are collinear. Taking all surface area elements of the

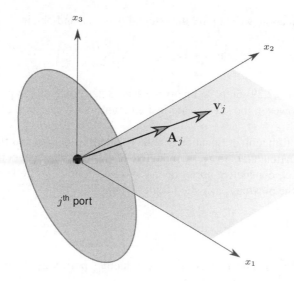

Figure 2.12 The area and velocity vectors through the jth port of material transfer

macroscopic system as vectors pointing perpendicularly outward from the body,

$$\mathbf{v}_j \cdot \hat{\mathbf{n}}_j A_j = \begin{cases} -v_j A_j & \text{at the entry port where } \mathbf{v}_j \text{ is opposite to } \mathbf{A}_j \\ v_j A_j & \text{at the exit port where } \mathbf{v}_j \text{ is along } \mathbf{A}_j \end{cases}$$

With this consideration, the second term of Equation (2.29) on the last page simplifies as follows:

$$\int_{A_p} -\rho \left(\frac{v^2}{2} + \hat{\Phi} + \hat{U} \right) \mathbf{v} \cdot \hat{\mathbf{n}} \, dA_p = \sum_{j=1}^{N_{\text{in}}} (\rho_j \hat{E}_j v_j A_j)_{\text{in}} - \sum_{j=1}^{N_{\text{out}}} (\rho_j \hat{E}_j v_j A_j)_{\text{out}}$$

$$= -\Delta \left[\rho \hat{E} v A \right] = -\Delta \left[\dot{m} \hat{E} \right] \qquad (2.30)$$

where $\dot{m}$ is the total mass flow rate through a port. Note that we have used the delta notation of Equation (2.2) on p. 20 in the last step.

The third term of Equation (2.29) involves momentum flow across all ports of material transfer as well as moving surfaces. Applying the divergence theorem [see Equation (2.55), p. 48], we get

$$\int_V -\nabla \cdot P\mathbf{v} \, dV = \int_A -P\mathbf{v} \cdot \hat{\mathbf{n}} \, dA = \int_{A_p} -P\mathbf{v} \cdot \hat{\mathbf{n}} \, dA_p + \underbrace{\int_{A_m} -P\mathbf{v} \cdot \hat{\mathbf{n}} \, dA_m}_{\equiv \dot{W}_s} \qquad (2.31)$$

where A_m is the area of all moving surfaces, and $\dot{W}_s$ denotes the rate of shaft work done on the system. Similar to the input terms in Equation (2.30) above, $\dot{W}_s$ is positive. Following the steps used to derive that equation, we simplify the integral over A_p, and obtain

$$\int_V -\nabla \cdot Pv \, dV \;=\; -\Delta[PvA] + \dot{W}_s \;=\; -\Delta\left[\dot{m}\frac{P}{\rho}\right] + \dot{W}_s \tag{2.32}$$

The fourth term of Equation (2.29) on p. 36 involves shear stresses across the ports as well as the moving surfaces. However, the stresses are insignificant relative to pressure, and that term may be ignored.

We are now left with the fifth term of Equation (2.29). This term is not much different from the second term of that equation, and involves conductive heat flux across and perpendicular to the parts of the wall (i.e., conductive ports) of the system. Following the steps used to derive Equation (2.30) on the previous page, the fifth term simplifies to

$$\int_A -\mathbf{q} \cdot \hat{\mathbf{n}} \, dA \;=\; -\Delta\left[\dot{Q}\right] \tag{2.33}$$

where $\dot{Q}$ is the rate of heat conduction across a conductive port. Based on the above simplifications, Equation (2.29) finally becomes

$$\frac{dE}{dt} \;=\; -\Delta\left[\dot{m}\hat{E}\right] - \Delta\left[\dot{m}\frac{P}{\rho}\right] - \Delta\left[\dot{Q}\right] + \dot{W}_s \tag{2.34}$$

which is the **macroscopic energy balance**.

Remarks

The microscopic energy balance obtained earlier is the starting point for the derivation of the equation of change for temperature, which is an important measurable property related to the energy content of the system, and is of immense practical interest. Next, we need to derive the equation of change for kinetic and potential energy combined. Subtraction of this equation from the energy balance will result in the equation of change for internal energy. This equation, in conjunction with the thermodynamics relation between temperature and internal energy, will eventually yield the equation of change for temperature.

2.6 Equation of Change for Kinetic and Potential Energy

We first derive the microscopic equation.

2.6.1 Microscopic Equation

The dot product of the equation of motion [Equation (2.23), p. 32] and $\mathbf{v}$ is given by

$$\mathbf{v} \cdot \frac{\partial(\rho\mathbf{v})}{\partial t} \;=\; -\mathbf{v} \cdot \nabla \cdot \phi + \mathbf{v} \cdot \rho\mathbf{g}$$

which is obtained by summing the following three equations:

$$v_i \frac{\partial(\rho v_i)}{\partial t} \;=\; -v_i \sum_{j=1}^{3} \frac{\partial \phi_{ji}}{\partial x_j} + v_i \rho g_i; \qquad i = 1, 2, 3 \tag{2.35}$$

We will now expand the terms of Equation (2.35) on the previous page, utilizing the product rule of differentiation as well as its rearrangement, i.e.,

$$a\frac{\partial b}{\partial y} = \frac{\partial(ab)}{\partial y} - b\frac{\partial a}{\partial y}$$

The first term of Equation (2.35) can be written as

$$v_i\frac{\partial(\rho v_i)}{\partial t} = v_i^2\frac{\partial\rho}{\partial t} + \underbrace{v_i\rho\frac{\partial v_i}{\partial t}}_{\frac{\partial(\rho v_i^2/2)}{\partial t} - \frac{v_i^2}{2}\frac{\partial\rho}{\partial t}} = \frac{v_i^2}{2}\underbrace{\frac{\partial\rho}{\partial t}}_{-\nabla\cdot\rho\mathbf{v}} + \frac{\partial}{\partial t}\left(\frac{\rho v_i^2}{2}\right)$$

[from Equation (2.8), p. 23]

$$= -\frac{v_i^2}{2}\sum_{j=1}^{3}\frac{\partial}{\partial x_j}(\rho v_j) + \frac{\partial}{\partial t}\left(\frac{\rho v_i^2}{2}\right) = -\sum_{j=1}^{3}\underbrace{\frac{v_i^2}{2}\frac{\partial}{\partial x_j}(\rho v_j)}_{\left[\frac{\partial}{\partial x_j}(\rho v_j v_i^2/2) - \rho v_j\frac{\partial}{\partial x_j}(v_i^2/2)\right]} + \frac{\partial}{\partial t}\left(\frac{\rho v_i^2}{2}\right)$$

$$= -\sum_{j=1}^{3}\left[\frac{\partial}{\partial x_j}\left(\rho v_j\frac{v_i^2}{2}\right) - \rho v_j v_i\frac{\partial v_i}{\partial x_j}\right] + \frac{\partial}{\partial t}\left(\frac{\rho v_i^2}{2}\right)$$

With the help of Equation (2.19) on p. 28, the second term of Equation (2.35) expands as

$$-v_i\sum_{j=1}^{3}\frac{\partial\phi_{ji}}{\partial x_j} = \underbrace{-v_i\frac{\partial P}{\partial x_i}}_{-\frac{\partial(Pv_i)}{\partial x_i} + P\frac{\partial v_i}{\partial x_i}} - \sum_{j=1}^{3}\left[\underbrace{v_i\frac{\partial\tau_{ji}}{\partial x_j}}_{\frac{\partial(v_i\tau_{ji})}{\partial x_j} - \tau_{ji}\frac{\partial v_i}{\partial x_j}} + \underbrace{v_i\frac{\partial}{\partial x_j}(\rho v_j v_i)}_{\frac{\partial(v_i\rho v_j v_i)}{\partial x_j} - \rho v_j v_i\frac{\partial v_i}{\partial x_j}}\right]$$

$$= -\frac{\partial}{\partial x_i}(Pv_i) + P\frac{\partial v_i}{\partial x_i} - \sum_{j=1}^{3}\left[\frac{\partial}{\partial x_j}(v_i\tau_{ji}) - \tau_{ji}\frac{\partial v_i}{\partial x_j} + \right.$$

$$\left.\frac{\partial}{\partial x_j}(\rho v_j v_i^2) - \rho v_j v_i\frac{\partial v_i}{\partial x_j}\right]$$

Finally, assuming that close to the earth's surface, gravity is equal to the gradient of the gravitational potential $\hat{\Phi}$, i.e.,

$$g_i = \frac{\partial\hat{\Phi}}{\partial x_i}$$

the last term of Equation (2.35) can be written as

$$v_i\rho g_i = -v_i\rho\frac{\partial\hat{\Phi}}{\partial x_i} = -\frac{\partial(\rho v_i\hat{\Phi})}{\partial x_i} + \hat{\Phi}\frac{\partial(v_i\rho)}{\partial x_i}$$

Substituting in Equation (2.35), the expressions obtained above for its three terms, we obtain upon simplification

$$\frac{\partial}{\partial t}\left(\frac{\rho v_i^2}{2}\right) = -\frac{\partial}{\partial x_i}(Pv_i) + P\frac{\partial v_i}{\partial x_i} - \sum_{j=1}^{3}\left[\frac{\partial}{\partial x_j}(v_i\tau_{ji}) - \tau_{ji}\frac{\partial v_i}{\partial x_j} + \frac{\partial}{\partial x_j}\left(\frac{\rho v_j v_i^2}{2}\right)\right]$$

$$- \frac{\partial(\rho v_i\hat{\Phi})}{\partial x_i} + \hat{\Phi}\frac{\partial(v_i\rho)}{\partial x_i}; \qquad i = 1,2,3$$

Summing the above set of three equations, we obtain the dot product of the equation of motion [Equation (2.23) on p. 32 and $\mathbf{v}$], i.e.,

$$\frac{\partial}{\partial t}\left(\frac{\rho}{2}\overbrace{\sum_{i=1}^{3}v_i^2}^{v^2}\right) = \underbrace{-\sum_{i=1}^{3}\frac{\partial}{\partial x_i}(Pv_i)}_{\nabla\cdot P\mathbf{v}} + \underbrace{P\sum_{i=1}^{3}\frac{\partial v_i}{\partial x_i}}_{\nabla\cdot\mathbf{v}} - \underbrace{\sum_{j=1}^{3}\frac{\partial}{\partial x_j}\left(\sum_{i=1}^{3}v_i\overbrace{\tau_{ji}}^{\tau_{ij}}\right)}_{\nabla\cdot\boldsymbol{\tau}\cdot\mathbf{v}}^{\boldsymbol{\tau}\cdot\mathbf{v}}$$

$$+\underbrace{\sum_{i=1}^{3}\sum_{j=1}^{3}\tau_{ji}\frac{\partial v_i}{\partial x_j}}_{\boldsymbol{\tau}:\nabla\mathbf{v}} - \underbrace{\sum_{j=1}^{3}\frac{\partial}{\partial x_j}\left(\frac{\rho v_j}{2}\overbrace{\sum_{i=1}^{3}v_i^2}^{v^2}\right)}_{\nabla\cdot\frac{\rho v^2}{2}\mathbf{v}}$$

$$-\underbrace{\sum_{i=1}^{3}\frac{\partial(\rho\hat{\Phi}v_i)}{\partial x_i}}_{\nabla\cdot\rho\hat{\Phi}\mathbf{v}} + \underbrace{\hat{\Phi}\sum_{i=1}^{3}\frac{\partial(\rho v_i)}{\partial x_i}}_{\nabla\cdot\rho\mathbf{v}}$$

where v is the magnitude of the velocity, and $\boldsymbol{\tau}$ is taken to be symmetric, i.e., $\tau_{ij}=\tau_{ji}$. In the matrix notation,

$$\frac{\partial}{\partial t}\left(\frac{\rho v^2}{2}\right) = -\nabla\cdot P\mathbf{v} + P\nabla\cdot\mathbf{v} - \nabla\cdot\boldsymbol{\tau}\cdot\mathbf{v} + \boldsymbol{\tau}:\nabla\mathbf{v} - \nabla\cdot\rho\left(\frac{v^2}{2}+\hat{\Phi}\right)\mathbf{v} + \hat{\Phi}\nabla\cdot\rho\mathbf{v}$$

where $\boldsymbol{\tau}:\nabla\mathbf{v}$ is the double dot product of $\boldsymbol{\tau}$ and $\nabla\mathbf{v}$ [see Section 8.6.6, p. 317]. The above equation is the **equation of change for kinetic energy** whose last term can be written as

$$\hat{\Phi}\nabla\cdot\rho\mathbf{v} = -\hat{\Phi}\frac{\partial\rho}{\partial t} = -\frac{\partial(\rho\hat{\Phi})}{\partial t}$$

based on the equation of continuity [Equation (2.8), p. 23], and time-independent $\hat{\Phi}$ sufficiently close to the earth's surface. With these considerations, we obtain

$$\frac{\partial}{\partial t}\left[\rho\left(\frac{v^2}{2}+\hat{\Phi}\right)\right] = -\nabla\cdot P\mathbf{v} + P\nabla\cdot\mathbf{v} - \nabla\cdot\boldsymbol{\tau}\cdot\mathbf{v} + \boldsymbol{\tau}:\nabla\mathbf{v} - \nabla\cdot\rho\left(\frac{v^2}{2}+\hat{\Phi}\right)\mathbf{v}$$

$$(2.36)$$

which is the **equation of change for kinetic and potential energy**.

2.6.2 Macroscopic Equation

The macroscopic equation for the combined kinetic and potential energy is obtained similar to Equation (2.34) on p. 38. The integral of $\nabla\cdot\boldsymbol{\tau}\cdot\mathbf{v}$ is discarded. Thus, the integration of Equation (2.36) on the previous page over the system volume yields the **microscopic equation of change for kinetic and potential energy,**

$$
\frac{\mathrm{d}}{\mathrm{d}t}\left[\rho\left(\frac{v^2}{2}+\hat{\Phi}\right)\right] = -\Delta\left[\dot{m}\left(\frac{v^2}{2}+\hat{\Phi}\right)\right] - \Delta\left[\dot{m}\frac{P}{\rho}\right] + \dot{W}_s
$$

$$
\underbrace{-\int_V -(P\nabla\cdot\mathbf{v})\,\mathrm{d}V}_{\equiv\,\dot{E}_c} \;\; \underbrace{-\int_V (-\boldsymbol{\tau}:\nabla\mathbf{v})\,\mathrm{d}V}_{\equiv\,\dot{E}_v} \tag{2.37}
$$

where the last two terms, $\dot{E}_c$ and $\dot{E}_v$, cannot be simplified, and are the rates of conversion of kinetic and potential energy, respectively, to thermal energy. While $\dot{E}_v$ is due to viscous effects and is positive, $\dot{E}_c$ is due to fluid compressibility and could be positive or negative. For incompressible fluids, E_c is zero.

2.7 Equation of Change for Temperature

We first obtain the microscopic equation of change for temperature.

2.7.1 Microscopic Equation

Subtracting the Equation (2.36) from Equation (2.28) on p. 35 yields

$$
\frac{\partial}{\partial t}(\rho\hat{U}) = -\nabla\cdot\rho\hat{U}\mathbf{v} - \nabla\cdot\mathbf{q} - P\nabla\cdot\mathbf{v} - \boldsymbol{\tau}:\nabla\mathbf{v} \tag{2.38}
$$

which is the equation of change for the internal energy of the system.

Next, we introduce specific enthalpy ($\hat{H}$) in the above equation. Since $\hat{U} = \hat{H} - P/\rho$, the first two terms of Equation (2.38) above expand as

$$
\frac{\partial(\rho\hat{U})}{\partial t} = \frac{\partial}{\partial t}\left[\rho\left(\hat{H}-\frac{P}{\rho}\right)\right] = \frac{\partial(\rho\hat{H})}{\partial t} - \frac{\partial P}{\partial t} \quad \text{and}
$$

$$
-\nabla\cdot\rho\hat{U}\mathbf{v} = -\nabla\cdot\rho\left(\hat{H}-\frac{P}{\rho}\right)\mathbf{v} = -\nabla\cdot\rho\hat{H}\mathbf{v} + \underbrace{\nabla\cdot P\mathbf{v}}_{\substack{(P\nabla\cdot\mathbf{v}+\nabla P\cdot\mathbf{v}) \\ \text{[see Equation (8.10), p. 317]}}}
$$

Substituting the above expansions in Equation (2.38) above, and expanding the time derivative yields

$$
\rho\frac{\partial\hat{H}}{\partial t} = -\hat{H}\frac{\partial\rho}{\partial t} - \nabla\cdot\rho\hat{H}\mathbf{v} - \nabla\cdot\mathbf{q} + \frac{\partial P}{\partial t} + \nabla P\cdot\mathbf{v} - \boldsymbol{\tau}:\nabla\mathbf{v} \tag{2.39}
$$

Assuming that the change in enthalpy is between successive states of system under equilibrium, $\partial \hat{H}/\partial t$ in the last equation is given by Equation (2.54) on p. 47 [see Appendix 2.A, p. 44]. Utilizing this result, we obtain

$$\rho \hat{C}_P \frac{\partial T}{\partial t} = -\nabla \cdot \rho \hat{H} \mathbf{v} - \nabla \cdot \mathbf{q} - \frac{T}{\rho} \frac{\partial \rho}{\partial T}\bigg|_{P,m} \frac{\partial P}{\partial t} - \underbrace{\hat{H}}_{\sum_{i=1}^{N_c} \omega_i \tilde{H}_i} \frac{\partial \rho}{\partial t} - \sum_{i=1}^{N_c} \tilde{H}_i \rho \frac{\partial \omega_i}{\partial t}$$

$$+ \nabla P \cdot \mathbf{v} - \boldsymbol{\tau} : \nabla \mathbf{v} \qquad \text{[see Equation (2.51), p. 47]}$$

where $\tilde{H}_i$ is the partial specific enthalpy of the ith species in a mixture of N_c species. In the above equation, substitution of $\partial \omega_i/\partial t$ from Equation (2.11) on p. 23 yields after simplification

$$\rho \hat{C}_P \frac{\partial T}{\partial t} = -\nabla \cdot \rho \hat{H} \mathbf{v} - \nabla \cdot \mathbf{q} - \frac{T}{\rho} \frac{\partial \rho}{\partial T}\bigg|_{P,m} \frac{\partial P}{\partial t} + \sum_{i=1}^{N_c} \tilde{H}_i \nabla \cdot \omega_i \rho \mathbf{v}$$

$$+ \sum_{i=1}^{N_c} \tilde{H}_i \nabla \cdot \mathbf{j}_i - \sum_{i=1}^{N_c} \sum_{j=1}^{N_r} \tilde{H}_i r_{\text{gen},ij} + \nabla P \cdot \mathbf{v} - \boldsymbol{\tau} : \nabla \mathbf{v} \qquad (2.40)$$

where the term involving $r_{\text{gen},ij}$ accounts for the enthalpy change due to the change in system composition because of chemical reactions. The above equation gets further simplified to [see Exercise 2.5, p. 58]

$$\frac{\partial T}{\partial t} = -\mathbf{v} \cdot \nabla T - \frac{1}{\rho \hat{C}_P} \left(\nabla \cdot \mathbf{q} + \frac{T}{\rho} \frac{\partial \rho}{\partial T}\bigg|_{P,m} \frac{\mathrm{D}P}{\mathrm{D}t} - \sum_{i=1}^{N_c} \tilde{H}_i \nabla \cdot \mathbf{j}_i \right.$$

$$\left. + \sum_{i=1}^{N_c} \sum_{j=1}^{N_r} \tilde{H}_i r_{\text{gen},ij} + \boldsymbol{\tau} : \nabla \mathbf{v} \right) \qquad (2.41)$$

which is the **microscopic equation of change for temperature**:

2.7.2 *Macroscopic Equation*

Let us consider a commonly encountered macroscopic system, which has the following attributes:

1. The material in the system is well-mixed so that T, ρ, $\hat{C}_P$ and $\tilde{H}_i$s are uniform, i.e., independent of space. Moreover, ρ does not change with time.

2. The system pressure is constant and uniform as well as viscous dissipation is negligible.

3. The mass and heat cross the system boundaries at relevant entry and exit ports. At any port, the applicable velocity, or heat flux is perpendicular to the cross-section area of the port.

4. Diffusive fluxes of species at a port are insignificant in comparison to the bulk flux.

Thus, for this system Equation (2.40) on the previous page simplifies to

$$\rho\hat{C}_\mathrm{P}\frac{\partial T}{\partial t} = -\nabla\cdot\rho\hat{H}\mathbf{v} - \nabla\cdot\mathbf{q} + \sum_{i=1}^{N_\mathrm{c}}\tilde{H}_i\nabla\cdot\omega_i\rho\mathbf{v} + \sum_{i=1}^{N_\mathrm{c}}\tilde{H}_i\nabla\cdot\mathbf{j}_i - \sum_{i=1}^{N_\mathrm{c}}\sum_{j=1}^{N_\mathrm{r}}\tilde{H}_i r_{\mathrm{gen},ij} - \boldsymbol{\tau}:\nabla\mathbf{v}$$

The volume integral of the above equation yields the macroscopic equation of change for temperature. The approach is similar to that used earlier for energy balances [refer Section 2.5.2, p. 35]. Following this approach, the volume integral is given by

$$\underbrace{\int_V \rho\hat{C}_\mathrm{P}\frac{\partial T}{\partial t}\,\mathrm{d}V}_{\text{first term}} = \underbrace{\int_V -\nabla\cdot\rho\hat{H}\mathbf{v}\,\mathrm{d}V}_{\text{second term}} + \underbrace{\int_V -\nabla\cdot\mathbf{q}\,\mathrm{d}V}_{\text{third term}} + \underbrace{\int_V \sum_{i=1}^{N_\mathrm{c}}\tilde{H}_i\nabla\cdot\omega_i\rho\mathbf{v}\,\mathrm{d}V}_{\text{fourth term}}$$
$$+ \underbrace{\int_V \sum_{i=1}^{N_\mathrm{c}}\tilde{H}_i\nabla\cdot\mathbf{j}_i\,\mathrm{d}V}_{\text{fifth term}} + \underbrace{\int_V \sum_{i=1}^{N_\mathrm{c}}\sum_{j=1}^{N_\mathrm{r}}(-\tilde{H}_i r_{\mathrm{gen},ij})\,\mathrm{d}V}_{\text{sixth term}}$$
$$+ \underbrace{\int_V (-\boldsymbol{\tau}:\nabla\mathbf{v})\,\mathrm{d}V}_{\text{seventh term}} \tag{2.42}$$

Since T, ρ and $\hat{C}_\mathrm{P}$ are uniform, the first term of the above equation simplifies as follows:

$$\int_V \rho\hat{C}_\mathrm{P}\frac{\partial T}{\partial t}\,\mathrm{d}V = \underbrace{\rho\hat{C}_P\int_V \frac{\partial T}{\partial t}\,\mathrm{d}V = \rho\hat{C}_\mathrm{P}\frac{\mathrm{d}}{\mathrm{d}t}\int_V T\,\mathrm{d}V}_{\substack{\text{from the general transport theorem}\\ \text{[Equation (2.65) on p. 53]}}} = \rho\hat{C}_\mathrm{P}\frac{\mathrm{d}T}{\mathrm{d}t}V = m\hat{C}_\mathrm{P}\frac{\mathrm{d}T}{\mathrm{d}t}$$

Note that we have applied the general transport theorem for the fixed volume V.

Applying the divergence theorem [see Equation (2.55), p. 48], the second term of Equation (2.42) above is equal to

$$\int_A -\rho\hat{H}\mathbf{v}\cdot\hat{\mathbf{n}}\,\mathrm{d}A = -\int_{A_\mathrm{p}} \rho\hat{H}\mathbf{v}\cdot\hat{\mathbf{n}}\,\mathrm{d}A_\mathrm{p} - \int_{A_\mathrm{m}} \rho\underbrace{\hat{H}}_{(\hat{U}+\frac{P}{\rho})}\mathbf{v}\cdot\hat{\mathbf{n}}\,\mathrm{d}A_\mathrm{m}$$
$$= \underbrace{-\int_{A_\mathrm{p}} \rho\hat{H}\mathbf{v}\cdot\hat{\mathbf{n}}\,\mathrm{d}A_\mathrm{p}}_{\substack{-\Delta[\dot{m}\hat{H}]\\ \text{[see Equation (2.30), p. 37]}}} - \underbrace{\int_{A_\mathrm{m}} \rho\hat{U}\mathbf{v}\cdot\hat{\mathbf{n}}\,\mathrm{d}A_\mathrm{m}}_{\substack{\text{0, since no material}\\ \text{flows through}}} + \underbrace{\int_{A_\mathrm{m}} -P\mathbf{v}\cdot\hat{\mathbf{n}}\,\mathrm{d}A_\mathrm{m}}_{\substack{\text{rate of shaft work, } \dot{W}_\mathrm{s}\\ \text{[see Equation (2.31), p. 37]}}}$$
$$= -\Delta[\dot{m}\hat{H}] + \dot{W}_\mathrm{s}$$

where A_p and A_m are, respectively, the areas of ports of material transfer and moving surfaces, and $\dot{m}$ is the overall bulk mass flow rate through a given port.

The third term of Equation (2.42) on the previous page is given by Equation (2.33) on p. 38. The fourth term of Equation (2.42) on the previous page gets simplified to [see Equation (2.30), p. 37]

$$
\int_{A_p} \sum_{i=1}^{N_c} \tilde{H}_i \omega_i \rho \mathbf{v} \cdot \hat{\mathbf{n}} \, \mathrm{d}A_p \;=\; \sum_{i=1}^{N_c} \tilde{H}_i \Delta \big[\underbrace{\omega_i \rho v A}_{\dot{m}_i} \big] \;=\; \sum_{i=1}^{N_c} \tilde{H}_i \Delta \big[\dot{m}_i \big]
$$

where (i) $\dot{m}_i$ is the bulk mass flow rate of the ith species through a given port, and (ii) $\tilde{H}_i$ is the property of the species *inside* the macroscopic system at system conditions.

Similarly, the fifth term of Equation (2.42) on the previous page gets simplified to

$$
\int_{A} \sum_{i=1}^{N_c} \tilde{H}_i \mathbf{j}_i \cdot \hat{\mathbf{n}} \, \mathrm{d}A \;=\; \sum_{i=1}^{N_c} \tilde{H}_i \Delta \big[\underbrace{j_i A}_{\underset{\sim}{m}_i} \big] \;=\; \sum_{i=1}^{N_c} \tilde{H}_i \Delta \big[\underset{\sim}{\dot{m}}_i \big] \;\approx\; 0
$$

where $\underset{\sim}{\dot{m}}_i$ is the mass flow rate of the ith species through a given port due to diffusion. Since $\underset{\sim}{\dot{m}}_i$s are considered insignificant relative to $\dot{m}_i$ at a port, the fifth term can be dropped.

The sixth term of Equation (2.42) simplifies as follows:

$$
\int_{V} \sum_{j=1}^{N_r} \sum_{i=1}^{N_c} (-\tilde{H}_i r_{\text{gen},ij}) \, \mathrm{d}V \;=\; V \sum_{j=1}^{N_r} \sum_{i=1}^{N_c} (-\tilde{H}_i r_{\text{gen},ij})
$$

where V is the volume of the macroscopic system.

Finally, the seventh term of Equation (2.42) on the previous page is $\dot{E}_v$ [see Equation (2.37), p. 41]. This term can be dropped since the viscous dissipation is assumed to be negligible. Thus, Equation (2.42) becomes

$$
m\hat{C}_P \frac{\mathrm{d}T}{\mathrm{d}t} \;=\; -\Delta\big[\dot{m}\hat{H}\big] - \Delta\big[\dot{Q}\big] + \sum_{i=1}^{N_c} \tilde{H}_i \Delta\big[\dot{m}_i\big] + V \sum_{j=1}^{N_r} \sum_{i=1}^{N_c} (-\tilde{H}_i r_{\text{gen},ij}) + \dot{W}_s
$$

$$(2.43)$$

which is the **macroscopic equation of change for temperature**.

2.A   Enthalpy Change from Thermodynamics

We will derive the equation for enthalpy change in three steps as follows.

Step 1

The differential changes in the internal energy (U), and entropy (S) of an open system are given by, respectively,

$$dU = \sum_{i=1}^{N_c} \tilde{H}_i dm_i + \delta Q - PdV \quad \text{and}$$

$$dS = \sum_{i=1}^{N_c} \tilde{S}_i dm_i + \frac{\delta Q}{T} + dS_{\text{gen}}$$

where (i) $\tilde{H}_i$ and $\tilde{S}_i$ are the partial specific enthalpy and partial specific entropy, respectively, of the ith species of mass m_i in the mixture of N_c species, (ii) δQ is the amount of heat transfer, and (iii) dS_{gen} is amount of entropy generated. Eliminating δQ from the last two equations, we obtain

$$dU = TdS - TdS_{\text{gen}} - PdV + \sum_{i=1}^{N_c} \underbrace{(\tilde{H}_i - T\tilde{S}_i)}_{\tilde{G}_i} dm_i$$

where $\tilde{G}_i$ is the partial specific Gibbs free energy. For a reversible process, dS_{gen} is zero, and therefore

$$dU = TdS - PdV + \sum_{i=1}^{N_c} \tilde{G}_i dm_i \qquad (2.44)$$

Since U is a path-independent property, dU given by the above equation is valid for *any* process for the same initial and final states.

Now from the definition of enthalpy, $H = U + PV$, so that

$$dH = dU + PdV + VdP = TdS + VdP + \sum_{i=1}^{N_c} \tilde{G}_i dm_i \qquad (2.45)$$

with the help of Equation (2.44) above.

Step 2

From $S = S(T, P, \mathbf{m})$ where $\mathbf{m}$ is the vector of all m_is, we obtain

$$dS = \frac{\partial S}{\partial T}\bigg|_{P,\mathbf{m}} dT + \frac{\partial S}{\partial P}\bigg|_{T,\mathbf{m}} dP + \sum_{i=1}^{N_c} \frac{\partial S}{\partial m_i}\bigg|_{T,P,\mathbf{m}_j} dm_i \qquad (2.46)$$

In the above equation $\mathbf{m}_j$ is $\mathbf{m}$ without the jth species. We will now get expressions for $\partial S/\partial T|_{P,\mathbf{m}}$ and $\partial S/\partial P|_{T,\mathbf{m}}$. In terms of heat capacity, $C_P \equiv \partial H/\partial T|_{P,\mathbf{m}}$, and the relation $\partial H/\partial S|_{P,\mathbf{m}} = T$ from Equation (2.45) above, we can express

$$\frac{\partial S}{\partial T}\bigg|_{P,\mathbf{m}} = \frac{\partial S}{\partial H}\bigg|_{P,\mathbf{m}} \times \frac{\partial H}{\partial T}\bigg|_{P,\mathbf{m}} = \frac{C_P}{T} \qquad (2.47)$$

From the definition of the Gibbs free energy, $G = H - TS$, so that

$$\mathrm{d}G = \mathrm{d}H - T\mathrm{d}S - S\mathrm{d}T = -S\mathrm{d}T + V\mathrm{d}P + \sum_{i=1}^{N_c} \tilde{G}_i \mathrm{d}m_i$$

where we have used Equation (2.45) on the previous page in the last step. From $G = G(T, P, \mathbf{m})$, we get

$$\mathrm{d}G = \left.\frac{\partial G}{\partial T}\right|_{P,\mathbf{m}} \mathrm{d}T + \left.\frac{\partial G}{\partial P}\right|_{T,\mathbf{m}} \mathrm{d}P + \sum_{i=1}^{N_c} \left.\frac{\partial G}{\partial m_i}\right|_{T,P,\mathbf{m}_j} \mathrm{d}m_i$$

Comparing the last two equations, we get the following relations:

$$\left.\frac{\partial G}{\partial T}\right|_{P,\mathbf{m}} = -S, \quad \left.\frac{\partial G}{\partial P}\right|_{T,\mathbf{m}} = V \quad \text{and}$$

$$\left.\frac{\partial G}{\partial m_i}\right|_{T,P,\mathbf{m}_j} = \tilde{G}_i; \quad i = 1, 2, \ldots, N_c \tag{2.48}$$

Since G is a path-independent property, using the first two of the above relations, we get

$$\left.\frac{\partial}{\partial P}\right|_{T,\mathbf{m}} \overbrace{(\partial G/\partial T)_{P,\mathbf{m}}}^{-S} = \left.\frac{\partial}{\partial T}\right|_{P,\mathbf{m}} \overbrace{(\partial G/\partial P)_{T,\mathbf{m}}}^{V} \quad \text{or} \quad \left.\frac{\partial S}{\partial P}\right|_{T,\mathbf{m}} = -\left.\frac{\partial V}{\partial T}\right|_{P,\mathbf{m}}$$

Substituting the above result and Equation (2.47) into Equation (2.46) on the previous page gives

$$\mathrm{d}S = \frac{C_{\mathrm{P}}}{T}\mathrm{d}T - \left.\frac{\partial V}{\partial T}\right|_{P,\mathbf{m}} \mathrm{d}P + \sum_{i=1}^{N_c} \left.\frac{\partial S}{\partial m_i}\right|_{T,P,\mathbf{m}_j} \mathrm{d}m_i$$

The above expression for $\mathrm{d}S$ when substituted into Equation (2.45) on the previous page yields

$$\mathrm{d}H = C_{\mathrm{P}}\,\mathrm{d}T + \left[V - T\left.\frac{\partial V}{\partial T}\right|_{P,\mathbf{m}}\right]\mathrm{d}P + \sum_{i=1}^{N_c} \Big(\underbrace{T\left.\frac{\partial S}{\partial m_i}\right|_{T,P,\mathbf{m}_j} + \tilde{G}_i}_{(\partial H/\partial m_i)_{T,P,\mathbf{m}_j}} \Big)\mathrm{d}m_i$$

$$\tag{2.49}$$

where the bracketed expression in the last term of the above equation gets simplified after applying Equation (2.48) above, and then utilizing the definition, $G \equiv H - TS$, as shown below.

$$T\frac{\partial S}{\partial m_i} + \tilde{G}_i = T\frac{\partial S}{\partial m_i} + \frac{\partial G}{\partial m_i} = \frac{\partial}{\partial m_i}(TS + G) = \frac{\partial H}{\partial m_i} \quad (\text{constant } T, P, \mathbf{m}_j)$$

Step 3

To obtain $\mathrm{d}H$ in terms of specific heat capacity ($\hat{C}_{\mathrm{P}}$), density (ρ), and mass fractions (ω_is), we express Equation (2.49) above as

$$\mathrm{d}(m\hat{H}) = m\hat{C}_{\mathrm{P}}\,\mathrm{d}T + \left[m\hat{V} - T\left.\frac{\partial(m\hat{V})}{\partial T}\right|_{P,\mathbf{m}}\right]\mathrm{d}P + \sum_{i=1}^{N_c} \left.\frac{\partial H}{\partial m_i}\right|_{T,P,\mathbf{m}_j} \mathrm{d}(m\omega_i)$$

where $\hat{H}$ is specific enthalpy, and $\hat{V}$ is specific volume. Upon expanding the differentials and derivatives,

$$m\,d\hat{H} + \hat{H}\,dm = m\hat{C}_{\mathrm{P}}\,dT + \left[m\hat{V} - T\left(m\frac{\partial \hat{V}}{\partial T}\Big|_{P,\mathbf{m}} + \underbrace{\hat{V}\frac{\partial m}{\partial T}\Big|_{P,\mathbf{m}}}_{\substack{= 0,\ \text{since} \\ \mathbf{m}\ \text{is constant}}} \right) \right] dP$$

$$+ \sum_{i=1}^{N_c} \frac{\partial H}{\partial m_i}\Big|_{T,P,\mathbf{m}_j} (m\,d\omega_i + \omega_i\,dm)$$

Further rearrangement of the above equation yields

$$\left[\hat{H} - \sum_{i=1}^{N_c} \omega_i \underbrace{\frac{\partial H}{\partial m_i}\Big|_{T,P,\mathbf{m}_j}}_{\tilde{H}_i} \right] dm + \left[d\hat{H} - \hat{C}_{\mathrm{P}}\,dT - \left(\hat{V} - T\frac{\partial \hat{V}}{\partial T}\Big|_{P,\mathbf{m}} \right) dP \right.$$

$$\left. - \sum_{i=1}^{N_c} \underbrace{\frac{\partial H}{\partial m_i}\Big|_{T,P,\mathbf{m}_j}}_{\tilde{H}_i} d\omega_i \right] m = 0 \tag{2.50}$$

where $\tilde{H}_i$ is called the partial specific enthalpy of the ith species in the mixture. Since dm and m are arbitrary, their respective coefficients must be zero. Thus,

$$\hat{H} = \sum_{i=1}^{N_c} \omega_i \tilde{H}_i \tag{2.51}$$

$$d\hat{H} = \hat{C}_{\mathrm{P}}\,dT + \left(\hat{V} - T\frac{\partial \hat{V}}{\partial T}\Big|_{P,\mathbf{m}} \right) dP + \sum_{i=1}^{N_c} \tilde{H}_i\,d\omega_i \tag{2.52}$$

Because $\hat{V} = 1/\rho$,

$$\frac{\partial \hat{V}}{\partial T} = \frac{\partial \hat{V}}{\partial \rho}\frac{\partial \rho}{\partial T} = -\frac{1}{\rho^2}\frac{\partial \rho}{\partial T}$$

Using the above result, Equation (2.52) above becomes

$$d\hat{H} = \hat{C}_{\mathrm{P}}\,dT + \frac{1}{\rho}\left(1 + \frac{T}{\rho}\frac{\partial \rho}{\partial T}\Big|_{P,\mathbf{m}} \right) dP + \sum_{i=1}^{N_c} \tilde{H}_i\,d\omega_i \tag{2.53}$$

Let $\hat{H}$ be a function of time as well as space. Then the partial differential of $\hat{H}$ is given by

$$\partial\hat{H} = \hat{C}_{\mathrm{P}}\,\partial T + \frac{1}{\rho}\left(1 + \frac{T}{\rho}\frac{\partial \rho}{\partial T}\Big|_{P,\mathbf{m}} \right) \partial P + \sum_{i=1}^{N_c} \tilde{H}_i\,\partial\omega_i$$

Dividing the above equation by the differential time, ∂t, we get **the equation of change for enthalpy** from thermodynamics.

$$\frac{\partial\hat{H}}{\partial t} = \hat{C}_{\mathrm{P}}\frac{\partial T}{\partial t} + \frac{1}{\rho}\left[1 + \frac{T}{\rho}\frac{\partial \rho}{\partial T}\Big|_{P,\mathbf{m}} \right]\frac{\partial P}{\partial t} + \sum_{i=1}^{N_c} \tilde{H}_i\frac{\partial\omega_i}{\partial t} \tag{2.54}$$

2.B   Divergence Theorem

According to this theorem, the integral of the divergence of a vector field over the volume of a system is equal to the net outward flow of the vector field through the boundary surface area of the system. Thus, given a vector field $\mathbf{f}$ in a system of volume V and boundary surface area A,

$$\int_V \nabla \cdot \mathbf{f}\, dV \;=\; \int_A \mathbf{f} \cdot \hat{\mathbf{n}}\, dA \qquad\qquad (2.55)$$

To prove this theorem, we split a system into upper and lower parts, as shown in Figure 2.13 below. The two parts are demarcated by a locus of points on the surface where the outward normal $\hat{\mathbf{n}}$ is perpendicular to the x_3-direction.

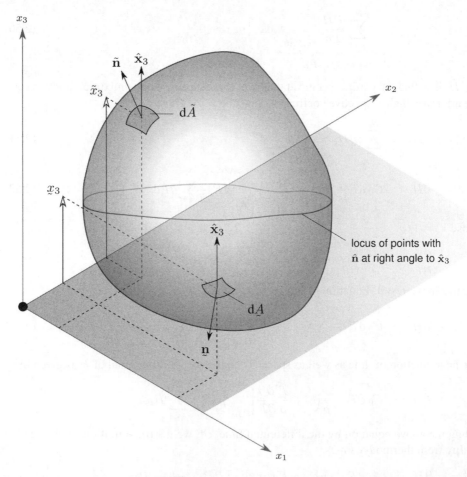

Figure 2.13   A system of volume V split into upper and lower parts by a locus of points with the outward normal $\hat{\mathbf{n}}$ perpendicular to the x_3-direction

Now from the definition of divergence of a vector,

$$\int_V \nabla \cdot \mathbf{f} \, dV = \int_V \frac{\partial f_1}{\partial x_1} \, dV + \int_V \frac{\partial f_2}{\partial x_2} \, dV + \int_V \frac{\partial f_3}{\partial x_3} \, dV$$

Let $\underset{\sim}{x}_3$ and $\tilde{x}_3$ be, respectively, the upper and lower limits of the system along the x_3-direction at any set of coordinates (x_1, x_2). Then we can write

$$\int_V \frac{\partial f_3}{\partial x_3} \, dV = \int_{x_1} \int_{x_2} \int_{\underset{\sim}{x}_3(x_1,x_2)}^{\tilde{x}_3(x_1,x_2)} \frac{\partial f_3}{\partial x_3} \, dx_3 \, dx_2 \, dx_1 = \int_{x_1} \int_{x_2} \Big[f_3 \Big]_{\underset{\sim}{x}_3(x_1,x_2)}^{\tilde{x}_3(x_1,x_2)} \, dx_2 \, dx_1$$

$$\text{or} \int_V \frac{\partial f_3}{\partial x_3} \, dV = \underbrace{\int_{x_1} \int_{x_2} f_3(\tilde{x}_3) \, dx_2 \, dx_1}_{\text{for upper surface area, } \tilde{A}} - \underbrace{\int_{x_1} \int_{x_2} f_3(\underset{\sim}{x}_3) \, dx_2 \, dx_1}_{\text{for lower surface area, } \underset{\sim}{A}} \qquad (2.56)$$

Observe that the differential area element in the upper part of the system is

$$d\tilde{\mathbf{A}} = (d\tilde{A})\tilde{\mathbf{n}}$$

where $d\tilde{A}$ is the magnitude, and $\tilde{\mathbf{n}}$ is the unit vector along $d\tilde{\mathbf{A}}$. The projection of $d\tilde{\mathbf{A}}$ on the $x_1 x_2$-plane is

$$d\tilde{A} \cos\theta = dx_1 dx_2$$

where θ is the angle between $\tilde{\mathbf{n}}$ and the unit vector $\hat{\mathbf{x}}_3$ along the x_3-direction. Thus, $\cos\theta = \hat{\mathbf{x}}_3 \cdot \tilde{\mathbf{n}}$ so that

$$dx_1 dx_2 = \hat{\mathbf{x}}_3 \cdot \tilde{\mathbf{n}} \, d\tilde{A} \qquad (2.57)$$

Similarly, the differential area element in the lower part of the system is

$$d\underset{\sim}{\mathbf{A}} = (d\underset{\sim}{A})\underset{\sim}{\mathbf{n}}$$

where $d\underset{\sim}{A}$ is the magnitude, and $\underset{\sim}{\mathbf{n}}$ is the unit vector along $d\underset{\sim}{\mathbf{A}}$. The projection of $d\underset{\sim}{\mathbf{A}}$ on the $x_1 x_2$-plane is

$$-d\underset{\sim}{A} \cos\theta = dx_1 dx_2$$

where θ is the angle between $\underset{\sim}{\mathbf{n}}$ and $\hat{\mathbf{x}}_3$. Since the magnitude of θ in the lower part is greater than the right angle, $\cos\theta$ is negative. To keep the projected area positive, we have multiplied the left-hand side of the above equation by -1. Since $\cos\theta = \hat{\mathbf{x}}_3 \cdot \underset{\sim}{\mathbf{n}}$,

$$dx_1 dx_2 = -\hat{\mathbf{x}}_3 \cdot \underset{\sim}{\mathbf{n}} \, d\underset{\sim}{A} \qquad (2.58)$$

Substituting Equations (2.57) and (2.58) above, respectively, in the integrals for the upper and lower surface areas of Equation (2.56) above, we get

$$\int_V \frac{\partial f_3}{\partial x_3} \, dV = \int_{\tilde{A}} f_3(\tilde{x}_3)\hat{\mathbf{x}}_3 \cdot \tilde{\mathbf{n}} \, d\tilde{A} + \int_{\underset{\sim}{A}} f_3(\underset{\sim}{x}_3)\hat{\mathbf{x}}_3 \cdot \underset{\sim}{\mathbf{n}} \, d\underset{\sim}{A} = \int_A f_3\hat{\mathbf{x}}_3 \cdot \hat{\mathbf{n}} \, dA$$

where $\hat{\mathbf{n}}$ is the unit vector perpendicular to an area element dA.

Proceeding in the same manner as above for the x_1- and x_2-directions, we obtain, respectively,

$$\int_V \frac{\partial f_1}{\partial x_1} \, dV \;=\; \int_A f_1 \hat{\mathbf{x}}_1 \cdot \hat{\mathbf{n}} \, dA \qquad \text{and}$$

$$\int_V \frac{\partial f_2}{\partial x_2} \, dV \;=\; \int_A f_2 \hat{\mathbf{x}}_2 \cdot \hat{\mathbf{n}} \, dA$$

Adding the last three equations results in the **divergence theorem**, i.e.,

$$\int_V \left(\frac{\partial f_1}{\partial x_1} + \frac{\partial f_2}{\partial x_2} + \frac{\partial f_3}{\partial x_3} \right) dV \;=\; \int_A (f_1 \hat{\mathbf{x}}_1 + f_2 \hat{\mathbf{x}}_2 + f_3 \hat{\mathbf{x}}_3) \cdot \hat{\mathbf{n}} \, dA$$

which can be written compactly as

$$\int_V \nabla \cdot \mathbf{f} \, dV \;=\; \int_A \mathbf{f} \cdot \hat{\mathbf{n}} \, dA \qquad\qquad (2.55)$$

2.C General Transport Theorem

This theorem provides a useful expression for the time derivative of the volume integral of a scalar function in a system whose volume depends on time.

To derive this theorem, we consider a system whose surface moves with velocity $\mathbf{w}$, which varies with space and time. As shown in Figure 2.14 on the next page, the volume changes in time Δt from the initial volume $V(t)$ to the final volume $V(t + \Delta t)$. The time derivative of the volume integral of a scalar function F, which depends on time as well as space, is given by

$$\frac{d}{dt} \int_{V(t)} F \, dV \;=\; \lim_{\Delta t \to 0} \frac{\displaystyle\int_{V(t+\Delta t)} F(t + \Delta t) \, dV - \int_{V(t)} F(t) \, dV}{\Delta t} \qquad (2.59)$$

The final volume can be written as

$$V(t + \Delta t) \;=\; V(t) + V_2(\Delta t) - V_1(\Delta t)$$

where, as shown in the figure, V_2 is the volume gained or swept forward, and V_1 is the volume lost or swept behind in time Δt by the system of initial volume $V(t)$. Utilizing the above equation, we can write

$$\int_{V(t+\Delta t)} F(t + \Delta t) \, dV \;=\; \int_{V(t)} F(t + \Delta t) \, dV + \int_{V_2(\Delta t)} F(t + \Delta t) \, dV_2 - \int_{V_1(\Delta t)} F(t + \Delta t) \, dV_1$$

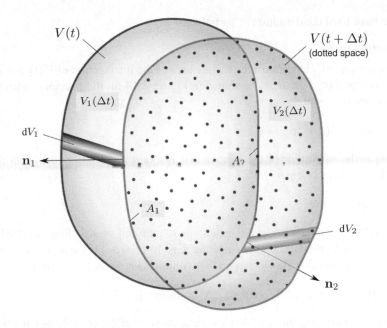

Figure 2.14 Snapshots of a moving volume V at times t and $(t + \Delta t)$. V_1 and V_2 are, respectively, the volumes left behind and swept forward in duration Δt

The above equation when substituted into Equation (2.59) on the previous page yields

$$\frac{\mathrm{d}}{\mathrm{d}t} \int_{V(t)} F \, \mathrm{d}V = \lim_{\Delta t \to 0} \underbrace{\frac{\displaystyle\int_{V(t)} F(t + \Delta t) \, \mathrm{d}V - \int_{V(t)} F(t) \, \mathrm{d}V}{\Delta t}}_{\equiv T_0}$$

$$+ \lim_{\Delta t \to 0} \underbrace{\frac{\displaystyle-\int_{V_1(\Delta t)} F(t + \Delta t) \, \mathrm{d}V_1}{\Delta t}}_{\equiv T_1} + \lim_{\Delta t \to 0} \underbrace{\frac{\displaystyle\int_{V_2(\Delta t)} F(t + \Delta t) \, \mathrm{d}V_2}{\Delta t}}_{\equiv T_2}$$

$$(2.60)$$

We now need to simplify the terms denoted by T_0, T_1 and T_2 in the above equation.

Simplification of T_0

The term T_0 in the above equation simplifies as follows:

$$T_0 = \lim_{\Delta t \to 0} \int_{V(t)} \frac{F(t + \Delta t) - F(t)}{\Delta t} \, \mathrm{d}V = \int_{V(t)} \frac{\partial F}{\partial t} \, \mathrm{d}V \qquad (2.61)$$

where we have used the definition of partial derivative.

Simplification of T_2

We first determine dV_2 in terms of the surface area of the intersection of $V(t)$ and $V(t + \Delta t)$. This area is made of A_1 and A_2 as shown in Figure 2.14 on the previous page. Consider a differential area element

$$d\mathbf{A}_2 = (dA_2)\hat{\mathbf{n}}_2$$

where dA_2 is the magnitude of the area, and $\hat{\mathbf{n}}_2$ is a unit vector along $d\mathbf{A}_2$. Let the velocity of the differential area element be

$$\mathbf{w} = w\hat{\mathbf{w}}$$

where w and $\hat{\mathbf{w}}$ are, respectively, the magnitude of and unit vector along $\mathbf{w}$. If θ is the angle between $\hat{\mathbf{w}}$ and $d\mathbf{A}_2$ then its component along $\hat{\mathbf{w}}$ is $dA_2 \cos\theta$. This area will sweep a distance $w\Delta t$ in time duration Δt. The corresponding volume swept along $\hat{\mathbf{w}}$ is then

$$dV_2 = (w\Delta t)\, dA_2 \cos\theta = \Delta t\, w \cos\theta\, dA_2$$

Note that on A_2, the magnitude of θ is less than the right angle so that $\cos\theta$ is positive. Now since $\cos\theta = \hat{\mathbf{w}} \cdot \hat{\mathbf{n}}_2$,

$$dV_2 = \Delta t\, w\hat{\mathbf{w}} \cdot \hat{\mathbf{n}}_2\, dA_2 = \Delta t\, \mathbf{w} \cdot \hat{\mathbf{n}}_2\, dA_2$$

Using the above expression for dV_2, we obtain

$$T_2 = \lim_{\Delta t \to 0} \int\limits_{A_2(t)} F(t + \Delta t)\mathbf{w} \cdot \hat{\mathbf{n}}_2\, dA_2 = \int\limits_{A_2(t)} F(t)\mathbf{w} \cdot \hat{\mathbf{n}}_2\, dA_2 \qquad (2.62)$$

Simplification of T_1

We proceed in a similar manner as above, and consider a differential area element [see Figure 2.14, previous page]

$$d\mathbf{A}_1 = (dA_1)\hat{\mathbf{n}}_1$$

where dA_1 is the magnitude of the area, and $\hat{\mathbf{n}}_1$ is a unit vector along $d\mathbf{A}_1$. The corresponding volume swept along $\hat{\mathbf{w}}$ is given by

$$dV_1 = -\Delta t\, w \cos\theta\, dA_1$$

where θ is the angle between $\hat{\mathbf{w}}$ and $d\mathbf{A}_1$, and the right-hand side is multiplied by -1 to keep the volume dV_1 positive. The reason is that on A_2, the magnitude of θ is greater than the right angle so that $\cos\theta$ is negative.

Since $\cos\theta = \hat{\mathbf{w}} \cdot \hat{\mathbf{n}}_1$ here, the above equation becomes

$$dV_1 = -\Delta t\, \mathbf{w} \cdot \hat{\mathbf{n}}_1\, dA_1$$

Thus, the term T_1, after substituting for dV_1 given by the above equation, is given by

$$T_1 = \int_{A_1(t)} F(t) \mathbf{w} \cdot \hat{\mathbf{n}}_1 \, dA_1 \tag{2.63}$$

The addition of Equations (2.62) and (2.63) above yields

$$T_1 + T_2 = \int_{A(t)} F(t) \mathbf{w} \cdot \hat{\mathbf{n}} \, dA \equiv \int_{A(t)} F \mathbf{w} \cdot \hat{\mathbf{n}} \, dA$$

where $A \equiv A_1 + A_2$, and $\hat{\mathbf{n}}$ denotes the unit vector along the differential element of area A. Substituting the above equation and Equation (2.61) into Equation (2.60) on p. 51, we finally obtain

$$\frac{d}{dt} \int_{V(t)} F \, dV = \int_{V(t)} \frac{\partial F}{\partial t} \, dV + \int_{A(t)} F \mathbf{w} \cdot \hat{\mathbf{n}} \, dA \tag{2.64}$$

The above equation is known as **general transport theorem**.

Special Case

For a system whose volume is fixed in space, $\mathbf{w}$ is zero. In this case the general transport theorem yields

$$\frac{d}{dt} \int_V F \, dV = \int_V \frac{\partial F}{\partial t} \, dV \tag{2.65}$$

2.D   Equations in Cartesian, Cylindrical and Spherical Coordinate Systems

In the following sections, we present the important equations derived in this chapter for the three most common coordinate systems – the Cartesian or rectangular, cylindrical or polar, and spherical coordinate systems. The equations in the last two systems are derivable from the equations in Cartesian coordinate system using suitable coordinate transformations. Coordinate transformations are presented in Chapter 5 on p. 139.

Coordinate Notation

In the equations below we use the following notation for the coordinates:

For Cartesian coordinates, x, y and z denote, respectively, the x_1, x_2 and x_3 coordinates that were used earlier. For cylindrical coordinates, r is the distance along the radial direction, θ is the azimuthal angle, and z is the distance along the axial direction. Lastly, for spherical coordinates, r is the distance along the radial direction, θ is the azimuthal angle, and ϕ is the polar angle.

2.D.1   Equations of Continuity

Cartesian coordinates

$$\frac{\partial \rho}{\partial t} = -\rho\left(\frac{\partial v_x}{\partial x} + \frac{\partial v_y}{\partial y} + \frac{\partial v_z}{\partial z}\right) - v_x\frac{\partial \rho}{\partial x} - v_y\frac{\partial \rho}{\partial y} - v_z\frac{\partial \rho}{\partial z}$$

Cylindrical coordinates

$$\frac{\partial \rho}{\partial t} = -\rho\left(\frac{\partial v_r}{\partial r} + \frac{v_r}{r} + \frac{1}{r}\frac{\partial v_\theta}{\partial \theta} + \frac{\partial v_z}{\partial z}\right) - v_r\frac{\partial \rho}{\partial r} - \frac{v_\theta}{r}\frac{\partial \rho}{\partial \theta} - v_z\frac{\partial \rho}{\partial z} \qquad (2.66)$$

Spherical coordinates

$$\frac{\partial \rho}{\partial t} = -\rho\left(\frac{\partial v_r}{\partial r} + \frac{2v_r}{r} + \frac{v_\theta}{r\tan\theta} + \frac{1}{r}\frac{\partial v_\theta}{\partial \theta} + \frac{1}{r\sin\theta}\frac{\partial v_\phi}{\partial \phi}\right)$$

$$- v_r\frac{\partial \rho}{\partial r} - \frac{v_\theta}{r}\frac{\partial \rho}{\partial \theta} - \frac{v_\phi}{r\sin\theta}\frac{\partial \rho}{\partial \phi}$$

2.D.2   Equations of Continuity for Individual Species

Cartesian Coordinates

$$\frac{\partial \omega_i}{\partial t} = -\frac{1}{\rho}\left(\frac{\partial j_{ix}}{\partial x} + \frac{\partial j_{iy}}{\partial y} + \frac{\partial j_{iz}}{\partial z} - \sum_{j=1}^{N_r} r_{\text{gen},ij}\right) - v_x\frac{\partial \omega_i}{\partial x} - v_y\frac{\partial \omega_i}{\partial y} - v_z\frac{\partial \omega_i}{\partial z}$$

Cylindrical Coordinates

$$\frac{\partial \omega_i}{\partial t} = -\frac{1}{\rho}\left(\frac{\partial j_{ir}}{\partial r} + \frac{j_{ir}}{r} + \frac{1}{r}\frac{\partial j_{i\theta}}{\partial \theta} + \frac{\partial j_{iz}}{\partial z} - \sum_{j=1}^{N_r} r_{\text{gen},ij}\right) - v_r\frac{\partial \omega_i}{\partial r} - \frac{v_\theta}{r}\frac{\partial \omega_i}{\partial \theta} - v_z\frac{\partial \omega_i}{\partial z}$$

Spherical Coordinates

$$\frac{\partial \omega_i}{\partial t} = -\frac{1}{\rho}\left(\frac{\partial j_{ir}}{\partial r} + \frac{2j_{ir}}{r} + \frac{1}{r}\frac{\partial j_{i\theta}}{\partial \theta} + \frac{j_{i\theta}}{r\tan\theta} + \frac{1}{r\sin\theta}\frac{\partial j_{i\phi}}{\partial \phi} - \sum_{j=1}^{N_r} r_{\text{gen},ij}\right)$$

$$- v_r\frac{\partial \omega_i}{\partial r} - \frac{v_\theta}{r}\frac{\partial \omega_i}{\partial \theta} - \frac{v_\phi}{r\sin\theta}\frac{\partial \omega_i}{\partial \phi}$$

2.D.3   Equations of Motion

The following equations of motions are for symmetric stress tensor τ:

Cartesian Coordinates

$$\frac{\partial v_x}{\partial t} = -v_x \frac{\partial v_x}{\partial x} - v_y \frac{\partial v_x}{\partial y} - v_z \frac{\partial v_x}{\partial z} - \frac{1}{\rho}\left(\frac{\partial \tau_{xx}}{\partial x} + \frac{\partial \tau_{yx}}{\partial y} + \frac{\partial \tau_{zx}}{\partial z} + \frac{\partial P}{\partial x} \right) + g_x$$

$$\frac{\partial v_y}{\partial t} = -v_x \frac{\partial v_y}{\partial x} - v_y \frac{\partial v_y}{\partial y} - v_z \frac{\partial v_y}{\partial z} - \frac{1}{\rho}\left(\frac{\partial \tau_{xy}}{\partial x} + \frac{\partial \tau_{yy}}{\partial y} + \frac{\partial \tau_{zy}}{\partial z} + \frac{\partial P}{\partial y} \right) + g_y$$

$$\frac{\partial v_z}{\partial t} = -v_x \frac{\partial v_z}{\partial x} - v_y \frac{\partial v_z}{\partial y} - v_z \frac{\partial v_z}{\partial z} - \frac{1}{\rho}\left(\frac{\partial \tau_{xz}}{\partial x} + \frac{\partial \tau_{yz}}{\partial y} + \frac{\partial \tau_{zz}}{\partial z} + \frac{\partial P}{\partial z} \right) + g_z$$

Cylindrical Coordinates

$$\frac{\partial v_r}{\partial t} = -v_r \frac{\partial v_r}{\partial r} - \frac{v_\theta}{r} \frac{\partial v_r}{\partial \theta} - v_z \frac{\partial v_r}{\partial z} - \frac{1}{\rho}\left(\frac{\partial \tau_{rr}}{\partial r} + \frac{1}{r}\frac{\partial \tau_{r\theta}}{\partial \theta} + \frac{\partial \tau_{rz}}{\partial z} \right.$$

$$\left. + \frac{\tau_{rr} - \tau_{\theta\theta}}{r} + \frac{\partial P}{\partial r} \right) + \frac{v_\theta^2}{r} + g_r \tag{2.67}$$

$$\frac{\partial v_\theta}{\partial t} = -v_r \frac{\partial v_\theta}{\partial r} - \frac{v_\theta}{r} \frac{\partial v_\theta}{\partial \theta} - v_z \frac{\partial v_\theta}{\partial z} - \frac{1}{\rho}\left(\frac{\partial \tau_{r\theta}}{\partial r} + \frac{1}{r}\frac{\partial \tau_{\theta\theta}}{\partial \theta} + \frac{1}{r}\frac{\partial \tau_{\theta z}}{\partial z} + \frac{2\tau_{r\theta}}{r} \right.$$

$$\left. + \frac{1}{r}\frac{\partial P}{\partial \theta} \right) - \frac{v_r v_\theta}{r} + g_\theta$$

$$\frac{\partial v_z}{\partial t} = -v_r \frac{\partial v_z}{\partial r} - \frac{v_\theta}{r} \frac{\partial v_z}{\partial \theta} - v_z \frac{\partial v_z}{\partial z} - \frac{1}{\rho}\left(\frac{\partial \tau_{rz}}{\partial r} + \frac{1}{r}\frac{\partial \tau_{\theta z}}{\partial \theta} + \frac{\partial \tau_{zz}}{\partial z} + \frac{\tau_{rz}}{r} \right.$$

$$\left. + \frac{\partial P}{\partial z} \right) + g_z$$

Spherical Coordinates

$$\frac{\partial v_r}{\partial t} = -v_r \frac{\partial v_r}{\partial r} - \frac{v_\theta}{r}\frac{\partial v_r}{\partial \theta} - \frac{v_\phi}{r\sin\theta}\frac{\partial v_r}{\partial \phi} - \frac{1}{\rho}\left(\frac{\partial \tau_{rr}}{\partial r} + \frac{1}{r}\frac{\partial \tau_{r\theta}}{\partial \theta} + \frac{1}{r\sin\theta}\frac{\partial \tau_{r\phi}}{\partial \phi} \right.$$

$$\left. + \frac{2\tau_{rr} - \tau_{\theta\theta} - \tau_{\phi\phi}}{r} + \frac{\tau_{r\theta}}{r\tan\theta} + \frac{\partial P}{\partial r} \right) + \frac{v_\theta^2 + v_\phi^2}{r} + g_r$$

$$\frac{\partial v_\theta}{\partial t} = -v_r \frac{\partial v_\theta}{\partial r} - \frac{v_\theta}{r}\frac{\partial v_\theta}{\partial \theta} - \frac{v_\phi}{r\sin\theta}\frac{\partial v_\theta}{\partial \phi} - \frac{1}{\rho}\left(\frac{\partial \tau_{r\theta}}{\partial r} + \frac{1}{r}\frac{\partial \tau_{\theta\theta}}{\partial \theta} + \frac{1}{r\sin\theta}\frac{\partial \tau_{\theta\phi}}{\partial \phi} \right.$$

$$\left. + \frac{3\tau_{r\theta}}{r} + \frac{\tau_{\theta\theta} - \tau_{\phi\phi}}{r\tan\theta} + \frac{1}{r}\frac{\partial P}{\partial \theta} \right) - \frac{v_r v_\theta}{r} + \frac{v_\phi^2}{r\tan\theta} + g_\theta$$

$$\frac{\partial v_\phi}{\partial t} = -v_r \frac{\partial v_\phi}{\partial r} - \frac{v_\theta}{r}\frac{\partial v_\phi}{\partial \theta} - \frac{v_\phi}{r\sin\theta}\frac{\partial v_\phi}{\partial \phi} - \frac{1}{\rho}\left(\frac{\partial \tau_{r\phi}}{\partial r} + \frac{1}{r}\frac{\partial \tau_{\theta\phi}}{\partial \theta} + \frac{1}{\sin\theta}\frac{\partial \tau_{\phi\phi}}{\partial \phi} \right.$$

$$\left. + \frac{3\tau_{r\phi}}{r} + \frac{2\tau_{\theta\phi}}{r\tan\theta} + \frac{1}{r\sin\theta}\frac{\partial P}{\partial \phi} \right) - \frac{v_r v_\phi}{r} - \frac{v_\theta v_\phi}{r\tan\theta} + g_\phi$$

2.D.4 *Equations of Change for Temperature*

Cartesian Coordinates

$$\frac{\partial T}{\partial t} = -v_x \frac{\partial T}{\partial x} - v_y \frac{\partial T}{\partial y} - v_z \frac{\partial T}{\partial z} - \frac{1}{\rho \hat{C}_{\mathrm{P}}}\left[\frac{\partial q_x}{\partial x} + \frac{\partial q_y}{\partial y} + \frac{\partial q_z}{\partial z} + \frac{T}{\rho}\frac{\partial \rho}{\partial T}\bigg|_{P,\mathbf{m}}\frac{\mathrm{D}P}{\mathrm{D}t} \right.$$

$$\left. - \sum_{i=1}^{N_c} \tilde{H}_i \left(\frac{\partial j_{ix}}{\partial x} + \frac{\partial j_{iy}}{\partial y} + \frac{\partial j_{iz}}{\partial z} \right) + \sum_{j=1}^{N_r}\sum_{i=1}^{N_c} \tilde{H}_i r_{\mathrm{gen},ij} + \boldsymbol{\tau} : \nabla\mathbf{v} \right]$$

Cylindrical Coordinates

$$\frac{\partial T}{\partial t} = -v_r \frac{\partial T}{\partial r} - \frac{v_\theta}{r}\frac{\partial T}{\partial \theta} - v_z \frac{\partial T}{\partial z} - \frac{1}{\rho \hat{C}_P}\left[\frac{\partial q_r}{\partial r} + \frac{q_r}{r} + \frac{1}{r}\frac{\partial q_\theta}{\partial \theta} + \frac{\partial q_z}{\partial z} + \frac{T}{\rho}\frac{\partial \rho}{\partial T}\bigg|_{P,\mathbf{m}}\frac{\mathrm{D}P}{\mathrm{D}t} \right.$$

$$\left. - \sum_{i=1}^{N_c} \tilde{H}_i \left(\frac{\partial j_{ir}}{\partial r} + \frac{j_{ir}}{r} + \frac{1}{r}\frac{\partial j_{i\theta}}{\partial \theta} + \frac{\partial j_{iz}}{\partial z} \right) + \sum_{j=1}^{N_r}\sum_{i=1}^{N_c} \tilde{H}_i r_{\mathrm{gen},ij} + \boldsymbol{\tau} : \nabla\mathbf{v} \right]$$

Spherical Coordinates

$$\frac{\partial T}{\partial t} = -v_r \frac{\partial T}{\partial r} - \frac{v_\theta}{r} \frac{\partial T}{\partial \theta} - \frac{v_\phi}{r \sin \theta} \frac{\partial T}{\partial \phi} - \frac{1}{\rho \hat{C}_\mathrm{P}} \left[\frac{\partial q_r}{\partial r} + \frac{2q_r}{r} + \frac{q_\theta}{r \tan \theta} + \frac{1}{r} \frac{\partial q_\theta}{\partial \theta} \right.$$

$$+ \frac{1}{r \sin \theta} \frac{\partial q_\phi}{\partial \phi} + \frac{T}{\rho} \frac{\partial \rho}{\partial T}\Big|_{P,\mathbf{m}} \frac{\mathrm{D}P}{\mathrm{D}t} - \sum_{i=1}^{N_c} \tilde{H}_i \left(\frac{\partial j_{ir}}{\partial r} + \frac{2j_{ir}}{r} + \frac{j_{i\theta}}{r \tan \theta} \right.$$

$$\left. + \frac{1}{r} \frac{\partial j_{i\theta}}{\partial \theta} + \frac{1}{r \sin \theta} \frac{\partial j_{i\phi}}{\partial \phi} \right) + \sum_{j=1}^{N_r} \sum_{i=1}^{N_c} \tilde{H}_i r_{\mathrm{gen},ij} + \boldsymbol{\tau} : \nabla \mathbf{v} \right]$$

Bibliography

[1] R.B. Bird, W.E. Stewart, and E.N. Lightfoot. *Transport Phenomena.* 2nd edition. New York: John Wiley & Sons, Inc., 2007.

[2] S. Whitaker. *Introduction to Fluid Mechanics.* New Jersey: Prentice-Hall, Inc., 1968.

[3] G.K. Batchelor. *An Introduction to Fluid Dynamics.* New York: Cambridge University Press, 2000.

[4] R. Aris. *Vectors, Tensors, and the Basic Equations of Fluid Mechanics.* New York: Dover Publications Inc., 1989.

[5] J.C. Slattery. *Advanced Transport Phenomena.* New York: Cambridge University Press, 2000.

[6] J.L. Plawsky. *Transport Phenomena Fundamentals.* 3rd edition. Boca Raton: CRC Press, 2014.

[7] S.I. Sandler. *Chemical, Biochemical, and Engineering Thermodynamics.* 4th edition. U.S.A.: John Wiley & Sons, Inc., 2006.

[8] R.J. Kee, M.E. Coltrin, and P. Glarborg. *Chemically Reacting Flow Theory and Practice.* New Jersey: Wiley-Interscience, 2003.

Exercises

2.1 Show that the following statement is false: The total number of moles of species in a system is always conserved.

2.2 Interpret the velocity of fluid under the continuum assumption.

2.3 Show that the equation of motion [see Equation (2.23), p. 32] is equivalent to

$$\rho \frac{\mathrm{D}\mathbf{v}}{\mathrm{D}t} = -\nabla \cdot \boldsymbol{\pi} + \rho \mathbf{g}$$

2.4 Obtain the microscopic and macroscopic equations of change of temperature in molar units.

2.5 Derive the microscopic equation of change for temperature [Equation (2.41), p. 42] from Equation (2.40).

2.6 For a fluid element of density ρ, and the surface moving with velocity $\mathbf{v}$, show that

$$\frac{\mathrm{d}}{\mathrm{d}t} \int_V \rho A \, \mathrm{d}V = \int_V \rho \frac{\mathrm{D}A}{\mathrm{D}t} \, \mathrm{d}V$$

where A is a scalar function. The above result is known as the *Reynolds Transport Theorem*.

3

Constitutive Relations

In this chapter, we present important constitutive relations involving mass, momentum, energy, chemical reactions, and equilibrium. These relations supplement fundamental relations by providing expressions for various quantities in terms of measurable intensive properties. This is what makes constitutive relations so important. Utilizing them, process models can account for various phenomena in terms of measurable properties.

We begin with constitutive relations for mass transport due to molecular diffusion of species.

3.1 Diffusion

Consider a mixture of two species, A and B. If their concentrations are not uniform in the mixture then either species has a propensity for diffusion, i.e., movement from a region of higher concentration to a region of lower concentration. The constitutive relation characterizing this phenomenon is **Fick's law of diffusion**. According to this law, the relative mass flux [see Equation (2.9), p. 23] of either of the species, say, A, at infinite dilution, and steady state is given by

$$\mathbf{j}_A = -\rho D_{AB} \nabla \omega_A$$

where ρ is the density of the mixture, D_{AB} is the (Fickian) diffusivity of A in the presence of B, and $\nabla \omega_A$ is the gradient of the mass fraction of A in the mixture. Likewise, for B,

$$\mathbf{j}_B = -\rho D_{BA} \nabla \omega_B$$

Using the expression for relative mass flux, and expressing ω_B as $(1 - \omega_A)$, it can be shown that

$$\mathbf{j}_A = -\mathbf{j}_B \quad \text{and} \quad D_{AB} = D_{BA}$$

Thus, for binary systems, single diffusivity describes the diffusion of both species.

In terms of the molar quantities, Fick's law of diffusion at infinite dilution, and steady state is given by

$$\mathbf{J}_A = -c D_{AB} \nabla y_A$$

Process Modeling and Simulation for Chemical Engineers: Theory and Practice, First Edition. Simant Ranjan Upreti.
© 2017 John Wiley & Sons Ltd. Published 2017 by John Wiley & Sons Ltd.
Companion website: www.wiley.com/go/upreti/pms_for_chemical_engineers

where $\mathbf{J_A}$ is the relative molar flux of A [see Equation (2.16), p. 25], c is the molar concentration of the mixture, and ∇y_A is the gradient of the mole fraction of A in the mixture.

It may be noted that Fick's law of diffusion is applicable to sufficiently dilute systems in which species are assumed to interact only with the bulk mixture. The law is an infinite dilution case of Maxwell–Stefan equations[1] of mass transfer in multicomponent systems. In systems with finite concentrations, the diffusivities defined by Fick's law can vary with composition.

3.1.1 Multicomponent Mixtures

For a mixture of N_c species where $N_c > 2$, Fick's law of diffusion can be generalized as

$$
\underbrace{\begin{bmatrix} \mathbf{j}_1 \\ \mathbf{j}_2 \\ \vdots \\ \mathbf{j}_{N_c-1} \end{bmatrix}}_{\bar{\mathbf{j}}} = -\rho \underbrace{\begin{bmatrix} D_{11} & D_{12} & \cdots & D_{1,N_c-1} \\ D_{21} & D_{22} & \cdots & D_{2,N_c-1} \\ \vdots & \vdots & \ddots & \vdots \\ D_{N_c-1,1} & D_{N_c-1,2} & \cdots & D_{N_c-1,N_c-1} \end{bmatrix}}_{\mathbf{D}} \underbrace{\begin{bmatrix} \nabla\omega_1 \\ \nabla\omega_2 \\ \vdots \\ \nabla\omega_{N_c-1} \end{bmatrix}}_{\nabla\omega}
$$

where D_{ij}s are the multicomponent Fickian diffusivities. In the matrix notation, the above set of equations can be written as

$$
\bar{\mathbf{j}} = -\rho \mathbf{D} \nabla \omega
$$

where $\bar{\mathbf{j}}$ is the vector of $\mathbf{j}_i$s, $\mathbf{D}$ is the matrix of D_{ij}s, and $\nabla\omega$ is the vector of $\nabla\omega_i$s. In the same manner, the relative molar fluxes are given by

$$
\bar{\mathbf{J}} = -c \mathbf{D} \nabla \mathbf{y}
$$

where $\bar{\mathbf{J}}$ is the vector of $\mathbf{J}_i$s, and $\nabla\mathbf{y}$ is the vector of ∇y_is.

Incorporating Fick's law of diffusion, Appendix 3.A.1 on p. 74 provides the equations of continuity for a binary system of constant density and diffusivity in Cartesian, cylindrical, and spherical coordinates.

Next, we present some important constitutive relations for momentum transport due to viscous motion.

3.2 Viscous Motion

Consider a moving material body (i.e., fluid) whose velocity $\mathbf{v}$ changes with the spatial location $\mathbf{x} \equiv \begin{bmatrix} x_1 & x_2 & x_3 \end{bmatrix}^\top$ in Cartesian coordinates. Using the first order Taylor expansion in the limit of $\Delta\mathbf{x}$ tending to zero vector [see Section 8.8.2, p. 323], we obtain

$$
\mathbf{v}(\mathbf{x} + \Delta\mathbf{x}) = \mathbf{v}(\mathbf{x}) + \nabla\mathbf{v}\Delta\mathbf{x}
$$

where $\nabla \mathbf{v}$ is the velocity gradient tensor given by

$$
\nabla \mathbf{v} =
\begin{bmatrix}
\dfrac{\partial v_1}{\partial x_1} & \dfrac{\partial v_1}{\partial x_2} & \dfrac{\partial v_1}{\partial x_3} \\[3mm]
\dfrac{\partial v_2}{\partial x_1} & \dfrac{\partial v_2}{\partial x_2} & \dfrac{\partial v_2}{\partial x_3} \\[3mm]
\dfrac{\partial v_3}{\partial x_1} & \dfrac{\partial v_3}{\partial x_2} & \dfrac{\partial v_3}{\partial x_3}
\end{bmatrix}
$$

This tensor can be expressed as

$$
\nabla \mathbf{v} = \frac{1}{2} \Big[\underbrace{\nabla \mathbf{v} + (\nabla \mathbf{v})^\top}_{\substack{\text{symmetric part} \\ (\dot{\gamma})}} + \underbrace{\nabla \mathbf{v} - (\nabla \mathbf{v})^\top}_{\substack{\text{antisymmetric part} \\ (\boldsymbol{\omega})}} \Big]
$$

In the above equation, the symmetric part $\dot{\gamma}$ is called the rate-of-strain tensor. The antisymmetric part $\boldsymbol{\omega}$ is the vorticity tensor, which represents pure rotation. The constitutive relations that follow characterize viscous motion based on $\dot{\gamma}$.

3.2.1　Newtonian Fluids

In Newtonian fluids, the viscous stress tensor $\boldsymbol{\tau}$ is a linear, isotropic (i.e., directionally independent) function of $\partial v_i / \partial x_j$s, but independent of $\boldsymbol{\omega}$. For these fluids, a constitutive relation known as **Newton's law of viscosity** relates $\boldsymbol{\tau}$ linearly to $\dot{\gamma}$ and $\nabla \cdot \mathbf{v}$ as[2,3]

$$
\boldsymbol{\tau} = -\mu \dot{\gamma} + \left(\frac{2}{3}\mu - \kappa \right)(\nabla \cdot \mathbf{v})\boldsymbol{\delta}
$$

where μ is viscosity is the viscosity of fluid, κ is its dilatational viscosity, and $\boldsymbol{\delta}$ is the unit dyadic [see Equation (8.8), p. 312].

In Cartesian coordinate system, the above equation expands to

$$
\boldsymbol{\tau} =
\begin{bmatrix}
-2\mu \dfrac{\partial v_1}{\partial x_1} + \left(\dfrac{2}{3}\mu - \kappa \right) \displaystyle\sum_{i=1}^{3} \dfrac{\partial v_i}{\partial x_i} & -\mu \left(\dfrac{\partial v_1}{\partial x_2} + \dfrac{\partial v_2}{\partial x_1} \right) & -\mu \left(\dfrac{\partial v_1}{\partial x_3} + \dfrac{\partial v_3}{\partial x_1} \right) \\[5mm]
-\mu \left(\dfrac{\partial v_2}{\partial x_1} + \dfrac{\partial v_1}{\partial x_2} \right) & -2\mu \dfrac{\partial v_2}{\partial x_2} + \left(\dfrac{2}{3}\mu - \kappa \right) \displaystyle\sum_{i=1}^{3} \dfrac{\partial v_i}{\partial x_i} & -\mu \left(\dfrac{\partial v_2}{\partial x_3} + \dfrac{\partial v_3}{\partial x_2} \right) \\[5mm]
-\mu \left(\dfrac{\partial v_3}{\partial x_1} + \dfrac{\partial v_1}{\partial x_3} \right) & -\mu \left(\dfrac{\partial v_3}{\partial x_2} + \dfrac{\partial v_2}{\partial x_3} \right) & -2\mu \dfrac{\partial v_3}{\partial x_3} + \left(\dfrac{2}{3}\mu - \kappa \right) \displaystyle\sum_{i=1}^{3} \dfrac{\partial v_i}{\partial x_i}
\end{bmatrix}
$$

Substituting the above equation in Equation (2.25) on p. 33 yields

$$
\frac{\partial \mathbf{v}}{\partial t} = -\mathbf{v} \cdot \nabla \mathbf{v} + \frac{1}{\rho} \Big[\underbrace{\mu \nabla \cdot \dot{\gamma}}_{\nabla^2 \mathbf{v}} - \left(\frac{2}{3}\mu - \kappa \right)\nabla(\nabla \cdot \mathbf{v}) - \nabla P \Big] + \mathbf{g}
$$

which is the equation of motion, or **Navier–Stokes equation**, for compressible Newtonian fluids.

Since $\nabla \cdot \mathbf{v} = 0$ when density is constant [see Equation (2.8), p. 23], the Newton's law of viscosity for an incompressible Newtonian fluid is given by $\boldsymbol{\tau} = -\mu\dot{\boldsymbol{\gamma}}$. The Navier–Stokes equation for that fluid having constant viscosity as well simplifies to

$$\frac{\partial \mathbf{v}}{\partial t} = -\mathbf{v} \cdot \nabla \mathbf{v} + \frac{1}{\rho}(\mu\nabla^2 \mathbf{v} - \nabla P) + \mathbf{g} \qquad (3.1)$$

Appendix 3.A.2 on p. 75 provides the equations of motion for Newtonian fluids of constant density and viscosity in Cartesian, cylindrical, and spherical coordinates.

3.2.2   Non-Newtonian Fluids

Many fluids such as molten polymers, and salt solutions have viscosities, which are not constant but vary, especially with shear rates. A general constitutive relation for momentum transport in such fluids is given by

$$\boldsymbol{\tau} = -\eta(\|\dot{\boldsymbol{\gamma}}\|)\,\dot{\boldsymbol{\gamma}}$$

where η is non-Newtonian viscosity, which is a function of $\|\dot{\boldsymbol{\gamma}}\|$, i.e., the norm of the rate-of-strain tensor,

$$\|\dot{\boldsymbol{\gamma}}\| = \sqrt{\frac{1}{2}\dot{\boldsymbol{\gamma}} : \dot{\boldsymbol{\gamma}}}$$

A simple characterization for η is given by the **power law model** according to which

$$\eta = m\|\dot{\boldsymbol{\gamma}}\|^{n-1}$$

where m is the consistency index, and n is the power law index of the fluid. The index m is sensitive to temperature, and is given by

$$m = m_0 e^{-a(T-T_0)}$$

where m_0 is the value of m at a reference temperature T_0, and a is an empirical constant.

In power law model, if $n > 1$ then the viscosity increases with the rate of shear, and the fluid is **shear-thickening**, or **dilatant**. If $n < 1$ then the viscosity decreases with the rate of shear, and the fluid is **shear-thinning**, or **pseudoplastic**. Note that the fluid is Newtonian if $n = 1$. The viscosity of many fluids is well-described by the power law model at high shear rates, i.e., large values of $\|\dot{\boldsymbol{\gamma}}\|$. However, for $\|\dot{\boldsymbol{\gamma}}\|$ tending to zero, the model predicts that η approaches

1. zero for dilatant fluids with $n > 1$, and

2. infinity for pseudoplastic fluids with $n < 1$.

In reality, however, η becomes a non-zero constant as $\|\dot{\boldsymbol{\gamma}}\|$ tends to zero.

A widely used relation that smoothly unifies the constant fluid viscosity at low shear rates with the power law trend at high shear rates is the **Cross–WLF model**. The Cross model[4] relates η to $\|\dot{\boldsymbol{\gamma}}\|$ as

$$\eta = \frac{\eta_0}{1 + \left(\dfrac{\eta_0 \|\dot{\boldsymbol{\gamma}}\|}{\tau^\star}\right)^{(1-n)}}$$

where η_0 is the limit of viscosity as $\|\dot{\gamma}\|$ approaches zero, and τ^* is the critical stress. This stress is determined when viscosity, which is constant at low shear rates, begins to change as $\|\dot{\gamma}\|$ increases. The WLF model[5] provides the temperature dependence through η_0, which is given by

$$\eta_0 = D_1 \exp\left[-\frac{A_1(T - T^\star)}{A_2 + (T - T^\star)}\right]$$

where D_1, A_1 and A_2 are empirical coefficients, T is the fluid temperature, and $T^\star$ is a reference temperature. For a polymer, $T^\star$ is the glass transition temperature.

The next constitutive relation is for energy transfer due to thermal conduction.

3.3 Thermal Conduction

The heat flux due to conduction in a medium is proportional to the temperature gradient. In an isotropic medium, the flux is given by **Fourier's law of heat conduction**, i.e.,

$$\mathbf{q} = -k\nabla T$$

where k is the thermal conductivity. It is not necessarily a constant but a function of temperature, pressure and composition of the medium.

Using Fourier's law, the microscopic equation of change for temperature [Equation (2.41), p. 42] can be written as

$$\rho\hat{C}_P\frac{\partial T}{\partial t} = -\rho\hat{C}_P\mathbf{v}\cdot\nabla T + \nabla\cdot(k\nabla T) + \sum_{i=1}^{N_c}\tilde{H}_i\nabla\cdot\mathbf{j}_i - \sum_{i=1}^{N_c}\sum_{j=1}^{N_r}\tilde{H}_i r_{\text{gen},ij}$$

$$-\frac{T}{\rho}\frac{\partial\rho}{\partial T}\frac{DP}{Dt} - \boldsymbol{\tau}:\nabla\mathbf{v} \tag{3.2}$$

For heat transfer in a non-reactive system with (i) constant density, or constant and uniform pressure, (ii) constant thermal conductivity, and (iii) negligible viscous dissipation, Equation (3.2) above becomes

$$\rho\hat{C}_P\frac{\partial T}{\partial t} = -\rho\hat{C}_P\mathbf{v}\cdot\nabla T + k\nabla^2 T$$

Appendix 3.A.3 on p. 76 provides the above equation in Cartesian, cylindrical, and spherical coordinates.

Constitutive relations that are described above associate the flux of mass, momentum, or energy to the gradients of measurable intensive properties. In the following section, we introduce an important constitutive relation, which relates the rate of reaction to the concentrations of species involved in a chemical reaction.

3.4 Chemical Reaction

We consider an elementary chemical reaction in which species react to yield products directly. The following representation of an elementary reaction

$$a\,\mathrm{A} + b\,\mathrm{B} \longrightarrow c\,\mathrm{C} + d\,\mathrm{D}$$

signifies that a moles of A and b moles of B disappear to produce c moles of C and d moles of D. While c and d are defined as **stoichiometric coefficients** of the products C and D, $-a$ and $-b$ are defined as stoichiometric coefficients of the reactants A and B.

Let us focus on two states, initial and final, of a mixture of the above species. In the initial state, the mixture has N_{A0}, N_{B0}, N_{C0} and N_{D0} mole numbers of A, B, C and D, respectively. The mole numbers change to N_A, N_B, N_C and N_D in the final state in which the reaction has progressed to a certain extent. Since, according to the reaction stoichiometry, c moles of C are produced when a moles of A are consumed,

$$\frac{\text{moles of C produced}}{\text{moles of A consumed}} = \frac{N_C - N_{C0}}{N_{A0} - N_A} = \frac{c}{a}$$

Let $\Delta N_Z \equiv (N_Z - N_{Z0})$ and ν_Z denote, respectively, the change in mole numbers, and the stoichiometric coefficient of species Z. Using this notation, the above relation becomes

$$\frac{\Delta N_C}{-\Delta N_A} = \frac{\nu_C}{-\nu_A}$$

Rearranging the above equation, we get

$$\frac{\Delta N_C}{\nu_C} = \frac{\Delta N_A}{\nu_A} \equiv X$$

where X is defined as the **extent of reaction**. We can similarly obtain

$$\frac{\Delta N_D}{\nu_D} = \frac{\Delta N_C}{\nu_C} = \frac{\Delta N_B}{\nu_B} = \frac{\Delta N_A}{\nu_A} = X$$

In general, the extent of a reaction (irreversible or reversible) is given by

$$X = \frac{\Delta N_Z}{\nu_Z} = \frac{N_Z - N_{Z0}}{\nu_Z} \tag{3.3}$$

where Z is *any* species involved in a reaction. Note that X is the same for all reaction-species. Once X is known, we can obtain the change in mole numbers of all the species from the reaction stoichiometry. Then given the initial mole numbers of a reaction-species, we can calculate its final mole numbers, and vice versa.

Example 3.4.1

Consider a mixture of species A, B, C and D, which undergo the following reaction:

$$4\,A + 3\,B \longrightarrow C + 2\,D$$

The mixture has initially 40, 30, 3 and 0 mol of A, B, C and D, respectively. We would like to calculate the final mole numbers of the species at 70% conversion of A.

Solution

At the initial state, $N_{A0} = 40$ mol. At the final state, $N_A = (1 - 0.70)N_{A0} = 12$ mol. Then from Equation (3.3) above, the extent of reaction at the final state is

$$X = \frac{\Delta N_A}{\nu_A} = \frac{N_A - N_{A0}}{\nu_A} = \frac{12 - 40}{-4} = 7$$

Application of Equation (3.3) on the previous page to species B yields

$$\Delta N_B \;=\; X\nu_B \;=\; 7 \times -3 \;=\; -21 \text{ mol}$$

Thus, the number of moles of B at the final state is

$$N_B \;=\; \Delta N_B + N_{B0} \;=\; -21 + 30 \;=\; 9 \text{ mol}$$

Likewise, the application of Equation (3.3) on the previous page to species C and D results in their number of moles at the final state: $N_C = 10$ mol and $N_D = 14$ mol, respectively.

❏

Generalization

For the general elementary chemical reaction

$$\alpha_1 A_1 + \alpha_2 A_2 + \ldots + \alpha_p A_p \longrightarrow \beta_1 B_1 + \beta_2 B_2 + \ldots + \beta_q B_q$$

the extent of reaction is given by

$$X \;=\; \frac{\Delta N_{A_1}}{\nu_{A_1}} = \frac{\Delta N_{A_2}}{\nu_{A_2}} = \cdots = \frac{\Delta N_{A_p}}{\nu_{A_p}} = \frac{\Delta N_{B_1}}{\nu_{B_1}} = \frac{\Delta N_{B_2}}{\nu_{B_2}} = \cdots = \frac{\Delta N_{B_q}}{\nu_{B_q}}$$

3.5   Rate of Reaction

Expressing Equation (3.3) on the previous page in terms of X per unit volume of the reaction mixture (i.e., $\tilde{X}$), and taking the derivative with respect to time, we get

$$\frac{d(\tilde{X}V)}{dt} \;=\; V\frac{d\tilde{X}}{dt} + \tilde{X}\frac{dV}{dt} \;=\; \frac{d}{dt}\left(\frac{N_Z - N_{Z0}}{\nu_Z}\right) \;=\; \frac{1}{\nu_Z}\frac{dN_Z}{dt}$$

Rearrangement of the above equation yields

$$\underbrace{\frac{d\tilde{X}}{dt}}_{r} \;=\; -\frac{\tilde{X}}{V}\frac{dV}{dt} + \frac{1}{\nu_Z V}\frac{dN_Z}{dt}$$

where r is called the **rate of reaction**. The molar rate of generation per unit volume of a reaction-species Z is given by

$$\underbrace{\frac{1}{V}\frac{dN_Z}{dt}}_{R_{\text{gen},Z}} \;=\; \nu_Z\left(r + \frac{\tilde{X}}{V}\frac{dV}{dt}\right)$$

Note that $R_{\text{gen},Z}$ would be negative if Z is consumed in the reaction. Similar to the extent of reaction (X), the rate of reaction (r) is the same for all species involved in a reaction.

For reaction mixtures with constant volume,

$$r = \frac{R_{\text{gen,Z}}}{\nu_{\text{Z}}} \quad \text{and} \quad R_{\text{gen,Z}} = \nu_{\text{Z}} r \tag{3.4}$$

Thus, from the last equation, we can determine the rate of change in mole numbers of all reaction-species with the help of r, which is furnished by a constitutive relation in terms of species concentrations.

The constitutive relation for the reaction rate of an elementary reaction, say,

$$a\,\mathrm{A} + b\,\mathrm{B} \longrightarrow c\,\mathrm{C} + d\,\mathrm{D}$$

is of the form

$$r = \underbrace{k_0 \exp\left(-\frac{E}{RT}\right)}_{k} c_{\mathrm{A}}^{a} c_{\mathrm{B}}^{b}$$

where (i) k is the reaction rate coefficient, (ii) c_{A} and c_{B} are, respectively, the concentrations of species A and B with a and b as stoichiometric coefficients, (iii) k_0 is the pre-exponential factor, (iv) E is the activation energy of the reaction (v) R is universal gas constant, and (vi) T is absolute temperature.

Reaction rates are included in the equations of change for moles, and temperature as follows.

3.5.1   Equations of Change for Moles

Let us consider a mixture of N_c species, which take part in N_r reactions. The volume of the mixture is constant. From Equation (3.4) above, the rate of generation per unit volume of the ith species of the mixture in the jth reaction is given by

$$R_{\text{gen},ij} = \nu_{ij} r_j \tag{3.5}$$

where ν_{ij} and r_j are, respectively, are the associated stoichiometric coefficient, and reaction rate. Substituting the above equation in Equation (2.13) on p. 24 and Equation (2.17) on p. 26, we obtain

$$\frac{dN_i}{dt} = -\Delta\left[\dot{N}_i\right] + V\sum_{j=1}^{N_r} \nu_{ij} r_j \quad \text{and} \tag{3.6}$$

$$\frac{\partial y_i}{\partial t} = -\tilde{\mathbf{v}} \cdot \nabla y_i - \frac{1}{c}\left(\nabla \cdot \mathbf{J}_i + y_i \sum_{k=1}^{N_c}\sum_{j=1}^{N_r} \nu_{kj} r_j - \sum_{j=1}^{N_r} \nu_{ij} r_j\right)$$

which are, respectively, the macroscopic mole balance, and the microscopic equation of change for the mole fraction of the ith species.

3.5.2 Equations of Change for Temperature

Recall the term involving enthalpy change due to the jth reaction in Equation (2.40) on p. 42. That term, with the help of Equation (3.5) on the previous page, can be expressed as

$$-\sum_{i=1}^{N_c} \overset{\bar{H}_i/M_i}{\overbrace{\bar{H}_i}} \times \underbrace{r_{\text{gen},ij}}_{R_{\text{gen},ij}M_i} = -\sum_{i=1}^{N_c} \bar{H}_i \times \underbrace{R_{\text{gen},ij}}_{\nu_{ij}r_j} = -r_j \sum_{i=1}^{N_c} \nu_{ij}\bar{H}_i = \underbrace{-\Delta H_{\text{r},j}r_j}_{\Delta H_{\text{r},j}}$$

where $\bar{H}_i$ and M_i are, respectively, the partial molar enthalpy, and molecular weight of the ith chemical species, and

$$\Delta H_{\text{r},j} \equiv \sum_{i=1}^{N_c} \nu_{ij}\bar{H}_i \tag{3.7}$$

is defined as the **heat of reaction** for the jth reaction. Thus, in terms of $\Delta H_{\text{r},j}$, the microscopic equation of change for temperature [Equation (3.2), p. 63] can be written as

$$\rho\hat{C}_{\text{P}}\frac{\partial T}{\partial t} = -\rho\hat{C}_{\text{P}}\mathbf{v}\cdot\nabla T + \nabla\cdot(k\nabla T) + \sum_{i=1}^{N_c}\tilde{H}_i\nabla\cdot\mathbf{j}_i + \sum_{j=1}^{N_r}(-\Delta H_{\text{r},j}r_j)$$

$$- \frac{T}{\rho}\frac{\partial\rho}{\partial T}\frac{\text{D}P}{\text{D}t} - \boldsymbol{\tau}:\nabla\mathbf{v} \tag{3.8}$$

Appendix 3.A.3 on p. 76 provides the above equation for a system of constant density (or constant and uniform pressure), and constant thermal conductivity in Cartesian, cylindrical, and spherical coordinates.

Incorporating $\Delta H_{\text{r},j}$, the macroscopic equation of change for temperature [Equation (2.43), p. 44] becomes

$$m\hat{C}_{\text{P}}\frac{\text{d}T}{\text{d}t} = -\Delta\left[\dot{m}\hat{H}\right] - \Delta\left[\dot{Q}\right] + \sum_{i=1}^{N_c}\tilde{H}_i\Delta\left[\dot{m}_i\right] + V\sum_{j=1}^{N_r}(-\Delta H_{\text{r},j}r_j) + \dot{W}_{\text{s}} \tag{3.9}$$

A more useful form of the above equation is obtained with the help of the heats of reaction. They can be expressed in terms of standard heats of reaction, which are typically available.

Standard Heat of Reaction

Consider the reaction
$$a\text{A} + b\text{B} \rightarrow c\text{C} + d\text{D}$$
under the standard condition, or state, which is at 25°C and 1 bar. Normally, $\bar{H}_i$ is assumed to be the same as molar enthalpy $\underline{H}_i$ of each pure component, i. In other words, $(\bar{H}_i - \underline{H}_i)$ is assumed negligible in comparison to energy changes due to chemical reaction, and temperature variations. Using this assumption in Equation (3.7) above, the standard heat of reaction is given by
$$\Delta H_{\text{r}}^{\circ} = c\underline{H}_{\text{C}}^{\circ} + d\underline{H}_{\text{D}}^{\circ} - a\underline{H}_{\text{A}}^{\circ} - b\underline{H}_{\text{B}}^{\circ} \tag{3.10}$$

where the superscript '°' denotes the standard state. Let Z stand for A, B, C, or D, which are produced from the stable atomic species, $\mathcal{A}_i$s, at the standard state in the formation reaction

$$z_1\mathcal{A}_1 + z_2\mathcal{A}_2 + \cdots + z_n\mathcal{A}_n \to Z$$

A stable atomic species is either a single atom, or a minimal cluster of identical atoms that is thermodynamically stable at the standard state. Examples of these species are He, H_2, N_2, O_2 (as gases), Fe (as alpha iron), C (as graphite), etc.

Note that if Z denotes A then z_is are a_is, if Z denotes B then z_is are b_is, and so on. Moreover, n is large enough to account for *all* $\mathcal{A}_i$s that are needed to form A, B, C and D. A z_i is zero in the formation reaction if the ith species is not involved. With this consideration, the standard heat of formation reaction (or, the heat of formation) for Z is given by

$$\Delta H_Z^\circ = \underline{H}_Z^\circ - \sum_{i=1}^n z_i \underline{H}_{\mathcal{A}_i}^\circ \quad \text{or} \quad \underline{H}_Z^\circ = \Delta H_Z^\circ + \sum_{i=1}^n z_i \underline{H}_{\mathcal{A}_i}^\circ$$

Substituting the last equation in Equation (3.10) on the previous page, we obtain

$$\Delta H_r^\circ = c\left(\Delta H_C^\circ + \sum_{i=1}^n c_i \underline{H}_{\mathcal{A}_i}^\circ\right) + d\left(\Delta H_D^\circ + \sum_{i=1}^n d_i \underline{H}_{\mathcal{A}_i}^\circ\right)$$
$$- a\left(\Delta H_A^\circ + \sum_{i=1}^n a_i \underline{H}_{\mathcal{A}_i}^\circ\right) - b\left(\Delta H_B^\circ + \sum_{i=1}^n b_i \underline{H}_{\mathcal{A}_i}^\circ\right)$$
$$= c\Delta H_C^\circ + d\Delta H_D^\circ - a\Delta H_A^\circ - b\Delta H_B^\circ + \sum_{i=1}^n \left[\underbrace{(\overbrace{cc_i + dd_i}^{\text{units of }\mathcal{A}_i\text{ in products}}) - (\overbrace{aa_i + bb_i}^{\text{units of }\mathcal{A}_i\text{ in reactants}})}_{=0}\right]\underline{H}_{\mathcal{A}_i}^\circ$$

The summation term in the last equation is zero since the number of atoms of the ith species (or, equivalently, the units of $\mathcal{A}_i$s) that are included in products is the same as that in reactants. The above result can be written for the jth elementary chemical reaction as

$$\Delta H_{r,j}^\circ = \sum_{i=1}^{N_c} \nu_{ij}\Delta H_i^\circ \tag{3.11}$$

where $\Delta H_{r,j}^\circ$ is the standard heat of that reaction, and ν_{ij} is the stoichiometric coefficient of the ith involved species having ΔH_i° as the heat of formation.

Enthalpy at Temperature above 25°C

Consider a species Z as a pure component, which is solid at the standard state but liquid at a higher temperature. At this temperature, the molar enthalpy of Z includes the energy needed to change itself from the solid to the liquid state. The molar enthalpy is then given by

$$\underline{H}_Z = \underbrace{\Delta H_Z^\circ}_{\text{heat of formation}} + \underbrace{\int_{T=25°C}^{T_{Z,melt}} \underline{C}_{P_Z,\text{solid}}\,dT}_{\substack{\text{energy intake in}\\\text{the solid state}}} + \Delta H_{Z,melt} + \underbrace{\int_{T_{Z,melt}}^{T<T_{Z,boil}} \underline{C}_{P_Z,\text{liquid}}\,dT}_{\substack{\text{energy intake in}\\\text{the liquid state}}}$$

where $T_{\text{Z,melt}}$, $\underline{C}_{\text{Pz,solid}}$, $\Delta H_{\text{Z,melt}}$, $T_{\text{Z,boil}}$, and $\underline{C}_{\text{Pz,liquid}}$, are, respectively, the melting point, solid-state molar specific heat capacity, molar latent heat of melting, boiling point, and liquid-state molar specific heat capacity of Z.

In general, for the ith species that undergoes N_π changes in states from the standard state, the molar enthalpy at the temperature of the final state can be written as

$$\underline{H}_i \;=\; \Delta\underline{H}_i^\circ + \underbrace{\int_{T=25°C}^{T<T_{i,1}} \underline{C}_{\text{P}_i,0}\,\mathrm{d}T + \sum_{k=1}^{N_\pi} u(T-T_{i,k})\!\left[\Delta\underline{H}_{i,k} + \int_{T_{i,k}}^{T<T_{i,k+1}} \underline{C}_{\text{P}_i,k}\,\mathrm{d}T\right]}_{\equiv \Delta\underline{H}_i}$$

In the above equation,

1. the second subscript denotes the state of the species,

2. $u(x)$ is a unit step function, which is zero for negative x but one otherwise,

3. $\Delta\underline{H}_{i,k}$ is the change in molar enthalpy of the ith species associated to its transition to the kth state from the previous one, and

4. $\Delta\underline{H}_i$ is the molar enthalpy in excess of the heat of formation.

Heat of Reaction at Elevated Temperature

Considering $\bar{H}_i$ the same as $\underline{H}_i$, and using the last equation in the definition of the heat of reaction [Equation (3.7), p. 67], we obtain

$$\Delta H_{\text{r},j} \;=\; \underbrace{\sum_{i=1}^{N_c} \nu_{ij}\Delta\underline{H}_i^\circ}_{=\Delta H_{\text{r},j}^\circ} \;+\; \sum_{i=1}^{N_c} \nu_{ij}\Delta\underline{H}_i \;=\; \Delta H_{\text{r},j}^\circ + \sum_{i=1}^{N_c} \nu_{ij}\Delta\underline{H}_i$$

[Equation (3.11), previous page]

which is the heat of reaction for the jth reaction at a temperature above 25°C.

With the help of the above results, we are now able to derive a practical equation of change for temperature in a macroscopic system.

3.5.3   *Macroscopic Equation of Change for Temperature*

Consider a macroscopic system similar to Figure 2.11 on p. 35. The system is fed through entry ports by fluid streams at different temperatures. The species in the streams react in the system to form products. There is no change of state, and the system is perfectly mixed so that the exit ports (usually one) the outgoing fluid streams are at system conditions. For this system, which is known as CSTR, we will use the above results to simplify Equation (3.9) on p. 67. Note that $\tilde{H}_i$ in that equation is *not* the property of the stream at a port, but of the system at system conditions – precisely at system temperature, pressure and composition [see Equation (2.50), p. 47].

The second term of Equation (3.9) on p. 67 is given by

$$
-\Delta\left[\dot{m}\hat{H}\right] = \sum_{j=1}^{N_{\text{in}}}(\dot{m}_j\hat{H}_j)_{\text{in}} - \sum_{j=1}^{N_{\text{out}}}(\dot{m}_j\hat{H}_j)_{\text{out}} = \sum_{j=1}^{N_{\text{in}}}\dot{m}_{jf}\underbrace{\hat{H}_{jf}}_{\sum\limits_{i=1}^{N_c}\omega_{ijf}\tilde{H}_{ijT_f}} - \sum_{j=1}^{N_{\text{out}}}\dot{m}_j\hat{H}_j
$$

where $\tilde{H}_{ijT_f}$ is the partial specific enthalpy of the ith species having mass fraction ω_{ijf} in the feed mixture of temperature T_f at the jth port of entry.

The fourth term of Equation (3.9) on p. 67 is given by,

$$
\sum_{i=1}^{N_c}\tilde{H}_i\Delta\left[\dot{m}_i\right] = -\sum_{j=1}^{N_{\text{in}}}\sum_{i=1}^{N_c}\underbrace{\dot{m}_{ijf}}_{\dot{m}_{jf}\omega_{ijf}}\tilde{H}_i + \sum_{j=1}^{N_{\text{out}}}\sum_{i=1}^{N_c}\underbrace{\dot{m}_{ij}}_{\dot{m}_j\omega_{ij}}\tilde{H}_i
$$

$$
= -\sum_{j=1}^{N_{\text{in}}}\dot{m}_{jf}\sum_{i=1}^{N_c}\omega_{ijf}\tilde{H}_i + \sum_{j=1}^{N_{\text{out}}}\dot{m}_j\underbrace{\sum_{i=1}^{N_c}\omega_{ij}\tilde{H}_i}_{\hat{H}_j}
$$

Substituting the above expressions of the second and fourth terms in Equation (3.9) on p. 67, and simplifying the result, we obtain

$$
m\hat{C}_P\frac{\mathrm{d}T}{\mathrm{d}t} = \sum_{j=1}^{N_{\text{in}}}\dot{m}_{jf}\sum_{i=1}^{N_c}\omega_{ijf}(\tilde{H}_{ijT_f} - \tilde{H}_i) - \Delta\left[\dot{Q}\right] + V\sum_{j=1}^{N_r}(-\Delta H_{r,j}r_j) + \dot{W}_s \quad (3.12)
$$

Relative to the system temperature T, and for no change in system state, we can write

$$
\tilde{H}_{ijT_f} = \tilde{H}_{ijT} + \int_{T}^{T_{jf}}\tilde{C}_{P,ij}\,\mathrm{d}T
$$

where $\tilde{H}_{ijT}$ is the partial specific enthalpy of the ith species in the jth port of feed mixture at temperature T, and $\tilde{C}_{P,ij}$ is the partial specific heat capacity of the species in that feed mixture. Substituting the above equation in Equation (3.12) above, we obtain after simplification

$$
m\hat{C}_P\frac{\mathrm{d}T}{\mathrm{d}t} = \sum_{j=1}^{N_{\text{in}}}\dot{m}_{jf}\sum_{i=1}^{N_c}\omega_{ijf}(\tilde{H}_{ijT} - \tilde{H}_i) + \sum_{j=1}^{N_{\text{in}}}\dot{m}_{jf}\sum_{i=1}^{N_c}\omega_{ijf}\int_{T}^{T_{jf}}\tilde{C}_{P,ij}\,\mathrm{d}T - \Delta\left[\dot{Q}\right]
$$

$$
+ V\sum_{j=1}^{N_r}(-\Delta H_{r,j}r_j) + \dot{W}_s
$$

For ideal mixtures, $\tilde{H}_{ijT}$ and $\tilde{H}_i$ are the same, and $\tilde{C}_{P,ij}$ is equal to $\hat{C}_{P,ij}$. Moreover, considering the change in $\hat{C}_{P,ij}$ to be negligible over the expected range of temperature variation in a process, the above equation simplifies to

$$m\hat{C}_\text{P}\frac{dT}{dt} = \sum_{j=1}^{N_\text{in}}\sum_{i=1}^{N_\text{c}}\dot{m}_{ijf}\hat{C}_{\text{P},ij}(T_{jf} - T) - \Delta\left[\dot{Q}\right] + V\sum_{j=1}^{N_\text{r}}(-\Delta H_{\text{r},j}r_j) + \dot{W}_\text{s} \qquad (3.13)$$

The above equation is the most widely used macroscopic equation of change for temperature in reactive chemical systems. For systems that do not involve fluid flow, and chemical reactions, the above equation simplifies to

$$m\hat{C}_\text{P}\frac{dT}{dt} = -\Delta\left[\dot{Q}\right] + \dot{W}_\text{s} \qquad\qquad (3.14)$$

3.6   Interphase Transfer

The constitutive relations that have been presented so far are applicable in a continuous medium, i.e., a phase. In these relations, the fluxes are, respectively, functions of the gradients (i.e., spatial derivatives) of concentration, velocity and temperature in the medium. However, the gradients are not defined at an interface, which is a discontinuity, between two phases. Consequently, the constitutive relations for interphase transfer involve property differences between the phases. The proportionality constants are coefficients, which are determined from experiments.

Mass Transfer

A constitutive relation for interphase molar flux of a species A is

$$\mathbf{J} = k_\text{c}\Delta c_\text{A}$$

where k_c is mass transfer coefficient, and Δc_A is the difference between the molar concentrations of the species in two phases across the interface.

Momentum Transfer

For interphase momentum transfer, the constitutive relation is

$$\frac{F}{A} = fE_\text{k}$$

where F is the force imparted at the interface, A is contact area, f is drag coefficient, and E_k is characteristic kinetic energy per unit volume. For fluid flow in conduits, $E_\text{k} = \rho\bar{v}^2/2$ where $\bar{v}$ is the average fluid velocity along the axis of the conduit. For fluid flow around objects, $E_\text{k} = \rho v_\infty^2/2$ where v_∞ is the velocity of fluid approaching the object from a large distance.

Heat Transfer

Similarly, for interphase heat transfer, the constitutive relation is

$$\frac{q}{A} = h\Delta T$$

where q is the rate of heat transfer across the area A, h is heat transfer coefficient, and ΔT is the temperature difference between two phases in thermal contact.

Finally, we present some important constitutive relations that associate properties of a system under equilibrium. These relations stem from thermodynamics.

3.7 Thermodynamic Relations

Some useful thermodynamic relations are as follows.

Dalton's Law

The partial pressure of a species in a gas mixture is the pressure that species would exert if it were to occupy the entire volume at the same temperature. According to Dalton's Law, the total pressure of a non-reacting, ideal gas mixture of N_c species is given by

$$P = \sum_{i=1}^{N_c} p_i$$

where p_i is the partial pressure of the ith species.

Lewis–Randall Rule

According to this rule, the fugacity* of the ith species in a gaseous mixture is given by

$$\bar{f}_i = y_i f_i$$

where y_i is the mole fraction of the species, and f_i is the fugacity of the pure species in the same physical state, temperature and pressure as that of the mixture.

Raoult's Law

Consider an ideal solution of a number of species in equilibrium with the vapor phase. Then according to Raoult's law, the partial pressure of the ith species in the vapor phase is given by

$$p_i = x_i p_i^{\text{v}}$$

where x_i is the mole fraction of the species in the solution, and p_i^{v} is the vapor pressure of the species in the pure state at the same temperature as that of the solution.

Henry's Law

According to this law, the fugacity of the ith species at close-to-zero concentrations in a liquid mixture is given by

$$\bar{f}_i = x_i H_i$$

where x_i is the mole fraction of the species in the mixture, and H_i is Henry's law constant for the species.

---

*It is the partial pressure of a species in an ideal gas mixture that has the same chemical potential as that in the given gas mixture.

If the mixture is under equilibrium with a vapor phase, and at very low pressure then the partial pressure of the ith species in the vapor phase is given by

$$p_i = x_i H_i$$

Antoine's Equation

This equation relates the vapor pressure (P^{v}) of a pure species to absolute temperature (T) as

$$\ln P^{\text{v}} = A - \frac{B}{C \mid T}$$

where A, B and C are constants depending on the species.

Vant Hoff's Relation

This is the relation between the equilibrium constant (K) of a chemical reaction, and absolute temperature (T). According to this relation

$$\frac{\mathrm{d} \ln K}{\mathrm{d} T} = \frac{\Delta H_{\text{r}}}{RT^2}$$

where ΔH_{r} is the heat of reaction, and R is universal gas constant. Integrating the above equation for K varying from $K_1(T_1)$ to $K_2(T_2)$, we get

$$\ln \left(\frac{K_2}{K_1} \right) = -\frac{\Delta H_{\text{r}}}{R} \left(\frac{1}{T_2} - \frac{1}{T_1} \right)$$

Given T_1, K_1 and T_2, the above equation yields K_2.

Equations of State

These are thermodynamic equations that relate two or more intensive variables of a system at equilibrium. Common, measurable intensive variables are pressure, temperature, specific volume, and mole fractions of species. From the equation of state for a system, all of its thermodynamic properties can be obtained. Equations of state are required for the determination of pressure in compressible systems under motion.

The simplest equation of state is the ideal gas law, which is applicable to systems at pressures close to zero. Another example is Soave–Redlich–Kwong equation of state[6],

$$P = \frac{RT}{V - b} - \frac{a\alpha}{V(V + b)}$$

where

$$a = \frac{0.42747 R^2 T_{\text{c}}^2}{P_{\text{c}}}, \qquad b = \frac{0.08664 R T_{\text{c}}}{P_{\text{c}}} \qquad \text{and}$$

$$\alpha = \left[1 + \left(0.48 + 1.574\omega - 0.176\omega^2 \right) \left(1 - \sqrt{\frac{T}{T_{\text{c}}}} \right) \right]^2$$

In the above equations, P is pressure, V is molar volume, R is universal gas constant, T is absolute temperature, T_{c} is critical temperature, P_{c} is critical pressure, and ω is the acentric factor of the species.

3.A   Equations in Cartesian, Cylindrical and Spherical Coordinate Systems

In the following sections, we present the important equations derived in this chapter for Cartesian, cylindrical and spherical coordinate systems. The equations in the last two systems are derivable from those in Cartesian coordinate system using suitable coordinate transformations [see Chapter 5, p. 139]. The coordinate notation is the same as that given on p. 53.

3.A.1   Equations of Continuity for Binary Systems of Constant Density and Diffusivity

For constant density systems comprising two chemical species of constant diffusivity, the equations of continuity of a species A are as follows:

Cartesian Coordinates

$$\frac{\partial \omega_A}{\partial t} = -v_x \frac{\partial \omega_A}{\partial x} - v_y \frac{\partial \omega_A}{\partial y} - v_z \frac{\partial \omega_A}{\partial z} + D_{AB}\left(\frac{\partial^2 \omega_A}{\partial x^2} + \frac{\partial^2 \omega_A}{\partial y^2} + \frac{\partial^2 \omega_A}{\partial z^2} \right)$$

$$+ \frac{1}{\rho} \sum_{j=1}^{N_r} r_{\mathrm{gen},Aj}$$

Cylindrical Coordinates

$$\frac{\partial \omega_A}{\partial t} = -v_r \frac{\partial \omega_A}{\partial r} - \frac{v_\theta}{r} \frac{\partial \omega_A}{\partial \theta} - v_z \frac{\partial \omega_A}{\partial z} + D_{AB}\left(\frac{\partial^2 \omega_A}{\partial r^2} + \frac{1}{r} \frac{\partial \omega_A}{\partial r} + \frac{1}{r^2} \frac{\partial^2 \omega_A}{\partial \theta^2} + \frac{\partial^2 \omega_A}{\partial z^2} \right)$$

$$+ \frac{1}{\rho} \sum_{j=1}^{N_r} r_{\mathrm{gen},Aj}$$

Spherical Coordinates

$$\frac{\partial \omega_A}{\partial t} = -v_r \frac{\partial \omega_A}{\partial r} - \frac{v_\theta}{r} \frac{\partial \omega_A}{\partial \theta} - \frac{v_\phi}{r \sin \theta} \frac{\partial \omega_A}{\partial \phi} + D_{AB}\left(\frac{\partial^2 \omega_A}{\partial r^2} + \frac{2}{r} \frac{\partial \omega_A}{\partial r} + \frac{1}{r^2} \frac{\partial^2 \omega_A}{\partial \theta^2} \right.$$

$$\left. + \frac{1}{r^2 \tan \theta} \frac{\partial \omega_A}{\partial \theta} + \frac{1}{r^2 \sin^2 \theta} \frac{\partial^2 \omega_A}{\partial \phi^2} \right) + \frac{1}{\rho} \sum_{j=1}^{N_r} r_{\mathrm{gen},Aj}$$

3.A.2 Equations of Motion for Newtonian Fluids of Constant Density and Viscosity

Cartesian Coordinates

$$\frac{\partial v_x}{\partial t} = -v_x \frac{\partial v_x}{\partial x} - v_y \frac{\partial v_x}{\partial y} - v_z \frac{\partial v_x}{\partial z} + \frac{\mu}{\rho} \left(\frac{\partial^2 v_x}{\partial x^2} + \frac{\partial^2 v_x}{\partial y^2} + \frac{\partial^2 v_x}{\partial z^2} \right) - \frac{1}{\rho} \frac{\partial P}{\partial x} + g_x$$

$$\frac{\partial v_y}{\partial t} = -v_x \frac{\partial v_y}{\partial x} - v_y \frac{\partial v_y}{\partial y} - v_z \frac{\partial v_y}{\partial z} + \frac{\mu}{\rho} \left(\frac{\partial^2 v_y}{\partial x^2} + \frac{\partial^2 v_y}{\partial y^2} + \frac{\partial^2 v_y}{\partial z^2} \right) - \frac{1}{\rho} \frac{\partial P}{\partial y} + g_y$$

$$\frac{\partial v_z}{\partial t} = -v_x \frac{\partial v_z}{\partial x} - v_y \frac{\partial v_z}{\partial y} - v_z \frac{\partial v_z}{\partial z} + \frac{\mu}{\rho} \left(\frac{\partial^2 v_z}{\partial x^2} + \frac{\partial^2 v_z}{\partial y^2} + \frac{\partial^2 v_z}{\partial z^2} \right) - \frac{1}{\rho} \frac{\partial P}{\partial z} + g_z$$

Cylindrical Coordinates

$$\frac{\partial v_r}{\partial t} = -v_r \frac{\partial v_r}{\partial r} - \frac{v_\theta}{r} \frac{\partial v_r}{\partial \theta} - v_z \frac{\partial v_r}{\partial z} + \frac{\mu}{\rho} \left(\frac{\partial^2 v_r}{\partial r^2} + \frac{1}{r} \frac{\partial v_r}{\partial r} + \frac{1}{r^2} \frac{\partial^2 v_r}{\partial \theta^2} \right.$$

$$\left. + \frac{\partial^2 v_r}{\partial z^2} - \frac{2}{r^2} \frac{\partial v_\theta}{\partial \theta} - \frac{v_r}{r^2} \right) + \frac{v_\theta^2}{r} - \frac{1}{\rho} \frac{\partial P}{\partial r} + g_r$$

$$\frac{\partial v_\theta}{\partial t} = -v_r \frac{\partial v_\theta}{\partial r} - \frac{v_\theta}{r} \frac{\partial v_\theta}{\partial \theta} - v_z \frac{\partial v_\theta}{\partial z} + \frac{\mu}{\rho} \left(\frac{\partial^2 v_\theta}{\partial r^2} + \frac{1}{r} \frac{\partial v_\theta}{\partial r} + \frac{1}{r^2} \frac{\partial^2 v_\theta}{\partial \theta^2} \right.$$

$$\left. + \frac{\partial^2 v_\theta}{\partial z^2} + \frac{2}{r^2} \frac{\partial v_r}{\partial \theta} - \frac{v_\theta}{r^2} \right) - \frac{v_r v_\theta}{r} - \frac{1}{r\rho} \frac{\partial P}{\partial \theta} + g_\theta$$

$$\frac{\partial v_z}{\partial t} = -v_r \frac{\partial v_z}{\partial r} - \frac{v_\theta}{r} \frac{\partial v_z}{\partial \theta} - v_z \frac{\partial v_z}{\partial z} + \frac{\mu}{\rho} \left(\frac{\partial^2 v_z}{\partial r^2} + \frac{1}{r} \frac{\partial v_z}{\partial r} + \frac{1}{r^2} \frac{\partial^2 v_z}{\partial \theta^2} \right.$$

$$\left. + \frac{\partial^2 v_z}{\partial z^2} \right) - \frac{1}{\rho} \frac{\partial P}{\partial z} + g_z$$

Spherical Coordinates

$$\frac{\partial v_r}{\partial t} = -v_r \frac{\partial v_r}{\partial r} - \frac{v_\theta}{r}\frac{\partial v_r}{\partial \theta} - \frac{v_\phi}{r\sin\theta}\frac{\partial v_r}{\partial \phi} + \frac{\mu}{\rho}\left[\frac{\partial^2 v_r}{\partial r^2} + \frac{4}{r}\frac{\partial v_r}{\partial r} + \frac{1}{r^2}\frac{\partial^2 v_r}{\partial \theta^2}\right.$$

$$\left. + \frac{1}{r^2\tan\theta}\frac{\partial v_r}{\partial \theta} + \frac{1}{r^2\sin^2\theta}\frac{\partial^2 v_r}{\partial \phi^2} + \frac{2v_r}{r^2}\right] + \frac{v_\theta^2 + v_\phi^2}{r} - \frac{1}{\rho}\frac{\partial P}{\partial r} + g_r$$

$$\frac{\partial v_\theta}{\partial t} = -v_r \frac{\partial v_\theta}{\partial r} - \frac{v_\theta}{r}\frac{\partial v_\theta}{\partial \theta} - \frac{v_\phi}{r\sin\theta}\frac{\partial v_\theta}{\partial \phi} + \frac{\mu}{\rho}\left[\frac{\partial^2 v_\theta}{\partial r^2} + \frac{4}{r}\frac{\partial v_\theta}{\partial r} + \frac{1}{r^2}\frac{\partial^2 v_\theta}{\partial \theta^2}\right.$$

$$\left. + \frac{1}{r^2\tan\theta}\frac{\partial v_\theta}{\partial \theta} + \frac{1}{r^2\sin^2\theta}\frac{\partial^2 v_\theta}{\partial \phi^2} - \frac{2\cos\theta}{r^2\sin^2\theta}\frac{\partial v_\phi}{\partial \phi} + \frac{2}{r^2}\frac{\partial v_r}{\partial \theta} - \frac{v_\theta}{r^2\sin^2\theta}\right.$$

$$\left. + \frac{2v_\theta}{r^2}\right] - \frac{v_r v_\theta}{r} + \frac{v_\phi^2}{r\tan\theta} - \frac{1}{r\rho}\frac{\partial P}{\partial \theta} + g_\theta$$

$$\frac{\partial v_\phi}{\partial t} = -v_r \frac{\partial v_\phi}{\partial r} - \frac{v_\theta}{r}\frac{\partial v_\phi}{\partial \theta} - \frac{v_\phi}{r\sin\theta}\frac{\partial v_\phi}{\partial \phi} + \frac{\mu}{\rho}\left[\frac{\partial^2 v_\phi}{\partial r^2} + \frac{4}{r}\frac{\partial v_\phi}{\partial r} + \frac{1}{r^2}\frac{\partial^2 v_\phi}{\partial \theta^2}\right.$$

$$\left. + \frac{1}{r^2\tan\theta}\frac{\partial v_\phi}{\partial \theta} + \frac{1}{r^2\sin^2\theta}\frac{\partial^2 v_\phi}{\partial \phi^2} + \frac{2}{r^2\sin\theta}\frac{\partial v_r}{\partial \phi} + \frac{1}{r^2\sin^2\theta}\left(2\cos\theta\frac{\partial v_\theta}{\partial \phi} - v_\phi\right)\right.$$

$$\left. + \frac{2v_\phi}{r^2}\right] - \frac{v_r v_\phi}{r} - \frac{v_\theta v_\phi}{r\tan\theta} - \frac{1}{r\rho\sin\theta}\frac{\partial P}{\partial \phi} + g_\phi$$

$$(3.15)$$

3.A.3 Equations of Change for Temperature in Non-Reactive, Non-Viscous Dissipative Systems of Constant Density and Thermal Conductivity

Cartesian Coordinates

$$\frac{\partial T}{\partial t} = -v_x \frac{\partial T}{\partial x} - v_y \frac{\partial T}{\partial y} - v_z \frac{\partial T}{\partial z} + \frac{k}{\rho \hat{C}_P}\left(\frac{\partial^2 T}{\partial x^2} + \frac{\partial^2 T}{\partial y^2} + \frac{\partial^2 T}{\partial z^2}\right)$$

Cylindrical Coordinates

$$\frac{\partial T}{\partial t} = -v_r \frac{\partial T}{\partial r} - \frac{v_\theta}{r} \frac{\partial T}{\partial \theta} - v_z \frac{\partial T}{\partial z} + \frac{k}{\rho \hat{C}_P} \left(\frac{\partial^2 T}{\partial r^2} + \frac{1}{r} \frac{\partial T}{\partial r} + \frac{1}{r^2} \frac{\partial^2 T}{\partial \theta^2} + \frac{\partial^2 T}{\partial z^2} \right)$$

Spherical Coordinates

$$\frac{\partial T}{\partial t} = -v_r \frac{\partial T}{\partial r} - \frac{v_\theta}{r} \frac{\partial T}{\partial \theta} - \frac{v_\phi}{r \sin \theta} \frac{\partial T}{\partial \phi} + \frac{k}{\rho \hat{C}_P} \left(\frac{\partial^2 T}{\partial r^2} + \frac{2}{r} \frac{\partial T}{\partial r} + \frac{1}{r^2} \frac{\partial^2 T}{\partial \theta^2} + \frac{1}{r^2 \tan \theta} \frac{\partial T}{\partial \theta} \right.$$
$$\left. + \frac{1}{r^2 \sin^2 \theta} \frac{\partial^2 T}{\partial \phi^2} \right)$$

References

[1] R. Taylor and R. Krishna. *Multicomponent Mass Transfer*. New York: John Wiley & Sons, 1993.

[2] R.B. Bird, W.E. Stewart, and E.N. Lightfoot. *Transport Phenomena*. 2nd. pp. 18–19. New York: John Wiley & Sons, Inc., 2007.

[3] G.K. Batchelor. *An Introduction to Fluid Dynamics*. pp. 142–147. New York: Cambridge University Press, 2000. Chap. 3.

[4] M.M. Cross. "Relation between Viscoelasticity and Shear-Thinning Behaviour in Liquids". In: *Rheologica Acta* 18.5 (1979), pp. 609–614.

[5] M.L. Williams, R.F. Landel, and J.D. Ferry. "The Temperature Dependence of Relaxation Mechanisms in Amorphous Polymers and Other Glass-Forming Liquids". In: *Journal of the American Chemical Society* 77.14 (1955), pp. 3701–3707.

[6] G. Soave. "Equilibrium constants from a modified Redlich–Kwong equation of state". In: *Chemical Engineering Science* 27 (1972), pp. 1197–1203.

Bibliography

[1] R. Aris. *Vectors, Tensors, and the Basic Equations of Fluid Mechanics*. New York: Dover Publications Inc., 1989.

[2] M.M. Denn. *Process Modeling*. Chapter 5. Massachusetts: Pitman Publishing Inc., 1986.

[3] S.R. de Groot and P. Mazur. *Non-Equilibrium Thermodynamics*. New York: Dover Publications Inc., 1984.

[4] J.M. Smith, H. Van Ness, and M. Abbott. *Introduction to Chemical Engineering Thermodynamics*. 7th edition. New York: McGraw-Hill, 2005.

Exercises

3.1 Why does the diffusivity matrix for the mixture of N_c species [see Section 3.1.1, p. 60] has $(N_c - 1)$ rows, and $(N_c - 1)$ columns?

3.2 Explain the significance of the negative sign in Fick's law of diffusion, Newton's law of viscosity, and Fourier's law of heat conduction.

3.3 Derive the equations of change in reactive systems with variable volume.

3.4 Do a literature search to find a constitutive relation for the determination of liquid flow rate in a porous medium.

4

Model Formulation

In this chapter, we present the formulation of process models, which involves carrying out the following steps:

1. description of objectives
2. specification of system
3. laying out of assumptions
4. interrelation of system properties
5. consolidation of resulting equations

Description of Objectives

Model formulation begins with the description of objectives that are motivated by what is required to be attained. These objectives are related to one or more processes taking place in a system, and eventually to measurable properties such as concentration, velocity and temperature that should be determined. The properties may be subject to changes with time and space. Consequently, the objectives implicitly include the development of equations by interconnecting system properties to capture their behavior.

Specification of System

Once the objectives are set in place, they help in a clear specification of a system as the basis for the process model. Depending on the expected behavior of the properties, the system could be uniform, or not. If the properties change spatially then the system is non-uniform. In this case, a differential element is specified according to the geometry of the system. Any possibility of the change in system size is identified. If a system is a combination of several systems then each one of them is specified analogously.

Laying out of Assumptions

In this step, we decide on the level of sophistication in the model to be formulated. Accordingly, we make assumptions that exclude one or more sub-processes from the model. This step is based on real-world observations, and information from previous experience. Note that assumptions are open to revision in the cycle of model development.

Process Modeling and Simulation for Chemical Engineers: Theory and Practice, First Edition. Simant Ranjan Upreti.
© 2017 John Wiley & Sons Ltd. Published 2017 by John Wiley & Sons Ltd.
Companion website: www.wiley.com/go/upreti/pms_for_chemical_engineers

Interrelation of System Properties

This is the most involved step of model formulation. In this step, fundamental and constitutive relations are used to describe system properties as per the objectives and assumptions. If the objective is to find mass (moles), species mass (mole) composition, or their changes then we use mass (mole) balances. To find velocity, or how it changes, we use momentum balance. To find temperature, or its change, we use the equation of change for temperature.

Macroscopic equations are used for a uniform system, or a differential element of the system having spatial variation in its size. In case of spatial variations in intensive properties, microscopic equations are applied on a differential element of the system. Chapter 2 on p. 17 provides fundamental macroscopic equations as well as microscopic ones in Cartesian, cylindrical and spherical coordinates. Based on these equations, Chapter 3 on p. 59 provides equations incorporating constitutive relations for common systems.

Consolidation of Resulting Equations

In this step, the equations obtained in the previous step are collated, and checked for dimensional consistency. Additional constitutive relations applicable to the system are obtained, and assimilated with existing equations. The initial and boundary conditions of differential equations [see Section 4.A, p. 131] are ascertained based on specifications and assumptions. Finally, all parameters are identified that are needed to solve the equations. The resulting set of equations constitutes the desired model.

In the remaining chapter, we illustrate the aforementioned steps of model formulation using several examples categorized under uniform, or lumped-parameter systems, and distributed-parameter systems. In the last category, we take up examples from Cartesian, cylindrical and spherical coordinate systems. Emphasis is on model formulation from basic principles as much as possible. All models are developed in a time-dependent scenario. Eliminating the time derivatives in the models will result in corresponding steady-state models.

4.1 Lumped-Parameter Systems

This section includes the formulation of models incorporating mass, momentum and energy balances in lumped-parameter systems.

4.1.1 Isothermal CSTR

This example involves macroscopic mass balance in the presence of chemical reactions in a continuous-flow stirred-tank reactor i.e., CSTR. Widely used in chemical industry, this reactor is typically used for species in liquid phase. Reactants are continuously fed to the reactor where they react in an agitated mixture to form products. The latter are simultaneously withdrawn from the reactor for further processing.

Consider the following set of elementary reactions with rate coefficients k_1 and k_2:

$$\text{A} + \text{B} \xrightarrow{k_1} \text{C} \qquad \text{and} \qquad \text{B} + \text{C} \xrightarrow{k_2} \text{D}$$

The above reactions are carried out in the liquid phase at a constant temperature in the CSTR shown in Figure 4.1 on the next page. The reactor is fed by liquid streams of reactants A

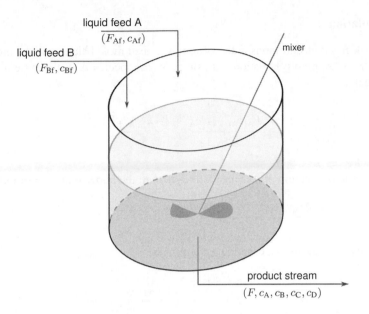

Figure 4.1   A CSTR carrying out liquid phase reactions at a constant temperature

and B, respectively, with (i) volumetric flow rates F_{Af} and F_{Bf}, and (ii) molar concentrations c_{Af} and c_{Bf}. The reactions result in the intermediate product C, and the final product D. The products along with residual reactants are withdrawn at a volumetric flow rate F from the reactor. The molar concentrations in the product stream are c_A, c_B, c_C, c_D, respectively, for A, B, C and D.

Objective

It is desired to develop a model that would enable the determination of species concentrations in the reactor as a function of time.

System

The system for this purpose is the reaction mixture inside the reactor.

Assumptions

We make the following assumptions for the model:

1. The temperature of the feed streams is the same as that of the reaction mixture, which is kept constant.

2. The reaction mixture is perfectly mixed so that its composition is the same as that of the product stream.

3. The volume of the reaction mixture is kept constant.

4. There is no vaporization loss of species.

Model Formulation

To determine species concentrations in the system, we need mole balances in the presence of chemical reactions. Applying Equation (3.6) on p. 66 for species A in the system (reaction mixture), we get

$$\underbrace{\frac{\mathrm{d}}{\mathrm{d}t}(V c_{\mathrm{A}})}_{\substack{\text{moles of A} \\ \text{in the system}}} = \underbrace{F_{\mathrm{Af}} c_{\mathrm{Af}} - F c_{\mathrm{A}}}_{-\Delta [\dot{N}_{\mathrm{A}}]} + \underbrace{(-k_1 c_{\mathrm{A}} c_{\mathrm{B}}) V}_{r_{\mathrm{A}}}$$

where r_{A} is the rate of reaction of A. Since V is constant, the above equation simplifies to

$$\frac{\mathrm{d} c_{\mathrm{A}}}{\mathrm{d}t} = \frac{F_{\mathrm{Af}} c_{\mathrm{Af}} - F c_{\mathrm{A}}}{V} - k_1 c_{\mathrm{A}} c_{\mathrm{B}} \tag{4.1}$$

The mole balances for species B, C and D are similarly

$$\frac{\mathrm{d} c_{\mathrm{B}}}{\mathrm{d}t} = \frac{F_{\mathrm{Bf}} c_{\mathrm{Bf}} - F c_{\mathrm{B}}}{V} + \underbrace{(-k_1 c_{\mathrm{A}} c_{\mathrm{B}} - k_2 c_{\mathrm{B}} c_{\mathrm{C}})}_{r_{\mathrm{B}}} \tag{4.2}$$

$$\frac{\mathrm{d} c_{\mathrm{C}}}{\mathrm{d}t} = -\frac{F c_{\mathrm{C}}}{V} + \underbrace{(k_1 c_{\mathrm{A}} c_{\mathrm{B}} - k_2 c_{\mathrm{B}} c_{\mathrm{C}})}_{r_{\mathrm{C}}} \tag{4.3}$$

$$\frac{\mathrm{d} c_{\mathrm{D}}}{\mathrm{d}t} = -\frac{F c_{\mathrm{D}}}{V} + \underbrace{k_1 c_{\mathrm{B}} c_{\mathrm{C}}}_{r_{\mathrm{D}}} \tag{4.4}$$

where r_{B}, r_{C} and r_{D} are, respectively, the reaction rates of B, C and D.

Each one of Equations (4.1)–(4.4) is a differential equation of first order with respect to time and, therefore, needs a condition for integration. The conditions are usually provided at the initial time, and are as follows.

Initial Conditions

With initial species concentrations in the reactor as c_{A0}, c_{B0}, c_{C0} and c_{D0}, the initial conditions for Equations (4.1)–(4.4) are

$$c_j(0) = c_{j0}; \qquad j = \mathrm{A, B, C, D} \tag{4.5}$$

Summary

Equations (4.1)–(4.5) constitute the dynamic model of the CSTR. Given the specifications for the parameter set

$$\{F_{\mathrm{Af}},\ F_{\mathrm{Bf}},\ F,\ c_{\mathrm{Af}},\ c_{\mathrm{Bf}},\ k_1,\ k_2,\ V,\ c_{\mathrm{A0}},\ c_{\mathrm{B0}},\ c_{\mathrm{C0}},\ c_{\mathrm{D0}}\}$$

the involved differential equations need to be integrated simultaneously to obtain

$$c_{\mathrm{A}} = c_{\mathrm{A}}(t), \quad c_{\mathrm{B}} = c_{\mathrm{B}}(t), \quad c_{\mathrm{C}} = c_{\mathrm{C}}(t) \quad \text{and} \quad c_{\mathrm{D}} = c_{\mathrm{D}}(t)$$

Steady State Model

At steady state, the time derivatives in the differential equations [Equations (4.1)–(4.4), previous page] become zero. The resulting algebraic equations,

$$F_{Af}c_{Af} - Fc_A - Vk_1c_Ac_B = 0$$

$$F_{Bf}c_{Bf} - Fc_B - V(k_1c_Ac_B + k_2c_Bc_C) = 0$$

$$-Fc_C + V(k_1c_Ac_B - k_2c_Bc_C) = 0$$

$$Fc_D + Vk_1c_Bc_C = 0$$

constitute the steady state model of the reactor. Using the aforementioned specifications, this model can be solved to obtain the steady state species concentrations.

4.1.2   Flow through Eccentric Reducer

This example involves macroscopic momentum balance in an eccentric reducer, which is a connector for two different diameter pipes that are not in a straight line. This reducer is commonly used to connect a pipe carrying liquid to a smaller diameter suction of a centrifugal pump.

Consider liquid, of density ρ, flowing through a reducer shown in Figure 4.2 below. From

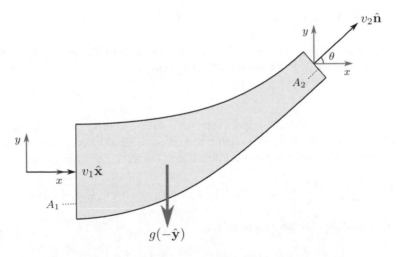

Figure 4.2   Liquid flow in and out of an eccentric reducer

the left end, the liquid enters the reducer along the horizontal x-direction with a constant average velocity $v_1\hat{\mathbf{x}}$ across the cross-section area $A_1\hat{\mathbf{x}}$. From the right end, the liquid exits the reducer with a constant average velocity $v_2\hat{\mathbf{n}}$ across the smaller cross-section area $A_2\hat{\mathbf{n}}$ where $\hat{\mathbf{n}}$ is the unit vector at an angle θ with $\hat{\mathbf{x}}$, the unit vector along the x-direction. The fluid is subject to gravity $g(-\hat{\mathbf{y}})$ where $\hat{\mathbf{y}}$ is the unit vector along the vertical y-direction.

Objective

It is desired to develop a model that would enable the determination of force exerted on the reducer by the liquid flow.

System

The system for this purpose is the fluid inside the reducer.

Assumptions

We assume that at the entrance as well as exit of the reducer, (i) the fluid velocity is perpendicular to the cross-section area, (ii) shear stresses are negligible, and (iii) pressures are identical.

Model Formulation

To determine the force, i.e., rate of change of momentum, we need momentum balance. Applying Equation (2.21) on p. 31 to the system, i.e., the fluid inside the reducer, we obtain the rate of change of momentum along the x- and y-directions as, respectively,

$$\frac{\mathrm{d}p_x}{\mathrm{d}t} = \rho v_1^2 A_1 - \rho v_2^2 A_2 \cos\theta \qquad \text{and} \tag{4.6}$$

$$\frac{\mathrm{d}p_y}{\mathrm{d}t} = -\rho v_2^2 A_2 \sin\theta - \rho V g \tag{4.7}$$

where (i) p_x and p_y are the components of the system momentum along, respectively, the x- and y-directions, and (ii) V is the volume of the system.

Summary

Equations (4.6) and (4.7) provide, respectively, the x- and y-components of the force exerted by the fluid on the reducer. To keep it secure, that force needs to be counteracted by an equal and opposite force. This force is provided by reducer supports, and is given by

$$\mathbf{F} = -\frac{\mathrm{d}p_x}{\mathrm{d}t}\hat{\mathbf{x}} - \frac{\mathrm{d}p_y}{\mathrm{d}t}\hat{\mathbf{y}} = \left(-\rho v_1^2 A_1 + \rho v_2^2 A_2 \cos\theta\right)\hat{\mathbf{x}} + \left(\rho v_2^2 A_2 \sin\theta + \rho V g\right)\hat{\mathbf{y}}$$

Given the specifications for the parameter set

$$\{\rho,\, v_1,\, v_2,\, A_1,\, A_2,\, \theta,\, V,\, g\}$$

the above equation can be used to obtain $\mathbf{F}$.

4.1.3 Liquid Preheater

This example involves a macroscopic mass balance, and the equation of change for temperature in a liquid preheater. It is a unit that is used to mix and heat different liquid streams. The mass balance accounts for any change in the liquid level inside the preheater.

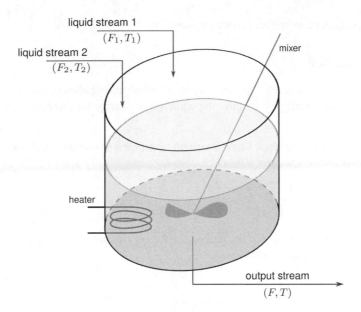

liquid stream 1
(F_1, T_1)

liquid stream 2
(F_2, T_2)

mixer

heater

output stream
(F, T)

Figure 4.3 A liquid preheater

Figure 4.3 above shows the preheater, which is fed by two liquid streams of the same composition but at different volumetric flow rates (F_1 and F_2), densities (ρ_1 and ρ_2), and temperatures (T_1 and T_2). The output stream from the preheater has a volumetric flow rate F, density ρ, and temperature T.

Objective

It is desired to develop a model that would enable the determination of liquid level as well as the temperature inside the preheater. We would like to include the possibility that the flow rates, and temperatures of the feed streams may change with time.

System

The system for this purpose is the liquid inside the preheater.

Assumptions

We make the following assumptions for the model:

1. There is no chemical reaction, or phase change in the system.

2. The preheater is well-insulated so that the heat loss to the surroundings is negligible.

3. The liquid in the preheater is perfectly mixed so that the system temperature is the same as that of the output stream.

4. The specific heat capacity of the liquid is constant in the range of temperature variation.

5. The preheater has uniform cross-section area.

Model Formulation

To determine the liquid level in the preheater, we need mass balance. For the determination of the system temperature, we will use the equation of change for temperature.

Mass Balance

Applying Equation (2.5) on p. 21 to our system (the liquid inside the preheater), we obtain

$$\frac{d}{dt}\underbrace{\left[Ah\rho \right]}_{\substack{\text{mass of the} \\ \text{system}}} = F_1\rho_1 + F_2\rho_2 - F\rho$$

where h is the liquid level inside the preheater of cross-section area A. The left-hand side of the above equation expands as follows:

$$A\frac{d}{dt}(h\rho) = A\left[h\frac{d\rho}{dt} + \rho\frac{dh}{dt} \right] = A\left[h\frac{d\rho}{dT}\frac{dT}{dt} + \rho\frac{dh}{dt} \right]$$

From the last two equations, we get after some rearrangement,

$$\frac{dh}{dt} = \frac{1}{\rho}\left[\frac{F_1\rho_1 + F_2\rho_2 - F\rho}{A} - h\frac{d\rho}{dT}\frac{dT}{dt} \right] \tag{4.8}$$

Equation of Change for Temperature

Applying Equation (3.13) on p. 71 to our system, we get

$$Ah\rho\hat{C}_P\frac{dT}{dt} = F_1\rho_1\hat{C}_P(T_1 - T) + F_2\rho_2\hat{C}_P(T_2 - T) + \underbrace{\dot{Q}_h}_{-\Delta[\dot{Q}]} + \dot{W}_s$$

where (i) $\hat{C}_P$ is the specific heat capacity of the liquid, which is at temperature T, (ii) $\dot{Q}_h$ is the rate of energy input by the heater, and (iii) $\dot{W}_s$ is the rate of shaft work equivalent to the power delivered by the mixer. Rearranging the above equation yields

$$\frac{dT}{dt} = \frac{1}{Ah\rho\hat{C}_P}\left[F_1\rho_1\hat{C}_P(T_1 - T) + F_2\rho_2\hat{C}_P(T_2 - T) + \dot{Q}_h + \dot{W}_s \right] \tag{4.9}$$

Equations (4.8) and (4.9) are differential equations of first order with respect to time. Each equation needs a condition for integration.

Initial Conditions

The initial conditions for Equations (4.8) and (4.9) are

$$h(0) = h_0 \quad \text{and} \quad T(0) = T_0 \tag{4.10}$$

where h_0 is the initial height of the liquid with the initial temperature, T_0.

Summary

Equations (4.8)–(4.10) constitute the dynamic model of the liquid preheater. Given the specifications for the parameter set

$$\{\rho(T),\ \rho_1,\ \rho_2,\ F_1,\ F_2,\ F,\ A,\ \hat{C}_P,\ T_1,\ T_2,\ \dot{Q}_h,\ \dot{W}_s,\ h_0,\ T_0\}$$

the involved differential equations need to be integrated simultaneously to obtain

$$h = h(t) \quad \text{and} \quad T = T(t)$$

At steady state, the time derivatives in the differential equations become zero, and the model becomes independent of A and h. With the remaining parameters, the resulting algebraic equations can be solved to obtain any two unknowns at the steady state. For example, to find F and T, the following set of parameters:

$$\{\rho(T),\ \rho_1,\ \rho_2,\ F_1,\ F_2,\ \hat{C}_P,\ T_1,\ T_2,\ \dot{Q}_h,\ \dot{W}_s\}$$

should be specified.

4.1.4 Non-Isothermal CSTR

This example involves macroscopic mass and energy balances in a water-cooled CSTR that is used to carry out an exothermic reaction in liquid phase. The following exothermic reaction takes place in the CSTR [see Figure 4.4, next page]:

$$a\,\mathrm{A} + b\,\mathrm{B} \xrightarrow{\ k\ } c\,\mathrm{C} + d\,\mathrm{D}$$

where k is the reaction rate coefficient, and a, b, c and d are stoichiometric coefficients. The rate of reaction is given by

$$r = k_0 \exp\left(-\frac{E}{RT}\right) c_\mathrm{A}^a c_\mathrm{B}^b \tag{4.11}$$

where k_0 is the pre-exponential factor, and E is the activation energy.

The reactor is fed by liquid streams of reactants A and B, respectively, with (i) volumetric flow rates F_Af and F_Bf, (ii) molar concentrations c_Af and c_Bf, and (iii) temperatures T_Af and T_Bf. The reaction results in products C and D. They are withdrawn, along with residual reactants at temperature T, and volumetric flow rate F from the reactor. The molar concentrations in the product stream are c_A, c_B, c_C, c_D, respectively, for A, B, C and D. The heat generated in the reaction is taken away by cooling water circulated at temperature T_J in the jacket around the reactor.

Objective

It is desired to develop a model that would enable the determination of molar concentrations of species, and the temperature of the reaction mixture as a function of time.

System

The system for this purpose is the reaction mixture inside the reactor.

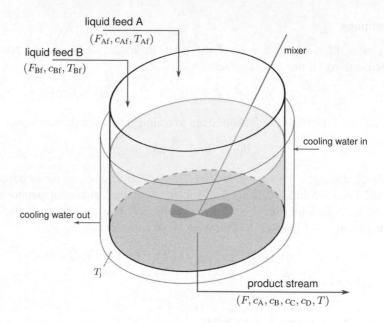

Figure 4.4 A CSTR, which carries out a reaction in liquid phase

Assumptions

We make the following assumptions for the model:

1. The reaction mixture is perfectly mixed so that its properties are the same as that of the product stream.

2. The volume of the reaction mixture is constant.

3. Inside the reactor jacket, cooling water is sufficiently agitated so that it is at uniform temperature. Moreover, the flow rate of cooling water is high enough so that its temperature stays constant.

4. The specific heat capacities of all streams are constant in the range of temperature variation.

5. There is no vaporization loss of any species.

Model Formulation

To determine species concentrations in the system, we need macroscopic mole balances in the presence of chemical reactions. To model system temperature, we require the macroscopic equation of change for temperature.

Mole Balances

Applying Equation (2.13) on p. 24 to our system (the reaction mixture inside the CSTR) for the ith species, we get

$$\underbrace{\frac{\text{d}}{\text{d}t}(Vc_i)}_{\substack{\text{moles of the } i^{\text{th}} \\ \text{species in the system}}} = F_{if}c_{if} - Fc_i + Vr_i; \qquad i = \text{A}, \text{B}, \text{C}, \text{D}$$

where (i) V is the reactor volume, (ii) c_i and c_{if} are, respectively, the molar concentrations of the ith species in the reactor, and the feed stream, (iii) $r_i = \nu_i r$ is the rate of change of the species due to the reaction, and (iv) F_{if} and F are the volumetric flow rates of, respectively, the inlet feed stream of the species, and the outlet stream. Note that $F_{if} = c_{if} = 0$ for the products C and D, and ν_i is the stoichiometric coefficient of the ith species as defined on p. 64.

Since V is constant, we obtain

$$\frac{\text{d}c_i}{\text{d}t} = \frac{F_{if}c_{if} - Fc_i}{V} + \nu_i r; \qquad i = \text{A}, \text{B}, \text{C}, \text{D} \tag{4.12}$$

Equation of Change for Temperature

Applying Equation (3.13) on p. 71 to our system, we obtain

$$V\rho\hat{C}_{\text{P}}\frac{\text{d}T}{\text{d}t} = \sum_{i=\text{A}}^{\text{B}} F_{if}c_{if}M_i\hat{C}_{\text{Pif}}(T_{if} - T) + \big[\underbrace{-UA(T - T_{\text{j}})}_{-\Delta[\dot{Q}]}\big] + V(-\Delta H_{\text{r}})r + \dot{W}_{\text{s}}$$
$$\tag{4.13}$$

where (i) ρ, $\hat{C}_{\text{P}}$ and T are, respectively, the density, specific heat capacity, and temperature of the system (reaction mixture), (ii) M_i is the molecular weight of the ith species having $\hat{C}_{\text{Pif}}$ and T_{if} as, respectively, the specific heat capacity, and temperature in the feed stream, (iii) U is the coefficient of heat transfer across the area of thermal contact A between the reaction mixture, and cooling water, (iv) $(-\Delta H_{\text{r}})$ is the heat of reaction, and (v) $\dot{W}_{\text{s}}$ is the rate of shaft work provided by the mixer.

Equations (4.12) and (4.13) are differential equations of first order with respect to time. Each of these equations needs a condition for integration.

Initial Conditions

The initial conditions for Equations (4.12) and (4.13) are, respectively,

$$c_i(0) = c_{i0}, \qquad i = \text{A}, \text{B}, \text{C}, \text{D}, \qquad \text{and} \qquad T(0) = T_0 \tag{4.14}$$

where c_{i0} is the initial concentration of the ith species in the system at initial temperature T_0.

Summary

Equations (4.11)–(4.14) constitute the dynamic model of the non-isothermal CSTR. Given the specifications for the parameter set

$$\{F_{\text{Af}}, F_{\text{Bf}}, c_{\text{Af}}, c_{\text{Bf}}, M_{\text{A}}, M_{\text{B}}, V, k_0, a, b, c, d, E, R, \rho, \hat{C}_{\text{P}},$$
$$\hat{C}_{\text{PAf}}, \hat{C}_{\text{PBf}}, T_{\text{Af}}, T_{\text{Bf}}, U, A, T_{\text{j}}, -\Delta H_{\text{r}}, \dot{W}_{\text{s}}, c_{\text{A0}}, c_{\text{B0}}, c_{\text{C0}}, c_{\text{D0}}, T_0\}$$

the involved differential equations need to be integrated simultaneously to obtain

$$c_A = c_A(t), \quad c_B = c_B(t), \quad c_C = c_C(t), \quad c_D = c_D(t) \quad \text{and} \quad T = T(t)$$

At steady state, the time derivatives in the differential equations become zero, and the model becomes independent of ρ and $\hat{C}_P$. With the remaining parameters, the resulting algebraic equations can be solved to obtain the species concentrations, and temperature at steady state.

4.2 Distributed-Parameter Systems

This section illustrates model formulation in distributed-parameter systems in Cartesian, cylindrical and spherical coordinates. Balance equations are derived on differential elements. The example on tapered fin [p. 96] illustrates model formulation when a system has a spatial variation in its size. Time-dependent system volumes with constant and variable density assumptions are handled, respectively, in examples from solvent induced heavy oil recovery [p. 108], and hydrogel tablet [p. 112]. The last two examples on Horton sphere [p. 119], and reactions around solid reactant [p. 122] deal with model formulation in complex interacting systems.

4.2.1 Nicotine Patch

This example involves microscopic mass balance of nicotine released slowly to the blood stream from a patch worn on the skin surface. The patch is used as a therapeutic device to help people quit smoking.

Figure 4.5 on the next page shows the transfer of nicotine from the patch on skin surface through the skin and fat layers into underlying blood vessels. The breadth and width of the patch are X and Z, respectively. The mass flux of nicotine is $\mathbf{f}$ along the depth, which is the y-direction across the xz-plane. The thicknesses of skin and fat layers are, respectively, Y_s and $(Y_f - Y_s)$.

Objective

It is desired to develop a model that would enable the determination of (i) nicotine mass fraction in the skin and fat layers at different depths and times, and (ii) the rate of nicotine delivery to the blood stream.

System

In this example, nicotine mass fraction is a function of the y-coordinate in the skin and fat layers. Hence, the system for nicotine mole balance in each layer comprises a rectangular differential element of thickness Δy along the y-direction. Figure 4.5 shows the differential element in the skin layer.

Assumptions

We make the following assumptions for the model:

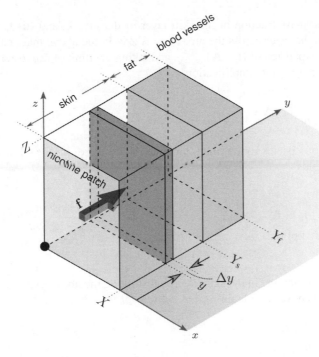

Figure 4.5   Nicotine flux through skin and fat into the blood vessels

1. Nicotine mass fraction in the skin and fat layers is a function of depth (i.e., the y-direction) and time.

2. The diffusivities of nicotine in the two layers are constant.

3. While the flux of nicotine is only along the y-direction, the flux of other species is zero.

4. Nicotine uptake by blood is fast enough so that the nicotine mass fraction is zero at the interface between fat layer, and blood vessels.

5. The densities and thicknesses of the skin and fat layers are constant.

Model Formulation

To determine nicotine mass fraction, and delivery rate, we need nicotine mass balance in differential elements in the skin and fat layers. Applying Equation (2.4) on p. 21 to the differential element in the skin layer, we get

$$\frac{\mathrm{d}}{\mathrm{d}t}\underbrace{(\Delta y X Z \rho_s \omega_s)}_{\substack{\text{nicotine mass} \\ \text{in the system}}} = \underbrace{(\mathbf{f_n} \cdot \mathbf{A})_y - (\mathbf{f_n} \cdot \mathbf{A})_{y+\Delta y}}_{-\Delta[\text{nicotine mass}]}$$

where (i) ω_s is the nicotine mass fraction in the skin layer of density ρ_s, and (ii) $\mathbf{f}_n$ is the net mass flux of nicotine in the layer across the area $\mathbf{A} = XZ\hat{\mathbf{y}}$. In the above mass balance, using the first order Taylor expansion of $(\mathbf{f}_n \cdot \mathbf{A})_{y+\Delta y}$, and taking the limit of Δy to zero [see Section 8.8.2, p. 323], we obtain after simplification

$$\frac{\partial}{\partial t}(dy\,XZ\rho_s\omega_s) = -\frac{\partial}{\partial y}(\mathbf{f}_n \cdot \mathbf{A})dy \qquad (4.15)$$

Considering the flux $\mathbf{f}_o$ of species other than nicotine to be zero in the skin layer, Equation (2.10) on p. 23 yields

$$\mathbf{f}_n = \omega_s(\underbrace{\mathbf{f}_n + \overbrace{\mathbf{f}_o}^{=0}}_{\text{total flux}}) + \underbrace{\mathbf{j}}_{\substack{-D_s\rho_s\frac{\partial\omega_s}{\partial y}\hat{\mathbf{y}} \\ \text{(from Fick's law)}}} \qquad\Rightarrow\qquad \mathbf{f}_n = -\frac{D_s\rho_s}{1-\omega_s}\frac{\partial\omega_s}{\partial y}\hat{\mathbf{y}}$$

where D_s is the diffusivity of nicotine in the skin layer. Substituting the above expressions of $\mathbf{A}$ and $\mathbf{f}_n$ in Equation (4.15) above, we obtain

$$\frac{\partial}{\partial t}(dy\,XZ\rho_s\omega_s) = -\frac{\partial}{\partial y}\left(-\frac{D_s\rho_s}{1-\omega_s}\frac{\partial\omega_s}{\partial y}XZ\right)dy$$

Since X and Z are constant, dy is independent of t, and D_s and ρ_s are assumed to be constant, the above equation simplifies to

$$\frac{\partial\omega_s}{\partial t} = D_s\frac{\partial}{\partial y}\left(\frac{1}{1-\omega_s}\frac{\partial\omega_s}{\partial y}\right) \qquad (4.16)$$

In the same manner, we obtain for the fat layer

$$\frac{\partial\omega_f}{\partial t} = D_f\frac{\partial}{\partial y}\left(\frac{1}{1-\omega_f}\frac{\partial\omega_f}{\partial y}\right) \qquad (4.17)$$

where ω_f and D_f are, respectively, the mass fraction, and diffusivity of nicotine in the fat layer.

Equations (4.16) and (4.17) are partial differential equations of (i) first order with respect to t, and (ii) second order with respect to y. Thus, for integration, each equation needs an initial condition, and two conditions on the y-axis. The conditions on a spatial axis are usually provided at the terminal points, and are called boundary conditions.

Initial Conditions
Initially, nicotine mass fraction is ω_0 at the outer skin surface, and zero elsewhere. Thus, the initial conditions are

$$\omega_s(0,0) = \omega_0, \qquad \omega_s(y,0) = 0 \ \forall \ 0 < y \leq Y_s \qquad (4.18)$$

$$\omega_f(y,0) = 0 \ \forall \ Y_s \leq y \leq Y_f \qquad (4.19)$$

Boundary Conditions

While nicotine mass fraction stays ω_0 at the outer skin surface, nicotine flux in the skin layer is equal to that in the fat layer at the skin–fat interface. Moreover, nicotine mass fraction is assumed to be zero at the fat–blood interface.

Hence, the boundary conditions for Equation (4.16) on the previous page at any time $t > 0$ are

$$\omega_s(0, t) = \omega_0 \quad \text{and} \quad -\frac{D_s \rho_s}{1 - \omega_s} \frac{\partial \omega_s}{\partial y} = -\frac{D_f \rho_f}{1 - \omega_f} \frac{\partial \omega_f}{\partial y} \quad \text{at} \quad y = Y_s \quad (4.20)$$

where ρ_f is the density of the fat layer. The boundary conditions for Equation (4.17) on the previous page at any time $t > 0$ are

$$-\frac{D_s \rho_s}{1 - \omega_s} \frac{\partial \omega_s}{\partial y} = -\frac{D_f \rho_f}{1 - \omega_f} \frac{\partial \omega_f}{\partial y} \quad \text{at} \quad y = Y_s \quad \text{and} \quad \omega_f(Y_f, t) = 0 \quad (4.21)$$

Summary

Equations (4.16)–(4.21) constitute the dynamic, distributed-parameter model of nicotine transfer through the skin and fat layers. Given the specifications for the parameter set

$$\left\{ D_s, \ D_f, \ Y_s, \ Y_f, \ \omega_0, \ \rho_s, \ \rho_f \right\}$$

the involved differential equations need to be integrated simultaneously to obtain

$$\omega_s = \omega_s(y, t) \quad \text{and} \quad \omega_f = \omega_f(y, t)$$

Steady State Model

At steady state, the time derivatives become zero in Equations (4.16) and (4.17). The resulting equations are

$$\left(\frac{1}{1 - \omega_i} \frac{\partial \omega_i}{\partial y} \right)^2 + \frac{1}{1 - \omega_i} \frac{\partial^2 \omega_i}{\partial y^2} = 0; \qquad i = s, f$$

which are independent of the diffusivities. With the remaining parameters, the above differential equations can be integrated simultaneously using the boundary conditions to obtain steady state mass fraction of nicotine in the skin and fat layers at different depths, namely, $\hat{\omega}_s = \hat{\omega}_s(y)$, and $\hat{\omega}_s = \hat{\omega}_s(y)$.

Nicotine Delivery Rate

With additional specifications for X and Z, the rate of nicotine delivery to the blood stream can be calculated from

$$\left(\mathbf{f}_n \cdot \mathbf{A} \right)_{Y_f} = -D_f \rho_f \left(\frac{1}{1 - \omega_f} \frac{\partial \omega_f}{\partial y} \right)_{Y_f} X Z$$

4.2.2 *Fluid Flow between Inclined Parallel Plates*

This example involves microscopic momentum balance in a fluid between two parallel plates separated by distance Z, and inclined at angle θ with the horizontal [Figure 4.6, next page].

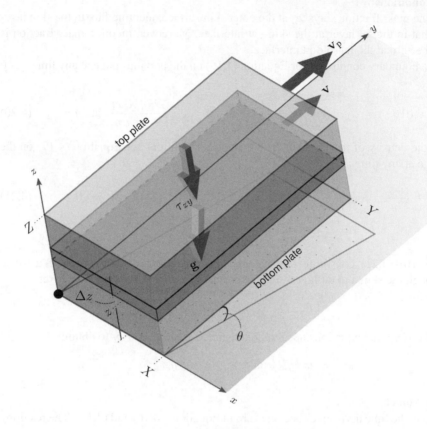

Figure 4.6 Fluid flow between inclined parallel plates

The width and length of the plates are X and Y, respectively. While the bottom plate is stationary, the top plate moves at a velocity $\mathbf{v}_p$ along the y-direction, and pulls the fluid along with velocity $\mathbf{v}$. The latter is dependent on the z-coordinate. The fluid does not flow in the x- and z-directions, and is subject to gravity $\mathbf{g}$ along the downward, vertical direction. It is at an angle $(\pi - \theta)$ to $\hat{\mathbf{z}}$, which is the unit vector along the z-direction.

Objective

It is desired to develop a model that would enable the determination of the fluid velocity between the parallel plates in the z-direction at different heights and times.

System

In this example, the fluid velocity is a function of the z-coordinate. Hence, the system for momentum balance is a rectangular differential element of thickness Δz along the z-direction, as shown in Figure 4.6 above.

Assumptions

We make the following assumptions for the model:

1. The fluid is Newtonian. It has constant density ρ and viscosity μ, and moves only along the y-direction.

2. Along the y-direction, the fluid velocity v_y is a function of height (the z-coordinate) and time.

3. Where the fluid comes in contact with a plate (i.e., at $z = 0$ and $z = Z$), the fluid velocity is equal to the plate velocity. This assumption is also known as the **no-slip** boundary condition.

4. The system pressure P is constant along the y-direction.

Model Formulation

To determine fluid velocity, we need momentum balance in our system, i.e., the differential fluid element. We will use Equation (2.1) on p. 18 with ϵ as the momentum along the y-direction, and $\dot{\epsilon}_{\text{in}}$ or $\dot{\epsilon}_{\text{out}}$ as the product of a momentum flux along that direction, and the area it crosses. The generation term $\dot{\epsilon}_{\text{gen}}$ is zero, and $\dot{\epsilon}_{\text{ext}}$ is due to gravity.

The momentum fluxes due to motion along the y-direction, and transmitted along the y-, x- and the *negative* z-direction are, respectively, [see Equation (2.19), p. 28]

$$\phi_{yy} = \underbrace{\tau_{yy}}_{=0} + \rho v_y^2 + P = \rho v_y^2 + P,$$

$$\phi_{xy} = \underbrace{\tau_{xy}}_{=0} + \rho \underbrace{v_x}_{=0} v_y = 0, \quad \text{and}$$

$$\phi_{zy} = -(\tau_{zy} + \rho \underbrace{v_z}_{=0} v_y) = -\tau_{zy}$$

because $v_y = v_y(z)$ and is the only non-zero velocity. Using Equation (2.1) on our system, we obtain,

$$\frac{\mathrm{d}}{\mathrm{d}t}\underbrace{(XY\Delta z\rho v_y)}_{\substack{\text{system momentum} \\ \text{in the } y\text{-direction}}} = (\phi_{zy}XY)_{z+\Delta z} - (\phi_{zy}XY)_z + \underbrace{(\phi_{yy}\Delta zX)_{y=0} - (\phi_{yy}\Delta zX)_{y=Y}}_{=0}$$
$$+ \Delta zXY\rho(-g\sin\theta)$$

In the above equation, the terms with ϕ_{yy} cancel each other out because v_y and P do not vary along the y-direction. Using the first order Taylor expansion of $(\phi_{zy}XY)_{z+\Delta z}$, taking the limit of Δz to zero, and substituting for ϕ_{zy}, we obtain after simplification

$$\frac{\partial}{\partial t}(XY\,\mathrm{d}z\rho v_y) = \frac{\partial}{\partial z}(-\tau_{zy}XY)\mathrm{d}z - \mathrm{d}zXY\rho g\sin\theta$$

Since ρ, X and Y are constants, and $\mathrm{d}z$ is independent of t, the above equation simplifies to

$$\frac{\partial v_y}{\partial t} = -\frac{1}{\rho}\frac{\partial \tau_{zy}}{\partial z} - g\sin\theta$$

Upon substituting for $\tau_{zy} = -\mu \partial v_y / \partial z$ with constant μ, we finally obtain

$$\frac{\partial v_y}{\partial t} = \frac{\mu}{\rho} \frac{\partial^2 v_y}{\partial z^2} - g \sin \theta \tag{4.22}$$

Equation (4.22) is a partial differential equation of (i) first order with respect to t, and (ii) second order with respect to z. Thus, for integration, that equation needs an initial condition, and two boundary conditions on the z-axis.

Initial Condition

Initially, the fluid velocity is zero throughout. Thus, the initial condition for Equation (4.22) above is

$$v_y(z, 0) = 0 \ \forall \ 0 \leq z \leq Z \tag{4.23}$$

Because of the no-slip condition, the fluid velocity is equal to the plate velocity where fluid and plate contact each other. Therefore, the boundary conditions at any time $t > 0$ are

$$v_y(z, t) = \begin{cases} 0 & \text{at} \ z = 0 \\ v_\text{p} & \text{at} \ z = Z \end{cases} \tag{4.24}$$

Summary

Equations (4.22)–(4.24) constitute the dynamic, distributed-parameter model of momentum transfer from a moving plate through a fluid layer to an underlying stationary plate. Given the specifications for the parameter set

$$\{\mu, \ \rho, \ g, \ \theta, \ Z, \ v_\text{p}\}$$

the involved differential equation needs to be integrated to obtain $v_y = v_y(z, t)$.

At steady state, the time derivative in Equation (4.22) becomes zero. The resulting differential equation can be integrated using the boundary conditions to obtain the steady state fluid velocity, $\hat{v}_y = \hat{v}_y(z)$.

4.2.3 Tapered Fin

This example involves the equation of change for temperature in a tapered fin designed to dissipate heat to the surroundings in tight spaces. Figure 4.7 on the next page shows the schematic of a tapered fin with the surroundings at temperature T_a. The left end of the fin is of thickness $2Z$, and is at the wall temperature T_w. The fin has width X, and is symmetrical across the xy-plane.

At a given y, the extent of the fin is $z(y)$ along both the positive and negative z-directions. The fin has heat flux $\mathbf{q}$ due to conduction along the length, which is the y-direction across the xz-plane. In addition, the fin loses heat to the surroundings from the top and bottom surfaces.

Objective

It is desired to develop a model that would enable the determination of temperature in the fin at different lengths and times.

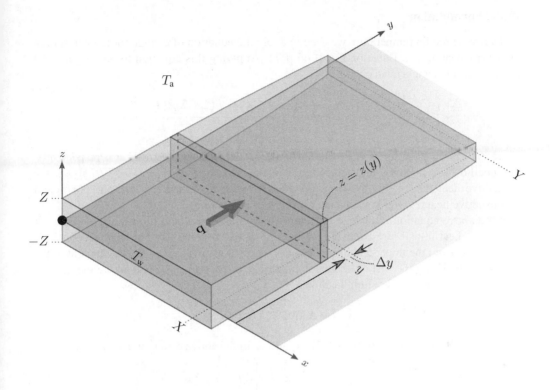

Figure 4.7   Heat transfer in a tapered fin

System

In this example, the fin temperature is a function of the y-coordinate. Hence, the system for the equation of change for temperature is a rectangular differential element of thickness Δy along the y-direction, as shown in Figure 4.7 above.

Assumptions

We make the following assumptions for the model:

1. Temperature in the fin is a function of length (the y-coordinate) and time.

2. There are no temperature gradients in the fin along the x- and z-axes.

3. The density ρ, and thermal conductivity k of the fin are constant in the operating range of temperature.

4. The fin is thin enough so that heat loss to the surroundings is negligible from the sides.

Model Formulation

To determine the fin temperature we need to apply the equation of change for temperature in a uniform system, i.e., Equation (3.14) on p. 71. Applying this equation to our system (the differential fin element), we obtain

$$\underbrace{\Delta V \rho \hat{C}_{\mathrm{P}}}_{\substack{\text{system} \\ \text{mass}}} \frac{\mathrm{d}T}{\mathrm{d}t} = \underbrace{(\mathbf{q} \cdot \mathbf{A})_y - (\mathbf{q} \cdot \mathbf{A})_{y+\Delta y} - h(2X\Delta y)(T - T_{\mathrm{a}})}_{-\Delta[\dot{Q}]}$$

where (i) $\Delta V = 2zX\Delta y$ is the volume of the system, (ii) $\hat{C}_{\mathrm{P}}$ and T are, respectively, the specific heat capacity, and temperature of the system, (iii) $\mathbf{q} = (-k\partial T/\partial y)\hat{\mathbf{y}}$ is the conductive heat flux in the fin across the area $\mathbf{A} = (2zX)\hat{\mathbf{y}}$, and (iv) h is the coefficient of convective heat transfer between the fin and the surroundings.

In the above equation of change for temperature, the second and third terms account for conductive heat transfer, respectively, in and out of the system. The last term accounts for the heat loss from the upper and lower surfaces (each of area $X\Delta y$) of the fin element to the surroundings. Using the first order Taylor expansion of $(\mathbf{q} \cdot \mathbf{A})_{y+\Delta y}$, and taking the limit of Δy to zero, we obtain after simplification

$$\mathrm{d}V \rho \hat{C}_{\mathrm{P}} \frac{\partial T}{\partial t} = -\frac{\partial}{\partial y}(\mathbf{q} \cdot \mathbf{A})\mathrm{d}y - 2hX\mathrm{d}y(T - T_{\mathrm{a}})$$

where $\mathrm{d}V = 2zX\mathrm{d}y$. Substituting for $\mathrm{d}V$, $\mathbf{q}$ and $\mathbf{A}$ in the above differential equation, we get

$$2zX\mathrm{d}y\rho \hat{C}_{\mathrm{P}} \frac{\partial T}{\partial t} = -\frac{\partial}{\partial y}\left[-k\frac{\partial T}{\partial y}(2zX)\right]\mathrm{d}y - h(2X\mathrm{d}y)(T - T_{\mathrm{a}})$$

Considering that $z = z(y)$ and X is constant, the above differential equation yields upon simplification

$$\frac{\partial T}{\partial t} = \frac{1}{z\rho \hat{C}_{\mathrm{P}}}\left[k\left(z\frac{\partial^2 T}{\partial y^2} + \frac{\partial T}{\partial y}\frac{\mathrm{d}z}{\mathrm{d}y}\right) - h(T - T_{\mathrm{a}})\right] \tag{4.25}$$

Equation (4.25) is a partial differential equation of (i) first order with respect to t, and (ii) second order with respect to y. Thus, for integration, that equation needs an initial condition, and two boundary conditions on the y-axis.

Initial Condition
The fin is assumed to be initially at uniform temperature T_0 except at the wall, which is always at temperature T_{w}. Hence, the initial condition for Equation (4.25) above is

$$T(y, 0) = \begin{cases} T_0 & \forall \ 0 < y \leq Y \\ T_{\mathrm{w}} & \text{at } y = 0 \end{cases} \tag{4.26}$$

Boundary Conditions
While the left end of the fin is at the wall temperature, the right end is assumed to have no heat transfer, i.e., zero temperature gradient. Hence, the boundary conditions for Equation (4.25) above at any time $t > 0$ are

$$T = T_{\mathrm{w}} \text{ at } y = 0 \quad \text{and} \quad \frac{\partial T}{\partial y} = 0 \text{ at } y = Y \tag{4.27}$$

Summary

Equations (4.25)–(4.27) constitute the dynamic, distributed-parameter model of heat transfer in a tapered fin. Given the specifications for the parameter set

$$\left\{ \rho,\ \hat{C}_{\mathrm{P}},\ z(y),\ k,\ h,\ T_{\mathrm{a}},\ T_0,\ T_{\mathrm{w}},\ Y \right\}$$

the involved differential equation needs to be integrated to obtain $T = T(y,t)$.

At steady state, the time derivative in Equation (4.25) becomes zero. The resulting differential equation is independent of ρ and $\hat{C}_{\mathrm{P}}$. With the remaining parameters, this equation can be integrated using the boundary conditions to obtain the steady state fin temperature, $\hat{T} = \hat{T}(y)$.

4.2.4 *Continuous Microchannel Reactor*

This example involves microscopic mass and momentum balances in a continuous microchannel reactor designed to grow cells for tissue engineering.[1] As shown in Figure 4.8 below, the reactor is a rectangular channel of width X and length Y, and has a cellular layer of height Z_1 at the bottom. A mixture of species, such as nutrients and metabolites needed for cell growth, flows along the length (i.e., the horizontal y-direction) in the space above

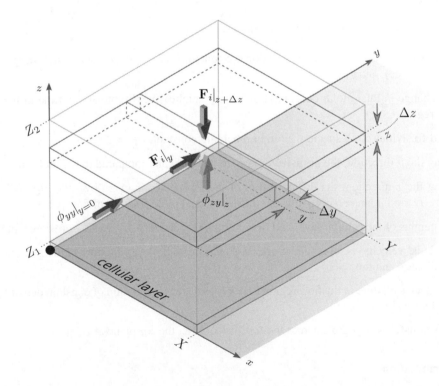

Figure 4.8   Schematic of a continuous microchannel reactor

the cellular layer up to the height Z_2. The species diffuse vertically down along the negative z-direction to feed the cells in the layer. At the same time, the metabolic products released by cells diffuse out from the layer, and leave the reactor.

Objective

It is desired to develop a model that would enable the determination of (i) species mole fractions at different heights, lengths and times, and (ii) the bulk velocity at different heights and times in the flow region above the cellular layer in the reactor.

System

In this example, species mole fractions are functions of the y- and z-coordinates. Hence, the system for mole balances is a rectangular differential element of thicknesses Δy and Δz along the y- and z-directions, respectively. On the other hand, the bulk velocity v_y along the y-direction is a function of only the z-coordinate. Thus, the system for momentum balance is a rectangular differential element of thickness Δz along the z-direction. Figure 4.8 on the previous page shows the two differential elements.

Assumptions

We make the following assumptions for the model:

1. Species mole fractions in the flow region are functions of length (y-coordinate), height (z-coordinate), and time.

2. The change in total molar concentration c with height is negligible at any time in the flow region.

3. The diffusivities of all species, N_c in number, are constant.

4. Along the y-direction, the diffusive fluxes are insignificant compared to the bulk flux.

5. Along the z-direction, the diffusive flux of the ith species is given by $-D_i c \partial y_i / \partial z$, where D_i and y_i are, respectively, the diffusivity and mole fraction of the species.

6. The heights of the cellular layer and flow region, respectively, Z_1 and Z_2, are constant.

7. The bulk velocity is only along the y-direction, and is a function of height (z-coordinate), and time.

8. The species mixture in the flow region behaves as a Newtonian fluid of constant density ρ and viscosity μ.

9. Mass transfer is negligible across the top surface, i.e., the xy-plane at Z_2.

Model Formulation

To determine the species concentrations, and bulk velocity, we need species mole balances, and momentum balance.

Species Mole Balances

Applying Equation (3.6) on p. 66 for the ith species in our system (the rectangular differential element of height Δz, width X, and length Δy in the gas phase), we get

$$\frac{d}{dt}\underbrace{(\Delta V c y_i)}_{\substack{\text{moles of the } i^{th} \\ \text{species in the system}}} = (\mathbf{F}_i \cdot \Delta\mathbf{A})_y - (\mathbf{F}_i \cdot \Delta\mathbf{A})_{y+\Delta y} + (\mathbf{F}_i \cdot \Delta\mathbf{A})_{z+\Delta z} - (\mathbf{F}_i \cdot \Delta\mathbf{A})_z$$

where y_i is the mole fraction of the ith species in the system of volume $\Delta V = X\Delta y\Delta z$, c is the molar concentration in the system, and $\mathbf{F}_i$ is the molar flux of the species across the area $\Delta\mathbf{A}$. Note that there is no reaction in the gas phase.

Along the y-direction, $\mathbf{F}_i = (v_y c y_i)\hat{\mathbf{y}}$, and $\Delta\mathbf{A} = (X\Delta z)\hat{\mathbf{y}}$. Along the z-direction, since there is no bulk movement, $\mathbf{F}_i = \mathbf{J}_i = -D_i c \partial y_i/\partial z(-\hat{\mathbf{z}})$ across the area $\Delta\mathbf{A} = (\Delta y X)\hat{\mathbf{z}}$.

In the above mole balance, we use the first order Taylor expansions of $(\mathbf{F}_i \cdot \Delta\mathbf{A})_{y+\Delta y}$ and $(\mathbf{F}_i \cdot \Delta\mathbf{A})_{z+\Delta z}$, take the limits of Δy and Δz to zero, and substitute the expressions for system volume, fluxes and areas. The result after simplification is

$$\frac{\partial}{\partial t}(X\mathrm{d}y\mathrm{d}z c y_i) = -\frac{\partial}{\partial y}(v_y c y_i X\mathrm{d}z)\mathrm{d}y + \frac{\partial}{\partial z}\left(D_i c \frac{\partial y_i}{\partial z}\mathrm{d}y X\right)\mathrm{d}z$$

Because $\mathrm{d}y$ is independent of t and z, X is constant, $\mathrm{d}z$ is independent of t and y, and c is assumed to be independent of z, the above equation simplifies to

$$c\frac{\partial y_i}{\partial t} = -y_i\frac{\partial c}{\partial t} - \frac{\partial}{\partial y}(v_y c y_i) + D_i c\frac{\partial^2 y_i}{\partial z^2} \tag{4.28}$$

Overall Mole Balance

The mole balance of all species in the system is given by

$$\frac{d}{dt}\underbrace{(\Delta V c)}_{\substack{\text{moles of all} \\ \text{species in the system}}} = (\mathbf{F}_y \cdot \Delta\mathbf{A}_y)_y - (\mathbf{F}_y \cdot \Delta\mathbf{A}_y)_{y+\Delta y} + (\mathbf{F}_z \cdot \Delta\mathbf{A}_z)_{z+\Delta z} - (\mathbf{F}_z \cdot \Delta\mathbf{A}_z)_z$$

where (i) $\mathbf{F}_y = (v_y c)\hat{\mathbf{y}}$ is the flux of all species along the y-direction, and across the area $\Delta\mathbf{A}_y = (X\Delta z)\hat{\mathbf{y}}$, and (ii) $\mathbf{F}_z$ is the diffusive flux of all species along the z-direction, and across the area $\Delta\mathbf{A}_z = (X\Delta y)\hat{\mathbf{z}}$.

In the above mole balance, we use the first order Taylor expansions of $(\mathbf{F}_y \cdot \Delta\mathbf{A}_y)_{y+\Delta y}$ and $(\mathbf{F}_z \cdot \Delta\mathbf{A}_z)_{z+\Delta z}$, take the limits of Δy and Δz to zero, and substitute the expressions for the fluxes and areas. The result after simplification is

$$\frac{\partial c}{\partial t} = -\frac{\partial}{\partial y}(v_y c) + c\sum_{j=1}^{N_c} D_j \frac{\partial^2 y_j}{\partial z^2}$$

Substituting the above equation in Equation (4.28) above and simplifying the result, we finally obtain

$$\frac{\partial y_i}{\partial t} = -v_y\frac{\partial y_i}{\partial y} + D_i\frac{\partial^2 y_i}{\partial z^2} - y_i\sum_{j=1}^{N_c} D_j\frac{\partial^2 y_j}{\partial z^2} \tag{4.29}$$

Momentum Balance

Keep in mind that there is only one non-zero velocity v_y, which is along the y-direction, and varies only along the z-direction. The relevant momentum fluxes are ϕ_{yy}, ϕ_{xy} and ϕ_{zy}, respectively. The first two fluxes, similar to those in Section 4.2.2 on p. 93, are

$$\phi_{yy} = \rho v_y^2 + P \quad \text{and} \quad \phi_{xy} = 0.$$

The last one is $\phi_{zy} = \tau_{zy}$, which is taken to be transmitted along the positive z-direction.

Applying Equation (2.1) on p. 18 with ϵ as the momentum of the system (the rectangular differential element of height Δz, width X, and length Y in the flow region) in the y-direction, we obtain

$$\frac{\mathrm{d}}{\mathrm{d}t} \underbrace{(XY\Delta z \rho v_y)}_{\substack{\text{system momentum} \\ \text{in the } y\text{-direction}}} = (\phi_{zy}XY)_z - (\phi_{zy}XY)_{z+\Delta z} + (\phi_{yy}X\Delta z)_{y=0} - (\phi_{yy}X\Delta z)_{y=Y}$$

In the above differential equation, using the first order Taylor expansion of $(\phi_{zy}XY)_{z+\Delta z}$, and taking the limit of Δz to zero, we obtain after simplification

$$\frac{\partial}{\partial t}(XY\mathrm{d}z \rho v_y) = -\frac{\partial}{\partial z}(\phi_{zy}XY)\mathrm{d}z + (\phi_{yy}X\mathrm{d}z)_{y=0} - (\phi_{yy}X\mathrm{d}z)_{y=Y}$$

Since ρ, X and Y are constant, $\mathrm{d}z$ is independent of t, and v_y is invariant along the y-direction, the above equation after substituting the fluxes simplifies to

$$\frac{\partial v_y}{\partial t} = -\frac{1}{\rho}\left(\frac{\partial \tau_{zy}}{\partial z} + \overbrace{\frac{P_{y=Y} - P_{y=0}}{Y}}^{\Delta P}\right)$$

where ΔP is the pressure difference across the reactor length. Since $\tau_{zy} = -\mu \partial v_y / \partial z$, and μ is constant,

$$\frac{\partial v_y}{\partial t} = \frac{1}{\rho}\left(\mu \frac{\partial^2 v_y}{\partial z^2} - \frac{\Delta P}{Y}\right) \tag{4.30}$$

Equation (4.29) on the previous page is a partial differential equation of (i) first order with respect to t as well as y, and (ii) second order with respect to z. Thus, for integration, that equation needs an initial condition, one boundary condition on the y-axis, and two boundary conditions on the z-axis. Equation (4.30) is a partial differential equation of (i) first order with respect to t, and (ii) second order with respect to z. Thus, for integration, that equation needs an initial condition, and two boundary conditions on the z-axis.

Initial Conditions

Initially, the concentration of the ith species is specified as y_{i0} at the reactor entrance but zero elsewhere. Moreover, there is no flow of species at the initial time. Hence, the initial conditions for Equation (4.29) on the previous page and Equation (4.30) above are

$$y_i(y,z,0) = \begin{cases} y_{i0} & \text{at} \ \ y = 0 \\ 0 & \forall \ \ 0 < y \le Y \end{cases} \quad \forall \ \ Z_1 \le z \le Z_2 \tag{4.31}$$

$$v_y(z,0) = 0 \ \ \forall \ \ Z_1 \le z \le Z_2 \tag{4.32}$$

Boundary Conditions

At all times, the concentration of the ith species is maintained as y_{i0} at the reactor entrance. Along the z-direction, and at the cellular-layer–gas interface, the rate of mass transfer of the ith species is considered equal to its rate of generation in the layer. Since mass transfer is negligible at the top surface, the mole fraction gradient over there is zero. As to the velocity v_y, it is zero at the cellular-layer–gas interface as well as at the top surface due to the no-slip condition. With these considerations, the boundary conditions at any time $t > 0$ are

$$y_i(y, z, t) = y_{i0} \quad \text{at} \quad y = 0 \ \forall \ Z_1 \leq z \leq Z_2$$

$$\left.\begin{array}{rll} -D_i c \dfrac{\partial y_i}{\partial z} &= R_{\text{gen},i} Z_1 & \text{at} \quad z = Z_1 \\[2mm] \dfrac{\partial y_i}{\partial z} &= 0 & \text{at} \quad z = Z_2 \end{array}\right\} \quad \forall \ 0 \leq y \leq Y \tag{4.33}$$

$$v_y(z, t) = 0 \quad \text{at} \quad z = Z_1, Z_2 \tag{4.34}$$

where $R_{\text{gen},i}$ is the rate of generation of the ith species in the cellular layer per unit volume.

Summary

Equations (4.29)–(4.34) constitute the dynamic, distributed-parameter model of mass and momentum transfer in a microchannel reactor. Given the specifications for the parameter set

$$\left\{ c, \ D_i\text{s}, \ y_{i0}\text{s}, \ R_{\text{gen},i}\text{s}, \ \mu, \ \rho, \ \Delta P, \ Y, \ Z_1, \ Z_2 \right\}$$

the involved differential equations need to be integrated simultaneously to obtain each $y_i = y_i(y, z, t)$, and $v_y = v_y(z, t)$.

At steady state, the time derivatives become zero in Equations (4.29) and (4.30). The resulting differential equations can be integrated using the boundary conditions to obtain at steady state, the species velocity $\hat{v}_y = \hat{v}_y(z)$, and mole fraction $\hat{y}_i = \hat{y}_i(y, z)$ of each species.

4.2.5 Oxygen Transport to Tissues

This example involves microscopic mass balance in a cylindrical tissue [see Figure 4.9, next page] surrounding a capillary that supplies oxygen.[2] From the capillary surface at R_1, oxygen penetrates the tissue whose outermost radius is R_2. The molar flux of oxygen in the tissue is **F** in the r-direction.

As shown in the figure, there are points of symmetry where tissue cylinders touch one other in a capillary cluster. At these points, the gradient of oxygen concentration is zero [see Appendix 4.B, p. 133]. Given that tissue cylinders have extremely small radii, the points of symmetry are so many that the entire surface of tissue at radius R_2 has almost zero gradient of oxygen concentration.

Objective

It is desired to develop a model that would enable the determination of oxygen mole fraction at different radii and times in the tissue around a capillary.

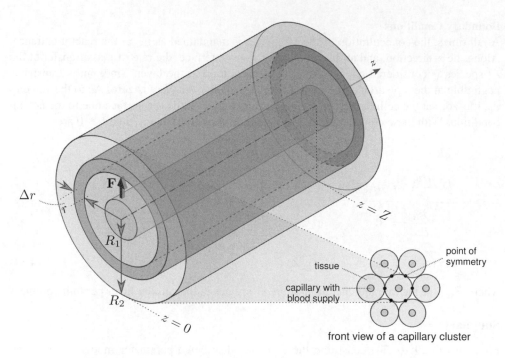

Figure 4.9   Oxygen transfer from blood to the surrounding tissue

System

In this example, the oxygen mole fraction in the tissue is a function of the r-coordinate. Hence, the system for oxygen mole balance is a cylindrical differential element (an annulus) of thickness Δr along the r-direction, as shown in Figure 4.9 above.

Assumptions

We make the following assumptions for the model:

1. Oxygen mole fraction in the tissue is a function of radial distance and time.

2. The diffusivity of oxygen in the tissue is constant.

3. The change in total molar concentration with radial distance is negligible at any time in the tissue.

4. The radial gradient of oxygen at the tissue surface can be neglected due to numerous points of symmetry of contiguous tissue cylinders of very small radii.

5. The thickness of the tissue is constant.

Model Formulation

To determine oxygen concentration in the tissue, we need mole balance. Applying Equation (3.6) on p. 66 for the moles of oxygen in the system (the annulus in the tissue), we get

$$\frac{d}{dt}\underbrace{(\Delta V c y)}_{\substack{\text{oxygen moles} \\ \text{in the system}}} = (\mathbf{F_{O_2}} \cdot \mathbf{A})_r - (\mathbf{F_{O_2}} \cdot \mathbf{A})_{r+\Delta r} - \Delta V r_{O_2}$$

where (i) y is the oxygen mole fraction in the system of volume $\Delta V = \pi[(r + \Delta r)^2 - r^2]Z$ with Z as capillary length, (ii) c is the molar concentration of all species in the system, (iii) $\mathbf{F_{O_2}}$ is the molar flux of oxygen across the area $\mathbf{A} = (2\pi r Z)\hat{\mathbf{r}}$, and (iv) r_{O_2} is the rate of oxygen consumption in the tissue per unit volume.

Since there is no bulk movement along the r-direction, $\mathbf{F_{O_2}} = (-Dc\partial y/\partial r)\hat{\mathbf{r}}$, where D is the diffusivity of oxygen in the tissue. In the above mole balance, we use the first order Taylor expansion of $(\mathbf{F_{O_2}} \cdot \mathbf{A})_{r+\Delta r}$, take the limit of Δr to zero, and substitute the expressions for the flux and area. The result after simplification is

$$\frac{\partial}{\partial t}(dV c y) = -\frac{\partial}{\partial r}\left[-Dc\frac{\partial y}{\partial r}(2\pi r Z)\right]dr - (dV)r_{O_2}$$

where $dV = 2\pi r dr Z$. Since r and dr are independent of t, and c, D and Z are constant, the above equation simplifies to

$$\frac{\partial y}{\partial t} = \frac{D}{r}\frac{\partial}{\partial r}\left(r\frac{\partial y}{\partial r}\right) - \frac{r_{O_2}}{c} \tag{4.35}$$

Equation (4.35) is a partial differential equation of (i) first order with respect to t, and (ii) second order with respect to r. Thus, for integration, that equation needs an initial condition, and two boundary conditions on the r-axis.

Initial Condition
Initially, oxygen mole fraction in the tissue is y_0 except at the blood–tissue interface, which is always at y_f. Hence, the initial condition for Equation (4.35) above is

$$y(r,0) = \begin{cases} y_0 \quad \forall \quad R_1 < r \le R_2 \\ y_f \quad \text{at} \quad r = R_1 \end{cases} \tag{4.36}$$

Boundary Conditions
While the blood–tissue interface has oxygen mole fraction y_f, the tissue periphery has no oxygen transfer, i.e., zero gradient of y. Hence, the boundary conditions for Equation (4.35) above at any time $t > 0$ are

$$y = y_f \text{ at } r = R_1 \quad \text{and} \quad \frac{\partial y}{\partial r} = 0 \text{ at } r = R_2 \tag{4.37}$$

Summary
Equations (4.35)–(4.37) constitute the dynamic, distributed-parameter model of oxygen transfer in a tissue around a blood capillary. Given the specifications for the parameter set

$$\{D, c, r_{O_2}, y_0, y_f, R_1, R_2\}$$

the involved differential equation needs to be integrated to obtain $y = y(r,t)$.

At steady state, the time derivative in Equation (4.35) becomes zero. The resulting differential equation can be integrated using the boundary conditions to obtain the steady state oxygen mole fraction in the tissue, $\hat{y} = \hat{y}(r)$.

4.2.6 Dermal Heat Transfer in Cylindrical Limb

This example involves the equation of change for temperature in the skin layer of a cylindrical limb.[3] As shown in Figure 4.10 below, the layer surrounds the core of the limb of radius R_1,

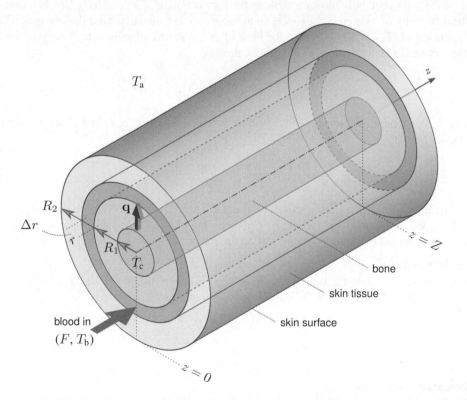

Figure 4.10   Heat transfer in the layer of skin of a cylindrical limb

and extends to the outer radius R_2. While the core is at temperature T_c, the layer of skin is fed by blood, which has volumetric flow rate F, and is at temperature T_b. The limb is exposed to the surroundings at temperature T_a.

Objective

It is desired to develop a model that would enable the determination of skin temperature at different radii and times.

System

In this example, the temperature in the skin layer is a function of the r-coordinate. Hence, the system for the equation of change for temperature is an annulus of thickness Δr along the r-direction, as shown in Figure 4.10 on the previous page.

Assumptions

We make the following assumptions for the model:

1. Temperature in the skin layer is a function of radial distance and time.

2. While heat conduction is only along the r-direction, blood flows only along the axial z-direction.

3. At any fixed radial distance, the temperature of blood in the tissue and at the exit is the same as the tissue temperature.

4. The density and thermal conductivity of the skin layer as well as blood are constant in the range of temperature variation.

Model Formulation

To determine the tissue temperature T in the skin layer, we need the equation of change for temperature. Applying Equation (3.13) on p. 71 to our system (the skin annulus), we obtain

$$\underbrace{\Delta V \rho \hat{C}_{\mathrm{P}}}_{\substack{\text{system} \\ \text{mass}}} \frac{dT}{dt} = \underbrace{(\Delta V \rho_{\mathrm{b}} \tilde{F})}_{\substack{\text{mass flow rate} \\ \text{of blood}}} \hat{C}_{\mathrm{Pb}}(T_{\mathrm{b}} - T) + \underbrace{(\mathbf{q} \cdot \mathbf{A})_r - (\mathbf{q} \cdot \mathbf{A})_{r+\Delta r}}_{-\Delta[\dot{Q}]} + \Delta V H_{\mathrm{m}}$$

where (i) $\Delta V = \pi[(r + \Delta r)^2 - r^2]Z$ is the volume of the annulus of length Z, (ii) ρ, $\hat{C}_{\mathrm{P}}$ and T are, respectively, the density, specific heat capacity, and temperature of the tissue (system), (iii) $\mathbf{q} = (-k\partial T/\partial r)\hat{\mathbf{r}}$ is the conductive heat flux, with k as the heat conductivity, across the area $\mathbf{A} = (2\pi r Z)\hat{\mathbf{r}}$, (iv) ρ_{b} is blood density, (v) $\tilde{F}$ is the rate of blood perfusion (volumetric blood flow rate per unit tissue volume), (vi) $\hat{C}_{\mathrm{Pb}}$ is the specific heat capacity of blood, and (vii) H_{m} is the rate of metabolic heat generation per unit volume in the tissue.

In the above equation of change for temperature, using the first order Taylor expansion of $(\mathbf{q} \cdot \mathbf{A})_{r+\Delta r}$, and taking the limit of Δr to zero, we obtain after simplification

$$dV \rho \hat{C}_{\mathrm{P}} \frac{\partial T}{\partial t} = dV \rho_{\mathrm{b}} \tilde{F} \hat{C}_{\mathrm{Pb}}(T_{\mathrm{b}} - T) - \frac{\partial}{\partial r}(\mathbf{q} \cdot \mathbf{A})dr + dV H_{\mathrm{m}}$$

where $dV = 2\pi r dr Z$. Substituting for dV, $\mathbf{q}$ and $\mathbf{A}$ in the above differential equation, and simplifying the result, we get

$$\frac{\partial T}{\partial t} = \frac{1}{\rho \hat{C}_{\mathrm{P}}} \left[\rho_{\mathrm{b}} \tilde{F} C_{\mathrm{Pb}}(T_{\mathrm{b}} - T) + k \left(\frac{\partial^2 T}{\partial r^2} + \frac{1}{r} \frac{\partial T}{\partial r} \right) + H_{\mathrm{m}} \right] \tag{4.38}$$

Equation (4.38) is a partial differential equation of (i) first order with respect to t, and (ii) second order with respect to r. Thus, for integration, that equation needs an initial condition, and two boundary conditions on the r-axis.

Initial Condition

Initially, the tissue temperature is T_0 except at the core–tissue interface, which is always at the core temperature T_c. Hence, the initial condition for Equation (4.38) on the previous page is

$$
T(r, 0) = \begin{cases} T_0 & \forall \ R_1 < r \le R_2 \\ T_c & \text{at} \ r = R_1 \end{cases} \tag{4.39}
$$

Boundary Conditions

While the temperature at core–tissue interface is T_c, the conductive heat flux in the tissue at the periphery is equal to the convective heat flux in the surroundings. Hence, the boundary conditions for Equation (4.38) on the previous page at any time $t > 0$ are

$$
T = T_c \ \text{at} \ r = R_1 \quad \text{and} \quad -k\frac{\partial T}{\partial r} = h(T - T_a) \ \text{at} \ r = R_2 \tag{4.40}
$$

where T_a is the temperature of the surroundings, and h is the convective heat transfer coefficient.

Summary

Equations (4.38)–(4.40) constitute the dynamic, distributed-parameter model of heat transfer in the skin tissue of a cylindrical limb. Given the specifications for the parameter set

$$
\left\{ \rho, \hat{C}_P, \rho_b, \tilde{F}, \hat{C}_{Pb}, T_b, k, H_m, T_0, T_c, R_1, R_2, T_a, h \right\}
$$

the involved differential equation needs to be integrated to obtain $T = T(r, t)$.

At steady state, the time derivative in Equation (4.38) becomes zero, and the model becomes independent of ρ and $\hat{C}_P$. With the remaining parameters, the resulting differential equation can be integrated using the boundary conditions to obtain the steady state oxygen mole fraction in the tissue, $\hat{T} = \hat{T}(r)$.

4.2.7   Solvent Induced Heavy Oil Recovery

This example involves microscopic and macroscopic mass balances in a lab-scale physical model of a heavy oil reservoir – a porous medium of heavy oil, and glass beads packed together as a cylinder of radius R, and initial height Z_0 [see Figure 4.11, next page].

The physical reservoir model is designed to investigate gravity-assisted oil recovery using a gas-phase solvent such as carbon dioxide, butane and propane.[4] During an experiment, the solvent penetrates the cylinder from the surroundings, and, upon absorption, brings down the viscosity of the oil assumed to be of constant density. This oil–solvent mixture flows due to gravity, and is produced at the bottom. As a result, the height of the physical reservoir model decreases with time. Hence, this is an example of a **moving boundary problem with constant density assumption**.

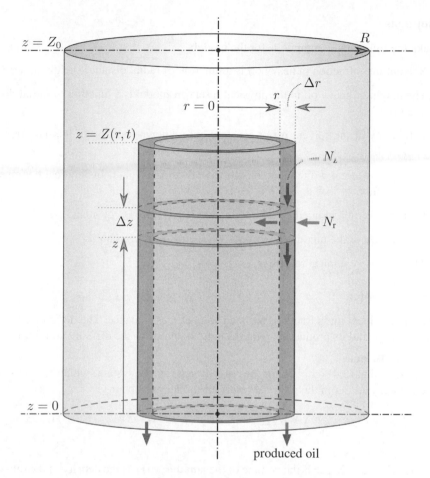

Figure 4.11   A lab-scale physical model of a heavy oil reservoir

Objective

It is desired to develop a model that would enable the determination of (i) solvent mass fraction in the physical reservoir model at different heights, radial distances, and times, (ii) the height of the oil–solvent mixture in the physical reservoir model at different radial distances and times, and (iii) rate of production of heavy oil, i.e., the oil–solvent mixture.

System

In this example, the solvent mass fraction in the physical reservoir model is a function of the r- and z-coordinates. Hence, the system for solvent mass balance is a cylindrical differential element (an annulus) of thicknesses Δr and Δz, respectively, along the r- and z-directions, as shown in Figure 4.11 above. Next, the height (and mass) of the oil is a function of time and r-coordinate. Hence, the system for the oil–solvent mass balance is the same annulus but of height Z, which is a function of time.

Assumptions

We make the following assumptions for the model:

1. Solvent mass fraction in heavy oil is a function of radial distance, height and time.

2. The height of heavy oil in the physical reservoir model is a function of radial distance, and time.

3. Diffusivity of the solvent is not constant but a known function of its mass fraction in heavy oil.

4. Along the vertical z-direction, the diffusive flux of solvent is insignificant compared to the bulk flux.

5. Bulk flux is only along the z-direction, and is given by Darcy's law.

6. The density of the mixture of solvent and heavy oil is constant.

7. There are no chemical reactions.

Model Formulation

To determine solvent mass fraction, we need solvent mass balance. The determination of the height of the oil, and its production requires mass balance for the oil–solvent mixture.

Solvent Mass Balance

Applying Equation (2.4) on p. 21 for the solvent mass in the system, which is an annulus of thicknesses Δr and Δz in the physical reservoir model, we obtain

$$\frac{d}{dt}\underbrace{(\Delta V \rho \phi \omega)}_{\substack{\text{solvent mass} \\ \text{in the system}}} = (\mathbf{f_r} \cdot \Delta \mathbf{A_r})_{r+\Delta r} - (\mathbf{f_r} \cdot \Delta \mathbf{A_r})_r + (\mathbf{f_z} \cdot \Delta \mathbf{A_z})_{z+\Delta z} - (\mathbf{f_z} \cdot \Delta \mathbf{A_z})_z$$

where (i) $\Delta V = 2\pi r \Delta r \Delta z$ is the volume of the annulus, (ii) ρ is the density of the oil–solvent mixture, (iii) ϕ is the porosity of the physical reservoir model, (iv) ω is solvent mass fraction, and (v) $\mathbf{f_r}$ and $\mathbf{f_z}$ are the solvent mass fluxes, respectively across areas $\Delta \mathbf{A_r} = 2\pi r \Delta z \hat{\mathbf{r}}$, and $\Delta \mathbf{A_z} = 2\pi r \Delta r \hat{\mathbf{z}}$.

Along the r-direction, since the total bulk flux is zero, the solvent flux is the diffusive flux, i.e., $\mathbf{f_r} = -D\rho\phi \partial\omega/\partial r (-\hat{\mathbf{r}})$ with D as solvent diffusivity. Along the z-direction, since diffusive flux is negligible, $\mathbf{f_z} = \omega\rho v_z (-\hat{\mathbf{z}})$ where v_z is downward, vertical velocity of the oil–solvent mixture.

Using the first order Taylor expansions of $(\mathbf{f_r} \cdot \Delta \mathbf{A_r})_{r+\Delta r}$ and $(\mathbf{f_z} \cdot \Delta \mathbf{A_z})_{z+\Delta z}$, taking the limits of Δr and Δz to zero, and substituting the expressions for the fluxes and areas, the solvent mass balance becomes

$$\frac{\partial}{\partial t}(dV \rho \phi \omega) = \frac{\partial}{\partial r}\left(D\rho\phi\frac{\partial \omega}{\partial r}2\pi r dz\right)dr - \frac{\partial}{\partial z}(\omega\rho v_z 2\pi r dr)dz$$

where $dV = 2\pi r dr dz$. Since r and dr are independent of t as well as z, dz is independent of t as well as r, and ρ and ϕ are constant, the above equation simplifies to

$$\frac{\partial \omega}{\partial t} = D\left(\frac{1}{r}\frac{\partial \omega}{\partial r} + \frac{\partial^2 \omega}{\partial r^2}\right) + \frac{\partial D}{\partial \omega}\left(\frac{\partial \omega}{\partial r}\right)^2 - \frac{1}{\phi}\left(v_z + \omega\frac{\partial v_z}{\partial \omega}\right)\frac{\partial \omega}{\partial z} \qquad (4.41)$$

In the last equation, v_z is given by Darcy's law, i.e.,

$$v_z = \frac{K_o \rho g}{\mu(\omega)} \tag{4.42}$$

where K_o is the effective permeability of the oil, g is gravity, and μ is the oil viscosity as a known function of solvent mass fraction ω. Note that $v_z = v_z[\omega(r, z, t)]$.

Oil–Solvent Mass Balance

Applying Equation (2.4) on p. 21 for the mass of the oil–solvent mixture in the system, which is an annulus of differential thickness Δr and height Z (see Figure 4.11 on p. 109), we get for Δr tending to zero,

$$\underbrace{\frac{\partial}{\partial t}(\rho \phi dV)}_{\substack{\text{oil–solvent mass} \\ \text{in the system}}} = \underbrace{-\rho v_0(r) \times 2\pi r dr}_{\substack{\text{rate of oil–solvent mass} \\ \text{out of the system}}}$$

In the above equation, $dV = 2\pi r dr Z$ is the annulus volume with Z as the annulus height, and $v_0(r)$ is v_z at $z = 0$ and a radial distance r. Note that $v_0(r)$ is obtained from Equation (4.42) above using $\omega(r, 0, t)$.

Simplifying the oil–solvent mass balance, we get

$$\frac{\partial Z}{\partial t} = -\frac{v_0(r)}{\phi} \tag{4.43}$$

Equation (4.41) is a partial differential equation of (i) first order with respect to t, (ii) second order with respect to r, and (iii) first order with respect to z. Thus, for integration, that equation needs an initial condition, two boundary conditions on the r-axis, and one boundary condition on the z-axis. Equation (4.43) above is of first order with respect to t, and thus needs an initial condition.

Initial Conditions

Initially, there is no solvent in the physical reservoir model except at the surface, where solvent mass fraction is the interfacial mass fraction ω_i. Also, the model height is Z_0. Hence, the initial condition for Equation (4.41) on the previous page is

$$\omega(r, z, 0) = \begin{cases} \omega_i \text{ at } \begin{cases} r = R & \forall \ 0 \leq z \leq Z_0 \\ z = 0, Z_0 & \forall \ 0 \leq r < R \end{cases} \\ 0 \quad \forall \ 0 < z < Z_0 \quad \text{and} \quad 0 \leq r < R \end{cases} \tag{4.44}$$

The initial condition for Equation (4.43) above is

$$Z(r, 0) = Z_0 \ \forall \ 0 \leq r \leq R \tag{4.45}$$

Boundary Conditions

While the solvent mass fraction at the surface of the physical reservoir model is ω_i, the mass fraction gradient at the vertical axis is zero due to symmetry [see Appendix 4.B, p. 133]. Hence, the boundary conditions for Equation (4.41) on p. 110 at any time $t > 0$ are

$$\omega(r, z, t) \;=\; \omega_i \;\; \text{at} \;\; \begin{cases} r = R & \forall \;\; 0 \leq z \leq Z(r, t) \\[2mm] z = 0, Z(r, t) & \forall \;\; 0 \leq r < R \end{cases} \tag{4.46}$$

$$\frac{\partial \omega}{\partial r} \;=\; 0 \;\; \text{at} \;\; r = 0 \;\; \forall \;\; 0 \leq z \leq Z(r, t) \tag{4.47}$$

Heavy Oil Production Rate

If Δm is the mass of the oil–solvent produced in time Δt from the annulus of thickness Δr, radius r, and height Z then

$$\left. \frac{\Delta m}{\Delta t} \right|_r \;=\; \rho \, \underbrace{v_0(r) \times \pi[(r + \Delta r)^2 - r^2]}_{\text{volumetric flow rate}}$$

In the limit of Δt and Δr tending to zero, the overall oil–solvent production rate from the physical reservoir model is then

$$\frac{dm}{dt} \;=\; 2\pi\rho \int\limits_0^R r v_0(r) \, dr$$

Summary

Equations (4.41)–(4.47) constitute the dynamic, distributed-parameter model of solvent transfer in the physical reservoir model. Given the specifications for the parameter set

$$\{D, \, \phi, \, K_o, \, g, \, \rho, \, \mu(\omega), \, \omega_i, \, R, \, Z_0\}$$

the involved partial differential equations need to be integrated simultaneously to obtain $\omega = \omega(r, z, t)$ and $Z = Z(r, t)$.

4.2.8 Hydrogel Tablet

This example involves microscopic mass balance of drug, polymer and water in a water-soluble hydrogel tablet.[5] As shown in Figure 4.12 on the next page, the tablet is a cylinder of radius R and height Z. When ingested, water in the stomach diffuses into the tablet. The polymer begins to diffuse out from the tablet.

From the surface, polymer and drug in the polymer matrix dissolve instantaneously into the bulk water. The water diffuses simultaneously into the tablet. During this process, the volume of the tablet changes with time. The tablet swells first but eventually dissolves in water. This is an example of a **moving boundary problem without constant density assumption**.

Objective

It is desired to develop a model that would enable the determination of the species mass concentrations in the tablet at different radial distances and times.

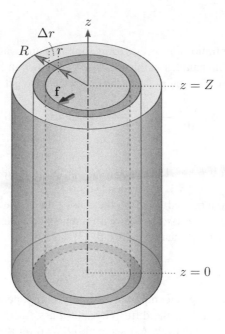

Figure 4.12   Hydrogel tablet in water

System

In this example, the species concentrations are functions of the r-coordinate. Hence, the system for mass balance is an annulus of thickness Δr along the r-direction, as shown in Figure 4.12 above.

Assumptions

We make the following assumptions for the model:

1. Species concentrations in the tablet are functions of the radial distance, and time.

2. Mass fluxes, and the change in tablet shape are along the radial direction only.

3. The fractional change in the length along the radial direction is always uniform (i.e., $\mathrm{d}r/r = \mathrm{d}R/R$).

4. The diffusion of the drug inside the tablet is insignificant.

5. The drug dissipates fast enough from the tablet surface so that its concentration at the surface, as well as outside, is negligible.

6. Outside the tablet, the mass concentration of water is not much different from its density.

Model Formulation

To determine the concentrations of species (drug, polymer and water), we need mass balances in our system, which is an annulus of thickness Δr, as shown in Figure 4.12 on the previous page. Applying Equation (2.4) on p. 21 for the ith species, we get

$$\underbrace{\frac{\mathrm{d}}{\mathrm{d}t}(\Delta V \rho_i)}_{\substack{\text{species mass} \\ \text{in the system}}} = \Delta V \frac{\mathrm{d}\rho_i}{\mathrm{d}t} + \rho_i \frac{\mathrm{d}(\Delta V)}{\mathrm{d}t} = (\mathbf{f}_i \cdot \mathbf{A})_r - (\mathbf{f}_i \cdot \mathbf{A})_{r+\Delta r}$$

where (i) the values of i are 1, 2 and 3 for drug, polymer and water, respectively, (ii) $\Delta V = \pi[(r + \Delta r)^2 - r^2]Z$ is the system volume, (iii) ρ_i is the mass concentration of the ith species in the system, and (iv) $\mathbf{f}_i$ is the mass flux of the ith species across the area $\mathbf{A} = (2\pi r Z)\hat{\mathbf{r}}$. Note that since Δr changes with time, so do V and ΔV.

The mass flux is given by [Equation (2.10), p. 23]

$$\mathbf{f}_i = \rho_i \mathbf{v} + \mathbf{j}_i = \rho_i v_r \hat{\mathbf{r}} - D_i \rho \frac{\partial \omega_i}{\partial r} \hat{\mathbf{r}}$$

where (i) $\mathbf{v}$ is the bulk velocity comprising the only component v_r along the r-direction, (ii) D_i and ω_i are, respectively, the diffusivity and mass fraction of the ith species in the tablet, and (iii) ρ is the tablet density. As assumed, $D_1 = 0$ for the drug.

In the above differential equation, we use the first order Taylor expansion of $(\mathbf{f}_i \cdot \mathbf{A})_{r+\Delta r}$, take the limit of Δr to zero, and substitute the expressions for the flux and area. The result after simplification is

$$2\pi r \mathrm{d}r Z \frac{\partial \rho_i}{\partial t} + \rho_i \lim_{\Delta r \to 0} \frac{\mathrm{d}(\Delta V)}{\mathrm{d}t}$$

$$= -\frac{\partial}{\partial r}(\mathbf{f}_i \cdot \mathbf{A})\mathrm{d}r = -2\pi Z \left[\frac{\partial}{\partial r}(\rho_i v_r r) - \frac{\partial}{\partial r}\left(r D_i \rho \frac{\partial \omega_i}{\partial r} \right) \right] \mathrm{d}r \tag{4.48}$$

Time Derivative of Annulus Volume

This term, which appears in above equation, is simplified as follows.

With $r_{\mathrm{s}} \equiv (r + \Delta r)$, we have

$$\frac{\mathrm{d}(\Delta V)}{\mathrm{d}t} = \frac{\mathrm{d}}{\mathrm{d}t}(\pi r_{\mathrm{s}}^2 Z - \pi r^2 Z) = \pi Z \frac{\mathrm{d}}{\mathrm{d}t}(r_{\mathrm{s}}^2 - r^2) = 2\pi Z \left(r_{\mathrm{s}} \frac{\mathrm{d}r_{\mathrm{s}}}{\mathrm{d}t} - r \frac{\mathrm{d}r}{\mathrm{d}t} \right)$$

Since the fractional change in radius is assumed to be the same at any radial distance,

$$\frac{\mathrm{d}r}{r} = \frac{\mathrm{d}r_{\mathrm{s}}}{r_{\mathrm{s}}} = \frac{\mathrm{d}R}{R} \quad \Rightarrow \quad \frac{\mathrm{d}r_{\mathrm{s}}}{\mathrm{d}t} = \frac{r_{\mathrm{s}}}{R}\frac{\mathrm{d}R}{\mathrm{d}t} \quad \text{and}$$

$$v_r = \frac{\mathrm{d}r}{\mathrm{d}t} = \frac{r}{R}\frac{\mathrm{d}R}{\mathrm{d}t} \tag{4.49}$$

With the help of the above expressions for $\mathrm{d}r_{\mathrm{s}}/\mathrm{d}t$ and $\mathrm{d}r/\mathrm{d}t$, we obtain

$$\frac{\mathrm{d}(\Delta V)}{\mathrm{d}t} = \frac{2\pi Z}{R}(r_{\mathrm{s}}^2 - r^2)\frac{\mathrm{d}R}{\mathrm{d}t} = \frac{2\pi Z}{R}\left[(r + \Delta r)^2 - r^2\right]\frac{\mathrm{d}R}{\mathrm{d}t}$$

Taking the limit of Δr to zero,

$$\lim_{\Delta r \to 0} \frac{d(\Delta V)}{dt} = \frac{2\pi Z}{R}(2r\,dr)\frac{dR}{dt}$$

With the help of the above result, and Equation (4.49) on the previous page, Equation (4.48) yields

$$\frac{\partial \rho_i}{\partial t} = -\frac{1}{R}\frac{dR}{dt}\left[3\rho_i + \frac{\partial}{\partial r}(r\rho_i)\right] + \frac{1}{r}\frac{\partial}{\partial r}\left(rD_i\rho\frac{\partial \omega_i}{\partial r}\right) \qquad (4.50)$$

where

$$\rho = \sum_{i=1}^{3}\rho_i, \qquad \text{and} \qquad \omega_i = \frac{\rho_i}{\rho}$$

Polymer Mass Balance

Applying Equation (2.4) on p. 21 for the polymer in the system, which is an annulus of differential thickness Δr, and height Z [see Figure 4.12, p. 113], we get for Δr tending to zero

$$\frac{d}{dt}\int_0^R 2\pi r Z\rho_2\,dr = -2\pi RZk\tilde{\rho}_2 \qquad \text{or} \qquad \frac{d}{dt}\underbrace{\int_0^R r\rho_2\,dr}_{\equiv I[R(t),t]} = -Rk\tilde{\rho}_2 \qquad (4.51)$$

$$\underbrace{\qquad\qquad}_{\substack{\text{polymer mass}\\\text{in the tablet}}} \qquad \underbrace{\qquad\qquad}_{\substack{\text{polymer mass lost}\\\text{from the tablet surface}}}$$

where k is the mass transfer coefficient of polymer in water, and $\tilde{\rho}_2$ is $\rho_2(R,t)$, i.e., the mass concentration of polymer at the tablet surface.

Using the chain rule of differentiation [see Section 8.7.3, p. 319], the left-hand side of the last equation simplifies as follows:

$$\frac{dI}{dt} = \frac{\partial I}{\partial R}\frac{dR}{dt} + \frac{\partial I}{\partial t} = R\tilde{\rho}_2\frac{dR}{dt} + \underbrace{\int_0^R \frac{\partial}{\partial t}(r\rho_2)\,dr}_{\equiv J}$$

where we have used the Leibniz's rule [see Section 8.10, p. 326] to evaluate $\partial I/\partial R$. The integral denoted above by J is given by

$$J = \int_0^R \left(r\frac{\partial \rho_2}{\partial t} + \rho_2\underbrace{\frac{dr}{dt}}_{=\frac{r\,dR}{R\,dt}}\right)dr = \int_0^R r\frac{\partial \rho_2}{\partial t}\,dr + \frac{1}{R}\frac{dR}{dt}\int_0^R r\rho_2\,dr$$

[from Equation (4.49), previous page]

Eliminating $\partial \rho_2/\partial t$ with the help of Equation (4.50) above for $i = 2$, we obtain after simplification

$$J = -\frac{1}{R}\frac{dR}{dt}\left(R^2\tilde{\rho}_2 + \int_0^R r\rho_2\,dr\right) + \int_0^R \left(rD_2\rho\frac{\partial \omega_2}{\partial r}\right)dr$$

Incorporating the expressions for J and $\mathrm{d}I/\mathrm{d}t$ in Equation (4.51) on the previous page, we obtain after simplification

$$\frac{\mathrm{d}R}{\mathrm{d}t} = \frac{k\tilde{\rho}_2 R + \int\limits_0^R \frac{\partial}{\partial r}\left(r D_2 \rho \frac{\partial \omega_2}{\partial r}\right) \mathrm{d}r}{\dfrac{1}{R} \int\limits_0^R r\rho_2 \, \mathrm{d}r} \tag{4.52}$$

Equation (4.50) on the previous page is a partial differential equation of (i) first order with respect to t, and (ii) second order with respect to r. Thus, for integration, that equation needs an initial condition, and two boundary conditions on the r-axis. On the other hand, Equation (4.52) above is a differential equation of first order with respect to t, and thus needs an initial condition.

Initial Conditions

The initial species concentrations ρ_{i0}s, and tablet radius R_0 are known. Hence, the initial conditions are

$$\rho_i(r,0) = \rho_{i0} \; \forall \; 0 \le r \le R_0; \quad i = 1, 2, 3; \quad \text{and} \quad R(0) = R_0 \tag{4.53}$$

Boundary Conditions

At the central axis of the tablet, the mass fraction gradients are zero due to symmetry [see Appendix 4.B, p. 133]. At the tablet surface, while the drug concentration is negligible, the diffusive flux of polymer as well as water is equal to the convective flux. Hence, the boundary conditions for Equation (4.50) on the previous page at any time $t > 0$ are

$$\left.\frac{\partial \rho_i}{\partial r}\right|_{r=0} = 0; \quad i = 1, 2, 3; \quad \rho_1(R) = 0 \tag{4.54}$$

$$-D_2 \rho \frac{\partial \omega_2}{\partial r} = k_2 \tilde{\rho}_2 \quad \text{and} \quad -D_3 \rho \frac{\partial \omega_3}{\partial r} = k_3(\tilde{\rho}_3 - \rho_{\mathrm{w}}) \quad \text{at} \quad r = R \tag{4.55}$$

where (i) k_2 and k_3 are, respectively, the mass transfer coefficients of polymer and water, (ii) $\tilde{\rho}_3$ is the mass concentration of water at the tablet surface, and (iii) ρ_{w} is the density of water.

Summary

Equations (4.50)–(4.55) constitute the dynamic, distributed-parameter model of mass transfer in a hydrogel tablet. Given the specifications for the parameter set

$$\left\{D_1 = 0, \, D_2, \, D_3, \, \rho_{10}, \, \rho_{20}, \, \rho_{30}, \, R_0, \, k_2, \, k_3, \, \rho_{\mathrm{w}}\right\}$$

the involved Equation (4.50) on the previous page for each species (drug, polymer and water), and Equation (4.52) above need to be integrated simultaneously to obtain $\rho_1 = \rho_1(r,t)$, $\rho_2 = \rho_2(r,t)$, $\rho_3 = \rho_3(r,t)$, and $R = R(t)$.

4.2.9 Neutron Diffusion

This example involves microscopic balance for neutrons in a spherical medium.[6] As shown in Figure 4.13 below, neutrons diffuse outward from a point source at the centre of the medium.

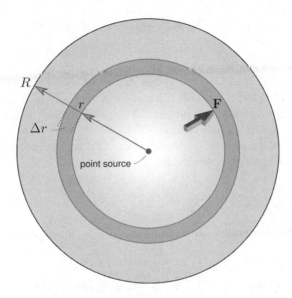

Figure 4.13   Neutron diffusion from a point source

Objective

It is desired to develop a model that would enable the determination of neutron concentration in the medium at different radial distances, and times.

System

In this example, the neutron concentration in the medium is a function of the r-coordinate. Hence, the system for neutron balance is a spherical differential element (or shell) of thickness Δr along the r-direction as shown in Figure 4.13 above.

Assumptions

We make the following assumptions for the model:

1. Neutron concentration in the medium is a function of radial distance, and time.

2. There is no bulk flux, and neutron flux is due to diffusion, which is only along the radial direction.

3. Diffusivity of neutrons in the medium is constant.

Model Formulation

To determine neutron concentration, i.e., the number of neutrons per unit volume, we need a balance of neutrons in our system, which is a spherical shell of thickness Δr. The balance is given by

$$\underbrace{\frac{d}{dt}(\Delta V n)}_{\substack{\text{number of} \\ \text{neutrons in the system}}} = (\mathbf{F} \cdot \mathbf{A})_r - (\mathbf{F} \cdot \mathbf{A})_{r+\Delta r} + \Delta V n_g$$

where (i) $\Delta V = 4\pi/3[(r + \Delta r)^3 - r^3]$ is the volume of the system (the spherical shell), (ii) n is the number of neutrons per unit volume, (iii) $\mathbf{F} = (-Ddn/dr)\hat{\mathbf{r}}$ is the neutron flux with D as neutron diffusivity across the area $\mathbf{A} = (4\pi r^2)\hat{\mathbf{r}}$, and (iv) n_g is the rate of the number of neutrons emitted per unit volume.

In the neutron balance above, using the first order Taylor expansion of $(\mathbf{F} \cdot \mathbf{A})_{r+\Delta r}$, and taking the limit of Δr to zero, we obtain after simplification

$$\frac{\partial}{\partial t}(dV n) = -\frac{\partial}{\partial r}(\mathbf{F} \cdot \mathbf{A}) + dV n_g$$

where $dV = 4\pi r^2 dr$. Substituting the expressions for dV, $\mathbf{F}$ and $\mathbf{A}$ in the above equation, we get

$$\frac{\partial}{\partial t}\left[(4\pi r^2 dr)n\right] = -\frac{\partial}{\partial r}\left[-D\frac{\partial n}{\partial r}(4\pi r^2)\right]dr + (4\pi r^2 dr)n_g$$

Since r as well as dr is independent of t and D is constant, the above equation simplifies to

$$\frac{\partial n}{\partial t} = D\left(\frac{2}{r}\frac{\partial n}{\partial r} + \frac{\partial^2 n}{\partial r^2}\right) + n_g \tag{4.56}$$

Equation (4.56) is a partial differential equation of (i) first order with respect to t, and (ii) second order with respect to r. Thus, for integration, that equation needs an initial condition, and two boundary conditions on the r-axis.

Initial Condition
Neutron concentration is zero everywhere in the beginning. Hence, the initial condition is

$$n(r, 0) = 0 \ \forall \ 0 \leq r \leq \infty \tag{4.57}$$

Boundary Conditions
At the centre, the gradient of neutron concentration is zero due to symmetry [see Appendix 4.B, p. 133]. At an infinite distance from the centre, the concentration is taken to be zero. Thus, the boundary conditions at any time $t > 0$ are

$$\frac{\partial n}{\partial r} = 0 \ \text{at} \ r = 0 \quad \text{and} \quad n = 0 \ \text{at} \ r = R \tag{4.58}$$

where R is a very large radius tending to infinity.

Summary

Equations (4.56)–(4.58) on the previous page constitute the dynamic, distributed-parameter model of neutron transfer from a point source in a spherical infinite medium. Given the specifications for D, n_g and R, Equation (4.56) on the previous page needs to be integrated to obtain $n = n(r, t)$.

At steady state, the time derivative in Equation (4.56) becomes zero. The resulting differential equation can be integrated using the boundary conditions to obtain the steady state neutron concentration, $\hat{n} = \hat{n}(r)$

4.2.10 Horton Sphere

This example involves the equations of change for temperature in a Horton sphere. It is a spherical vessel that is commonly used to store liquified gases. Figure 4.14 below shows the schematic of the sphere. The core of the sphere is a hollow metal sphere of inside and outside

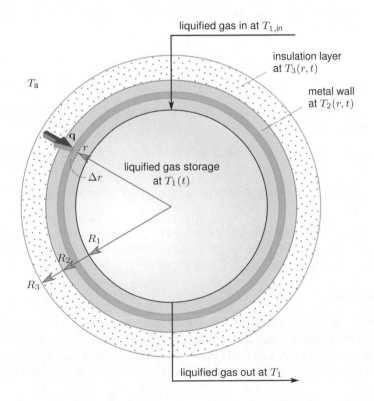

Figure 4.14 Schematic of a Horton sphere used to store liquified gas

radii, R_1 and R_2, respectively. The core is covered externally by a spherical shell of outside radius R_3. From the surroundings, heat flows radially inward across the insulation and the metal wall to the pool of well-mixed, liquified gas at a uniform temperature T_1 in the core. To discard this heat, a small amount of the liquified gas is continuously withdrawn, cooled and fed back to the core.

Objective

It is desired to develop a model that would enable the determination of (i) the temperature of the liquified gas stored in the core at different times, (ii) the temperature of the metal wall at different radial distances, and times, and (iii) the temperature of the insulation layer at different radial distances, and times.

System

In this example, the liquified gas temperature is uniform. Thus, the system for the equation of change for the gas temperature is the sphere of radius R_1. On the other hand, the temperatures of the metal wall, and insulation layer are functions of the r-coordinate. Therefore, the system for the equation of change for the temperature in either medium (wall, or insulation) is a spherical shell of thickness Δr along the r-direction. Figure 4.14 on the previous page shows the shell in the metal wall.

Assumptions

We make the following assumptions for the model:

1. The liquified gas in the core is sufficiently mixed to have a uniform temperature, which is a function of time.

2. Temperatures in the insulation layer, and metal wall are functions of radial distance, and time.

3. Heat conduction is only along the radially inward direction through the insulation layer, and the metal wall.

4. The density as well as specific heat capacity of liquified gas, and the densities as well as thermal conductivities of the insulation layer, and the metal wall are constant in the range of temperature variation.

Model Formulation

We will apply the equation of change for temperature to determine the temperature of liquified gas, metal wall, and insulation layer. In the following equations, we use subscripts '1', '2' and '3' for the properties of the gas, metal and insulation, respectively. Thus, ρ_2, $\hat{C}_{P2}$, k_2 and T_2 denote, respectively, the density, specific heat capacity, thermal conductivity, and temperature of the metal wall.

Temperature of Liquified Gas in the Core
Applying Equation (3.13) on p. 71 to the liquified gas in the core, we obtain

$$\underbrace{\frac{4}{3}\pi R_1^3 \rho_1}_{\substack{\text{system} \\ \text{mass}}} \hat{C}_{P1} \frac{dT_1}{dt} = wĈ_{P1}(T_{1\text{in}} - T_1) + \underbrace{h_1(4\pi R_1^2)[T_2(R_1) - T_1]}_{-\Delta[\dot{Q}]}$$

In the last equation, (i) R_1 is the radius of the core, (ii) w is the rate of gas mass that is withdrawn from the core, and fed back at temperature $T_{1\text{in}}$, and (iii) h_1 is the coefficient of heat transfer between the gas and metal wall. Simplifying the last equation, we get

$$\frac{dT_1}{dt} = \frac{3}{\rho_1 \hat{C}_{P1} R_1} \left\{ \frac{w \hat{C}_{P1}(T_{1\text{in}} - T_1)}{4\pi R_1^2} + h_1[T_2(R_1) - T_1] \right\} \tag{4.59}$$

Temperature of Metal Wall and Insulation Layer

Applying Equation (3.14) on p. 71 to spherical shells of thickness Δr in the metal wall, and in the insulation layer, we obtain

$$\underbrace{\Delta V \rho_i \hat{C}_{Pi}}_{\substack{\text{system} \\ \text{mass}}} \frac{dT_i}{dt} = \underbrace{(\mathbf{q}_i \cdot \mathbf{A})_{r+\Delta r} - (\mathbf{q}_i \cdot \mathbf{A})_r}_{-\Delta[\dot{Q}]}; \quad i = 2, 3$$

where (i) $\Delta V = 4\pi/3[(r + \Delta r)^3 - r^3]$ is the volume of the system (spherical shell), and (ii) $\mathbf{q}_i = (-k_i \partial T_i/\partial r)(-\hat{\mathbf{r}})$ is the conductive heat flux across the area $A = 4\pi r^2(\hat{\mathbf{r}})$.

In the above differential equation, using the first order Taylor expansion of $(\mathbf{q} \cdot \mathbf{A})_{r+\Delta r}$, and taking the limit of Δr to zero, we obtain after simplification

$$dV \rho_i \hat{C}_{Pi} \frac{\partial T_i}{\partial t} = \frac{\partial}{\partial r}(\mathbf{q}_i \cdot \mathbf{A})dr; \quad i = 2, 3$$

where $dV = 4\pi r^2 dr$. Substituting for dV, $\mathbf{q}_i$ and $\mathbf{A}$ in the above differential equation, and simplifying the result yields

$$\frac{\partial T_i}{\partial t} = \frac{k_i}{\rho_i \hat{C}_{Pi} r^2} \frac{\partial}{\partial r}\left(r^2 \frac{dT_i}{dr}\right); \quad i = 2, 3 \tag{4.60}$$

Equations (4.59) and (4.60) are differential equations of first order with respect to t, and each needs one initial condition. The last set of equations is additionally of second order with respect to r, and needs two boundary conditions on the r-axis.

Initial Conditions

The initial temperatures for liquified gas, metal wall, and insulation layer are specified as T_{10}, T_{20} and T_{30}, respectively. Hence, the initial conditions are

$$T_1(0) = T_{10}, \quad T_2(r, 0) = T_{20} \;\; \forall \;\; R_1 \leq r \leq R_2 \quad \text{and}$$

$$T_3(r, 0) = T_{30} \;\; \forall \;\; R_2 \leq r \leq R_3 \tag{4.61}$$

Boundary Conditions

At the inner surface of the metal wall, the conductive heat flux is equal to the convective heat flux in the liquified gas. At the outer surface of metal wall, the conductive flux is equal to that in the insulation layer. At the outer surface of insulation layer, the conductive heat flux is equal to the convective heat flux in surrounding air. Accordingly, the boundary conditions at any time $t > 0$ for Equation (4.60) on the previous page in the metal shell, and insulation layer are, respectively,

$$k_2 \left.\frac{\partial T_2}{\partial r}\right|_{R_1} = h_1[T_2(R_1) - T_1], \qquad k_2 \left.\frac{\partial T_2}{\partial r}\right|_{R_2} = k_3 \left.\frac{\partial T_3}{\partial r}\right|_{R_2} \qquad \text{and}$$

$$k_3 \left.\frac{\partial T_3}{\partial r}\right|_{R_2} = k_2 \left.\frac{\partial T_2}{\partial r}\right|_{R_2}, \qquad k_3 \left.\frac{\partial T_3}{\partial r}\right|_{R_3} = h_a[T_a - T_3(R_3)] \qquad (4.62)$$

where h_a is the coefficient of heat transfer in the surroundings.

Summary

Equations (4.59)–(4.62) constitute the dynamic heat transfer model for the Horton sphere. Given the specifications for the parameter set

$$\left\{\rho_1, \rho_2, \rho_3, \hat{C}_{P1}, \hat{C}_{P2}, \hat{C}_{P3}, R_1, R_2, R_3, w, T_{1\text{in}}, T_{10}, T_{20}, T_{30}, h_1, h_a, k_2, k_3, T_a\right\}$$

the involved differential equations need to be integrated to obtain $T_1 = T_1(t)$, $T_2 = T_2(r, t)$, and $T_3 = T_3(r, t)$.

At steady state, the time derivatives in the differential equations become zero, and the model becomes independent of ρ_1, ρ_2, ρ_3, $\hat{C}_{P2}$ and $\hat{C}_{P3}$. With the remaining parameters, resulting differential equations can be integrated simultaneously using the boundary conditions to obtain the steady state temperatures, $\hat{T}_1$, $\hat{T}_2 = \hat{T}_2(r)$, and $\hat{T}_3 = \hat{T}_3(r)$.

4.2.11   Reactions around Solid Reactant

This example involves macroscopic and microscopic mass balances with chemical reactions, and multicomponent diffusion in spherical geometry. As shown in Figure 4.15 on the next page, gas A diffuses from the surroundings through a porous layer of intermediate product C to the inside core of solid B. Both B and C have of constant molar densities. The following non-elementary reactions take place:

$$a_1 \underset{\text{gas}}{\text{A}} + b \underset{\text{solid}}{\text{B}} \xrightarrow{k_1} c_1 \underset{\text{solid}}{\text{C}} \qquad \text{and} \qquad a_2 \underset{\text{gas}}{\text{A}} + c_2 \underset{\text{solid}}{\text{C}} \xrightarrow{k_2} d \underset{\text{gas}}{\text{D}}$$

respectively, with reaction rates $k_1 c_A^{a_1}$ (based on per unit area) at the core surface, and $k_2 c_A^{a_2}$ (based on per unit volume as usual) in the porous layer. Gas D is produced in the layer, and diffuses out to the surroundings. The core radius, and porous layer thickness change with time as the reactions proceed. Hence, this is an example of a **moving boundary problem with constant density assumption**.

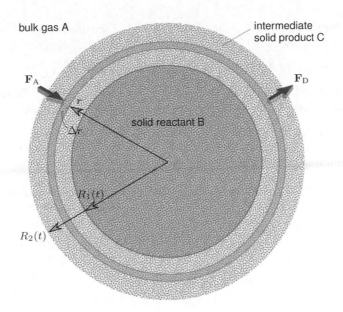

Figure 4.15 Solid reactant B covered with a porous layer of intermediate solid product C. Through the layer, reactant gas A diffuses in while the product gas D diffuses out

Objective

It is desired to develop a model that would enable the determination of (i) the radius of the core of solid reactant B at different times, (ii) the outer radius of the porous layer of intermediate product C at different times, (iii) the concentration of the reactant gas A in the porous layer at different radial distances, and times, and (iv) the concentration of product gas D in the porous layer at different radial distances, and times.

System

In this example, the core is made of pure solid B. Thus, the system for the mole balance of B is the core, i.e., the sphere of radius R_1. The porous layer has constant molar density of C. Hence, its mole balance has the layer as the system. On the other hand, concentrations of A and D in the layer are functions of the r-coordinate. Therefore, the system for the mole balances of A and D is a spherical shell of thickness Δr along the r-direction, as shown in Figure 4.15 above.

Assumptions

We make the following assumptions for the model:

1. Mole fractions of the gases A and D in the porous layer (made of solid C) are functions of radial distance, and time.

2. At the surface of the core, A is consumed instantaneously in the chemical reaction.

3. In the porous layer, only A and D are mobile. Their mass transfer is solely due to diffusion along the radial inward and outward directions, respectively.

4. The diffusivities of A and D in the porous layer are constant.

5. At the outer surface of the porous layer, D is dispersed immediately in the surroundings. Thus, the concentration of D at the outer surface of the layer is zero.

6. The molar densities of solids B and C are constant.

Model Formulation

To determine species concentrations in the system, we need mole balances in the presence of chemical reactions.

Mole Balance of B in the Core

Applying Equation (3.6) on p. 66 for the moles of B in the system, which is the core, we get

$$\frac{d}{dt}\underbrace{(V_B c_B)}_{\substack{\text{moles of B} \\ \text{in the system}}} = (4\pi R_1^2) r_B$$

where (i) $V_B = 4\pi R_1^3/3$ is the core volume, c_B is the molar density of B in the core, $R_1 = R_1(t)$ is the radius of the core, and (ii) $r_B = -bk_1[c_A^{a_1}]_{R_1}$ is the molar rate of change of B due to reaction with gas A per unit core surface area. While b and a_1 are stoichiometric coefficients, k_1 is the rate coefficient of this reaction.

In the above mole balance, we substitute for V_B and r_B, and simplify the result considering c_B as constant to obtain

$$\frac{dR_1}{dt} = -\frac{k_1 b}{c_B}[c_A^{a_1}]_{R_1} \tag{4.63}$$

Mole Balances in Porous Layer

Our second system is the porous layer, which is made of C, and has thickness $(R_2 - R_1)$ around the core. Gases A and D diffuse, respectively, into and out of this system.

Mole Balance of C Applying Equation (3.6) on p. 66 for the moles of C in the system, we get

$$\frac{d}{dt}\underbrace{(V_C c_C)}_{\substack{\text{moles of C} \\ \text{in the system}}} = (4\pi R_1^2) r_{1C} + \dot{R}_{\text{gen,C}}$$

where (i) $V_C = 4\pi(R_2^3 - R_1^3)/3$ is the layer volume with $R_2 = R_2(t)$ as the outer radius of the layer, (ii) c_C is the molar density of C in the layer, (iii) $r_{1C} = c_1 k_1[c_A^{a_1}]_{R_1}$ is the molar rate of change of C due to reaction between A and B per unit area at the core surface, and (iv) $\dot{R}_{\text{gen,C}}$ is the molar rate of generation of C due to its reaction with A in the layer, and is given by

$$\dot{R}_{\text{gen,C}} = -4\pi c_2 k_2 \int_{R_1}^{R_2} c_A^{a_2} r^2 \, dr$$

Since c_C is constant, the mole balance of C becomes

$$\frac{4\pi c_C}{3}\frac{\mathrm{d}}{\mathrm{d}t}(R_2^3 - R_1^3) = (4\pi R_1^2)c_1 k_1 [c_A^{a_1}]_{R_1} - 4\pi c_2 k_2 \int_{R_1}^{R_2} c_A^{a_2} r^2 \,\mathrm{d}r$$

Simplifying the above equation, and using Equation (4.63) on the previous page for $\mathrm{d}R_1/\mathrm{d}t$, we obtain

$$\frac{\mathrm{d}R_2}{\mathrm{d}t} = k_1[c_A^{a_1}]_{R_1}\left(\frac{R_1}{R_2}\right)^2\left(\frac{c_1}{c_C} - \frac{b}{c_B}\right) - \frac{c_2 k_2}{c_C R_2^2}\int_{R_1}^{R_2} c_A^{a_2} r^2 \,\mathrm{d}r \qquad (4.64)$$

Mole Balance of A For this balance, the system is a spherical shell of thickness Δr in the porous layer. Applying Equation (3.6) on p. 66 for the moles of A in this system, we get

$$\frac{\mathrm{d}}{\mathrm{d}t}\underbrace{(\Delta V c_A)}_{\substack{\text{moles of A} \\ \text{in the system}}} = (\mathbf{F}_A \cdot \mathbf{A})_{r+\Delta r} - (\mathbf{F}_A \cdot \mathbf{A})_r + \Delta V r_A$$

where (i) $\Delta V = 4\pi/3[(r + \Delta r)^3 - r^3]$ is the system volume, (ii) $\mathbf{F}_A = F_A(-\hat{\mathbf{r}})$ is the molar flux of A of magnitude F_A, and is across the area $\mathbf{A} = 4\pi r^2\hat{\mathbf{r}}$, and (iii) $r_A = -a_2 k_2 c_A^{a_2}$ is the molar rate of change of A per unit volume due to reaction in the porous layer.

In the above mole balance, we use the first order Taylor expansion of $(\mathbf{F}_A \cdot \mathbf{A})_{r+\Delta r}$, take the limit of Δr to zero, and substitute the expressions for system volume, flux and area. The result after simplification is

$$\frac{\partial}{\partial t}(4\pi r^2 \mathrm{d}r c_A) = -\frac{\partial}{\partial r}\left[F_A(4\pi r^2)\right]\mathrm{d}r - (4\pi r^2 \mathrm{d}r)a_2 k_2 c_A^{a_2}$$

Since r and $\mathrm{d}r$ are independent of t, the above equation simplifies to

$$\frac{\partial c_A}{\partial t} = -\frac{1}{r^2}\frac{\partial}{\partial r}(r^2 F_A) - a_2 k_2 c_A^{a_2} \qquad (4.65)$$

Mole Balance of D Applying Equation (3.6) on p. 66 for the moles of D in the same system as above (the porous layer), we get

$$\frac{\mathrm{d}}{\mathrm{d}t}\underbrace{(\Delta V c_D)}_{\substack{\text{moles of D} \\ \text{in the system}}} = (\mathbf{F}_D \cdot \mathbf{A})_r - (\mathbf{F}_D \cdot \mathbf{A})_{r+\Delta r} + \Delta V r_D$$

where (i) y_D mole fraction of D in the system, (ii) $\mathbf{F}_D = F_D\hat{\mathbf{r}}$ is the molar flux of D of magnitude F_D, and is across the area $\mathbf{A}$, and (iii) $r_D = dk_2 c_A^{a_2}$ is the molar rate of change of D per unit volume due to reaction in the porous layer.

In the above mole balance, we use the first order Taylor expansion of $(\mathbf{F}_D \cdot \mathbf{A})_{r+\Delta r}$, take the limit of Δr to zero, and substitute the expressions for system volume, flux and area. Simplifying as before, we obtain

$$\frac{\partial c_D}{\partial t} = -\frac{1}{r^2}\frac{\partial}{\partial r}(r^2 F_D) + dk_2 c_A^{a_2} \qquad (4.66)$$

For mole balances in the porous layer, we need expressions for F_A and F_D. They are derived next.

Expressions for F_A and F_D

The molar fluxes of A and D in the porous layer are given by, respectively,

$$\mathbf{F_A} = \frac{c_A}{c}\underbrace{(\mathbf{F_A} + \mathbf{F_D})}_{\text{total flux}} + \mathbf{J_A} \quad \text{and} \quad \mathbf{F_D} = \frac{c_D}{c}(\mathbf{F_A} + \mathbf{F_D}) + \mathbf{J_D}$$

where (i) c is the sum of the concentrations of A, D and C, and (ii) $\mathbf{J_A}$ and $\mathbf{J_D}$ are, respectively, the diffusive fluxes of A and D. Note that the total flux does not involve C, which is stationary.

Eliminating either $\mathbf{F_A}$ or $\mathbf{F_D}$ from the above equations, we obtain

$$\mathbf{F}_i = \frac{(c - c_j)\mathbf{J}_i + c_i\mathbf{J}_j}{c - c_i - c_j} \equiv F_i\hat{\mathbf{r}}; \quad i, j = \text{A, D} \text{ and } i \neq j$$

The diffusive fluxes in the above equations are given by

$$\mathbf{J}_i = -\left(D_{ii}c\frac{dy_i}{dr} + D_{ij}c\frac{dy_j}{dr}\right)\hat{\mathbf{r}}; \quad i, j = \text{A, D} \text{ and } i \neq j$$

where (i) D_{ij}s (namely, D_{AA}, D_{AD}, D_{DD} and D_{DA}) are the multicomponent diffusivities in the porous layer, and (ii) y_A and y_D are, respectively, the mole fractions of A and D. Since

$$\frac{\partial c_i}{\partial r} = \frac{\partial}{\partial r}(y_i c) = \underbrace{y_i}_{\frac{c_i}{c}}\frac{\partial c}{\partial r} + c\frac{\partial y_i}{\partial r} \quad \Rightarrow \quad c\frac{\partial y_i}{\partial r} = \frac{\partial c_i}{\partial r} - \frac{c_i}{c}\frac{\partial c}{\partial r}; \quad i = \text{A, D}$$

the diffusive fluxes simplify to

$$\mathbf{J}_i = -\left[D_{ii}\left(\frac{\partial c_i}{\partial r} - \frac{c_i}{c}\frac{\partial c}{\partial r}\right) + D_{ij}\left(\frac{\partial c_j}{\partial r} - \frac{c_j}{c}\frac{\partial c}{\partial r}\right)\right]\hat{\mathbf{r}}; \quad i, j = \text{A, D} \text{ and } i \neq j$$

Substituting the above equation in the above expression of $\mathbf{F}_i$ leads to

$$F_i = \frac{1}{c - c_i - c_j}\left\{(c_j - c)\left[D_{ii}\left(\frac{\partial c_i}{\partial r} - \frac{c_i}{c}\frac{\partial c}{\partial r}\right) + D_{ij}\left(\frac{\partial c_j}{\partial r} - \frac{c_j}{c}\frac{\partial c}{\partial r}\right)\right]\right.$$

$$\left. - c_i\left[D_{jj}\left(\frac{\partial c_j}{\partial r} - \frac{c_j}{c}\frac{\partial c}{\partial r}\right) + D_{ji}\left(\frac{\partial c_i}{\partial r} - \frac{c_i}{c}\frac{\partial c}{\partial r}\right)\right]\right\}$$

$$i, j = \text{A, D} \text{ and } i \neq j \tag{4.67}$$

In the above equation, since c_C is constant,

$$\frac{\partial c}{\partial r} = \frac{\partial c_A}{\partial r} + \frac{\partial c_D}{\partial r} \quad \text{where} \quad c = c_A + c_C + c_D \tag{4.68}$$

Equation (4.63) on p. 124, and Equation (4.64) on p. 125 are differential equations of first order with respect to t, and each equation needs one initial condition. Equations (4.65) and (4.66) on p. 125 are partial differential equations of (i) first order with respect to t, and (ii) second order with respect to r. Each of the last two equations needs an initial condition, and two boundary conditions on the r-axis.

Initial Conditions
Initially, the core radius is known as R_{10}, and there is no porous layer. The bulk concentration of A is specified as $\tilde{c}_A$. In the incipient porous layer, while the concentration of A is $\tilde{c}_A$, the concentration of D is zero. Hence, the initial conditions are

$$R_1(0) \;=\; R_2(0) \;=\; R_{10} \tag{4.69}$$

$$\left.\begin{aligned} c_A(r,0) &= \tilde{c}_A \\ c_D(r,0) &= 0 \end{aligned}\right\} \quad \forall \;\; R_1(0) \le r \le R_2(0) \tag{4.70}$$

Boundary Conditions
At R_1, the core–porous-layer interface, there is no D. Moreover, the concentration of A over there is zero, since A is consumed instantaneously at the core surface. At the outer surface of the porous layer, while the concentration of A is its bulk concentration in the surroundings, the concentration of D is zero. Hence, the boundary conditions for Equations (4.65) and (4.66) at any time $t > 0$ are, respectively,

$$c_A(R_1,t) \;=\; 0, \quad c_A(R_2,t) \;=\; \tilde{c}_A \qquad \text{and} \tag{4.71}$$

$$c_D(R_1,t) \;=\; 0, \quad c_D(R_2,t) \;=\; 0 \tag{4.72}$$

Summary

Equations (4.63)–(4.72) constitute the dynamic model for the solid reactant covered by a porous layer of an intermediate product. Given the specifications for the parameter set

$$\{a_1, \, a_2, \, b, \, c_1, \, c_2, \, d, \, k_1, \, k_2, \, \tilde{c}_A, \, c_B, \, c_C, \, D_{AA}, \, D_{AD}, \, D_{DD}, \, D_{DA}, \, R_{10}\}$$

the involved differential equations need to be integrated to obtain $R_1 = R_1(t)$, $R_2 = R_2(t)$, $c_A = c_A(r,t)$, and $c_A = c_A(r,t)$.

4.3 Fluxes along Non-Linear Directions

The above examples on distributed-parameter systems involved fluxes only along linear directions, namely, the directions in Cartesian coordinates, and the radial direction in the cylindrical or spherical coordinates. In those examples, it was straightforward to apply fundamental equations on appropriate differential elements, and derive model equations incorporating suitable constitutive relations. This approach is known as the shell-balance method, and is an important model development tool.

However, the shell-balance method gets complicated in the presence of fluxes along non-linear directions, e.g., the θ-direction in cylindrical coordinates. The reason is that such a flux may make a contribution in some other direction. Appendix 4.C on p. 134 illustrates this subtlety by deriving the equation of motion along the radial direction in cylindrical coordinates. In the involved balance, the momentum flux $\phi_{\theta\theta}$, which is along the non-linear θ-direction, contributes to the momentum along the radially inward direction.

Therefore, when dealing with fluxes along non-linear directions, e.g., in cylindrical and spherical coordinate systems, it is convenient to directly use the applicable equations that are readily available [see Appendix 2.D, p. 53]. As a matter of fact, the available equations in Cartesian, cylindrical and spherical coordinate systems can be directly applied regardless of the nature of flux path. For any other coordinate system, say, helical, an additional step of coordinate transformation is required [see Chapter 5, p. 139].

The following example illustrates the direct use of momentum balance in spherical coordinates.

4.3.1 Saccadic Movement of an Eye

This example involves the microscopic momentum balance in the vitreous humor[*] of an eye during a saccade, i.e., a quick simultaneous movement of both eyes between two points. This movement influences drug dispersion in eyes.[7]

Figure 4.16 on the next page shows the schematic of an eye ball turning about the vertical z-axis in spherical coordinates.

The angular movement during a saccade can be expressed as

$$\alpha = -\alpha_0 \cos(\omega_0 t)$$

where α_0 is a small magnitude, ω_0 is the frequency, and t is time. The velocity components of the vitreous humor along the r- and θ-directions are zero. Moreover, these are the only directions along which the ϕ-component of the velocity changes.

Objective

It is desired to develop a model that would enable the determination of the vitreous humor velocity at different radial distances, θ-positions, and times due to the saccadic movement of an eye.

Assumptions

We make the following assumptions for the model:

1. The vitreous humor in the eye is a Newtonian fluid of constant density, and moves only along the ϕ-direction.

2. Along the ϕ-direction, the fluid velocity is a function of radial distance, and time.

3. There is a no-slip boundary condition at $r = R$ where the fluid comes in contact with the inner eye surface. Over there, the fluid velocity is equal to that of the eye velocity.

4. Pressure is constant along the ϕ-direction.

---

[*]transparent liquid that occupies the space between the lens and the retina of an eye

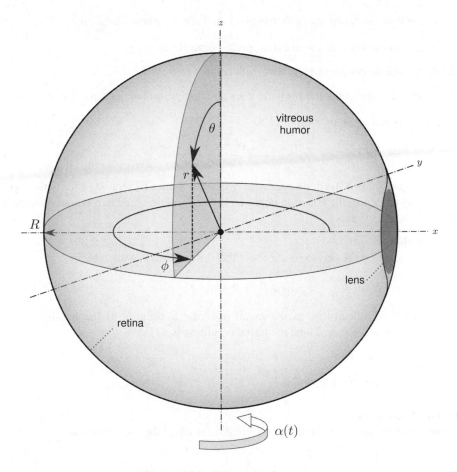

Figure 4.16   Schematic of an eye

Model Formulation

Since the involved momentum flux is along non-linear path, i.e., the curvilinear ϕ-axis, it is convenient to utilize the microscopic momentum balance available for spherical coordinates.

The applicable equation is Equation (3.15) on p. 76, which is

$$
\begin{aligned}
\frac{\partial v_\phi}{\partial t} ={} & -v_r \frac{\partial v_\phi}{\partial r} - \frac{v_\theta}{r}\frac{\partial v_\phi}{\partial \theta} - \frac{v_\phi}{r\sin\theta}\frac{\partial v_\phi}{\partial \phi} + \frac{\mu}{\rho}\left[\frac{\partial^2 v_\phi}{\partial r^2} + \frac{2}{r}\frac{\partial v_\phi}{\partial r} + \frac{1}{r^2}\frac{\partial^2 v_\phi}{\partial \theta^2} \right. \\
& + \frac{1}{r^2\tan\theta}\frac{\partial v_\phi}{\partial \theta} + \frac{1}{r^2\sin^2\theta}\frac{\partial^2 v_\phi}{\partial \phi^2} + \frac{2}{r^2\sin\theta}\frac{\partial v_r}{\partial \phi} \\
& \left. + \frac{1}{r^2\sin^2\theta}\left(2\cos\theta\frac{\partial v_\theta}{\partial \phi} - v_\phi\right) \right] - \frac{v_r v_\phi}{r} - \frac{v_\theta v_\phi}{r\tan\theta} \\
& - \frac{1}{r\rho\sin\theta}\frac{\partial P}{\partial \phi} + \rho g_\phi
\end{aligned}
$$

where the grayed out terms are zero, based on the following considerations:

1. v_r and v_θ are zero, and so are their partial derivatives

2. the partial derivative of v_ϕ is zero with respect to ϕ

3. along the ϕ-direction, the pressure does not change, and the gravity is zero

As a result, the last equation simplifies to

$$\frac{\partial v_\phi}{\partial t} = \frac{\mu}{\rho}\left(\frac{\partial^2 v_\phi}{\partial r^2} + \frac{2}{r}\frac{\partial v_\phi}{\partial r} + \frac{1}{r^2}\frac{\partial^2 v_\phi}{\partial \theta^2} + \frac{1}{r^2\tan\theta}\frac{\partial v_\phi}{\partial \theta} - \frac{v_\phi}{r^2\sin^2\theta}\right) \qquad (4.73)$$

The above equation is of (i) first order with respect to t, and (ii) second order with respect to each of r and θ. Hence, the equation needs an initial condition, and two boundary conditions on the r- as well as θ-axis.

Initial Condition
At time $t = 0$, the eye begins to rotate about the vertical z-axis with velocity*

$$v = R\sin\theta\frac{d\alpha}{dt} = \alpha_0\omega_0 R\sin\theta\sin(\omega_0 t) \qquad (4.74)$$

which is the fluid velocity v_ϕ at the radial distance R because of the no-slip condition. Everywhere else, v_ϕ is zero. Thus, the initial condition is

$$v_\phi = \begin{cases} v & \text{at } r = R \\ 0 & \forall \ 0 \leq r < R \end{cases} \quad \forall \ 0 \leq \theta \leq \pi \qquad (4.75)$$

Boundary Conditions
For times greater than zero, the boundary conditions along the r-direction are

$$v_\phi = \begin{cases} v & \text{at } r = R \\ \dfrac{\partial v_\phi}{\partial r} = 0 & \text{at } r = 0 \end{cases} \quad \forall \ 0 \leq \theta \leq \pi \qquad (4.76)$$

The last condition arises from the radial symmetry [see Appendix 4.B, p. 133].
Along the θ-direction, the boundary conditions are

$$v_\phi = 0 \ \text{at} \ \theta = 0, \pi \ \forall \ 0 \leq r \leq R$$

which means zero velocity at the vertical z-axis.

Summary

Equations (4.73)–(4.76) constitute the dynamic, distributed-parameter model of momentum transfer to vitreous humor in an eye during its saccadic movement. Given the specifications for the parameter set

$$\{\mu, \rho, \alpha_0, \omega_0, R\}$$

the involved differential equation needs to be integrated to obtain $v_\phi = v_\phi(r, \theta, t)$ in the θ-interval $[0, \pi]$. This velocity is mirrored in the other θ-interval $[\pi, 2\pi]$ due to symmetry.

---

*This is the tangential velocity, which is equal to radial distance times angular velocity.

4.A　Initial and Boundary Conditions

These conditions are equations associated with the dependent variables of differential equations in process models.

4.A.1　Initial Condition

This condition specifies the value of the dependent variable at the initial value of the independent variable, typically time. Thus, for the differential equation

$$\frac{dT}{dt} = g(t) \tag{4.77}$$

describing a heat transfer process,

$$T = T_0 \ \text{ at } \ t = 0$$

is the initial condition where T_0 is the specified initial temperature.

4.A.2　Boundary Condition

This condition is the requirement that must be satisfied by the dependent variable at a terminal value of the independent variable.

Boundary conditions can be classified into the following:

Dirichlet Boundary Condition

Also known as the boundary condition of the first type, this condition prescribes the value of the dependent variable at a terminal value of the independent variable. Given an equation of change for temperature (T) along the x-direction in a fin as,

$$\frac{d^2 T}{dx^2} = f(x), \qquad 0 \leq x \leq x_{\text{f}} \tag{4.78}$$

the Dirichlet boundary conditions, e.g., are

$$T = \begin{cases} T_0 & \text{at } \ x = 0 \\ T_{\text{f}} & \text{at } \ x = x_{\text{f}} \end{cases}$$

These conditions fix the temperature at the terminal values of x.

Neumann Boundary Condition

Also known as the boundary condition of the second type, this condition prescribes the value of the derivative of the dependent variable at a terminal value of the independent variable. An example is the condition

$$\frac{dT}{dx} = 0 \ \text{ at } \ x = x_{\text{f}}$$

for Equation (4.78) on the previous page. The above condition implies that there is no temperature gradient, and thus no heat transfer at the final x.

Robin Boundary Condition

Also known as the boundary condition of the third type, this condition prescribes the value of a linear combination of dependent variables, and their derivatives at a terminal value of the independent variable. An example pertaining to Equation (4.78) on the previous page is

$$-k\frac{\mathrm{d}T}{\mathrm{d}x} - h(T - T_\infty) = 0 \quad \text{at} \quad x = x_\mathrm{f}$$

where k, h and T_∞ denote, respectively, thermal conductivity, convective heat transfer coefficient, and ambient temperature. The above condition basically equates conductive heat transfer at the final x in the fin to convective heat transfer in the surroundings.

Mixed Boundary Condition

This boundary condition is a set of two non-overlapping boundary conditions, one being Dirichlet and the other Neumann. An example of a mixed boundary condition for Equation (4.78) is

$$T = T_0 \quad \text{at} \quad x = 0$$

$$\frac{\mathrm{d}T}{\mathrm{d}x} = 0 \quad \text{at} \quad x = x_\mathrm{f}$$

Cauchy Boundary Condition

This boundary condition comprises one Dirichlet, and one Neumann boundary condition, both at the same location. An example of this condition for Equation (4.78) is

$$\left.\begin{aligned} T &= T_0 \\ \frac{\mathrm{d}T}{\mathrm{d}x} &= -\frac{q_0}{k} \end{aligned}\right\} \quad \text{at} \quad x = 0$$

where q_0 is a known value of heat flux. This condition is posed at a boundary when it is difficult to prescribe conditions at the other boundaries. The above condition could mean that the fin at x_f is not accessible for temperature measurements due to physical constraints, or extreme temperatures.

4.A.3   Periodic Condition

This condition appears in a periodic process in which the value of a dependent variable at a boundary repeats after a time period. Thus, if Equation (4.77) on the previous page describes a periodic process then

$$T(0) = T(\tau)$$

is a periodic boundary condition with time period τ.

4.B   Zero Derivative at the Point of Symmetry

Consider a property $y(x)$ across a point of symmetry x_2 on the x-axis, as shown in Figure 4.17 below. In such a case, the variable value is the same at equidistant locations from the point of

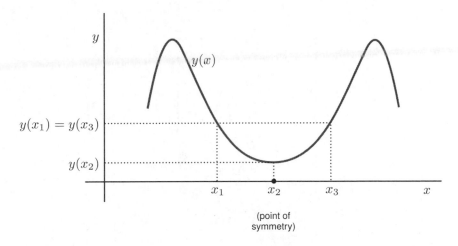

Figure 4.17   A property that is a symmetrical function, $y(x)$

symmetry. Thus, given any two locations x_1 and x_3, each at a distance Δx from the point of symmetry x_2,

$$\underbrace{y(x_2) + \int_{x_2}^{x_1} \frac{\mathrm{d}y}{\mathrm{d}x}\,\mathrm{d}x}_{y(x_1)} = \underbrace{y(x_2) + \int_{x_2}^{x_3} \frac{\mathrm{d}y}{\mathrm{d}x}\,\mathrm{d}x}_{y(x_3)}$$

where it is assumed that y is differentiable. Simplifying the above equation, and taking the limit of Δx to zero, we obtain

$$\frac{\mathrm{d}y}{\mathrm{d}x}\bigg|_{x_2} \underbrace{(x_1 - x_2)}_{-\Delta x} = \frac{\mathrm{d}y}{\mathrm{d}x}\bigg|_{x_2} \underbrace{(x_3 - x_2)}_{\Delta x}$$

The above equation implies that by necessity

$$\frac{\mathrm{d}y}{\mathrm{d}x}\bigg|_{x_2} = 0$$

Thus, the derivative of a property with respect to an independent variable at the point of symmetry is zero.

4.C   Equation of Motion along the Radial Direction in Cylindrical Coordinates

To obtain this equation, we apply momentum balance along the r-direction on the differential element in cylindrical coordinates [see Figure 4.18 below].

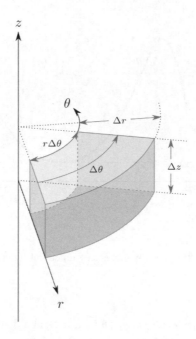

Figure 4.18   Differential element in cylindrical coordinates

As shown in Figure 4.19 on the next page, this differential element has faces of areas $\mathbf{A_r} = r\Delta\theta\Delta z\hat{\mathbf{r}}$, $\mathbf{A_\theta} = \Delta r\Delta z\hat{\boldsymbol{\theta}}$, and $\mathbf{A_z} = r\Delta\theta\Delta r\hat{\mathbf{z}}$, respectively, along the r-, θ- and z-directions.

The momentum balance yields

$$\frac{\mathrm{d}}{\mathrm{d}t}(\rho\Delta V v_r) = (\phi_{rr}\cdot\mathbf{A_r})_r - (\phi_{rr}\cdot\mathbf{A_r})_{r+\Delta r} + (\phi_{\theta r}\cdot\mathbf{A_\theta})_\theta - (\phi_{\theta r}\cdot\mathbf{A_\theta})_{\theta+\Delta\theta}$$

$$+ (\phi_{zr}\cdot\mathbf{A_z})_z - (\phi_{zr}\cdot\mathbf{A_z})_{z+\Delta z} + \rho\Delta V g_r$$

$$- \left[-\phi_{\theta\theta}(\theta+\Delta\theta)\cdot\mathbf{A_\theta}\sin\Delta\theta \right]$$

where $\Delta V = r\Delta r\Delta\theta\Delta z$.

Note that the last term in the above equation is an output term at the $(\theta + \Delta\theta)$-face, and stems from the momentum flux $\phi_{\theta\theta}$, which is along the non-linear θ-direction. As shown in Figure 4.20 on p. 136, the term enclosed in square brackets is the component of the momentum rate $\phi_{\theta\theta}(\theta + \Delta\theta)\cdot\mathbf{A_\theta}$, and is opposite to the r-direction.

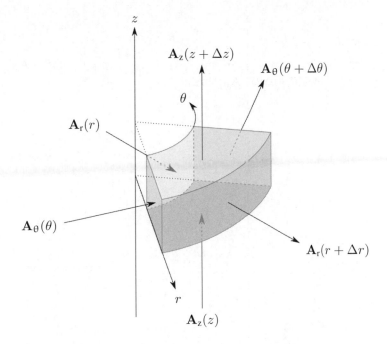

Figure 4.19 Area vectors along the r-, θ- and z-directions

In the differential equation on the previous page, using the first order Taylor expansion for the third, fifth, seventh and ninth terms, and taking the limit of Δr, $\Delta\theta$ and Δz to zero yields

$$\frac{\partial}{\partial t}\left[\rho(rdrd\theta dz)v_r\right] = -\frac{\partial}{\partial r}(\phi_{rr}rd\theta dz)dr - \frac{\partial}{\partial\theta}(\phi_{\theta r}drdz)d\theta - \frac{\partial}{\partial z}(\phi_{zr}rd\theta dr)dz +$$

$$\rho(rdrd\theta dz)g_r + \left(\phi_{\theta\theta} + \frac{\partial\phi_{\theta\theta}}{\partial\theta}d\theta\right)d\theta(drdz)$$

where we have applied following result for the last term:

$$\lim_{\theta\to 0}\sin\Delta\theta = d\theta$$

Simplifying the above differential equation for fixed system volume, and discarding the term that has $d\theta^2$, we obtain

$$\rho\frac{\partial v_r}{\partial t} + v_r\frac{\partial\rho}{\partial t} = -\frac{1}{r}\left[\frac{\partial}{\partial r}(r\phi_{rr}) + \frac{\partial\phi_{\theta r}}{\partial\theta}\right] - \frac{\partial\phi_{zr}}{\partial z} + \rho g_r + \frac{\phi_{\theta\theta}}{r}$$

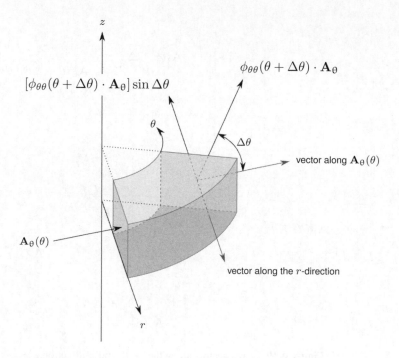

Figure 4.20 Momentum rate $\phi_{\theta\theta}(\theta + \Delta\theta) \cdot \mathbf{A}_\theta$ and its component $[\phi_{\theta\theta}(\theta + \Delta\theta) \cdot \mathbf{A}_\theta]\sin\Delta\theta$, which is opposite to the r-direction

In the last equation, after substituting for

$$\phi_{rr} = P + \tau_{rr} + \rho v_r v_r \qquad \phi_{\theta r} = \tau_{\theta r} + \rho v_\theta v_r$$

$$\phi_{zr} = \tau_{zr} + \rho v_z v_r \qquad \phi_{\theta\theta} = P + \tau_{\theta\theta} + \rho v_\theta v_\theta$$

and the continuity equation, Equation (2.66) on p. 54,

$$\frac{\partial \rho}{\partial t} = -\rho\left(\frac{\partial v_r}{\partial r} + \frac{v_r}{r} + \frac{1}{r}\frac{\partial v_\theta}{\partial \theta} + \frac{\partial v_z}{\partial z}\right) - v_r\frac{\partial \rho}{\partial r} - \frac{v_\theta}{r}\frac{\partial \rho}{\partial \theta} - v_z\frac{\partial \rho}{\partial z}$$

we finally obtain after simplification

$$\frac{\partial v_r}{\partial t} = -v_r\frac{\partial v_r}{\partial r} - \frac{v_\theta}{r}\frac{\partial v_r}{\partial \theta} - v_z\frac{\partial v_r}{\partial z} - \frac{1}{\rho}\left(\frac{\partial \tau_{rr}}{\partial r} + \frac{1}{r}\frac{\partial \tau_{r\theta}}{\partial \theta} + \frac{\partial \tau_{rz}}{\partial z}\right.$$

$$\left. + \frac{\tau_{rr} - \tau_{\theta\theta}}{r} + \frac{\partial P}{\partial r}\right) + \frac{v_\theta^2}{r} + g_r$$

which is the desired equation of motion along the r-direction in cylindrical coordinates for symmetric stress tensor $\boldsymbol{\tau}$ i.e., Equation (2.67) on p. 55.

References

[1] K. Mehta and J.J. Linderman. "Model-Based Analysis and Design of a Microchannel Reactor for Tissue Engineering". In: *Biotechnology and Bioengineering* 94.3 (2006), pp. 596–609.

[2] A. Krogh. "The number and distribution of capillaries in muscles with calculations of the oxygen pressure head necessary for supplying the tissue". In: *The Journal of Physiology* 52.6 (1919), pp. 409–415.

[3] H.H. Pennes. "Analysis of tissue and arterial blood temperatures in the resting human forearm". In: *Journal of Applied Physiology* 1.2 (1948), pp. 93–122.

[4] H. Muhamad et al. "Optimal control study to enhance oil production in labscale Vapex by varying solvent injection pressure with time". In: *Optimal Control Applications and Methods* 37.2 (2016), pp. 424–440.

[5] R.T.C Ju et al. "Drug Release From Hydrophilic Matrices. 2. A Mathematical Model Based on the Polymer Disentanglement Concentration and the Diffusion Layer". In: *Journal of Pharmaceutical Sciences* 84.12 (1995), pp. 1464–1477.

[6] J.R. Lamarsh and A.J. Baratta. *Introduction to Nuclear Engineering*. 3rd edition. Chapter 5. New Jersey: Prentice-Hall, Inc., 2001.

[7] R. Repetto, J.H. Siggers, and A. Stocchino. "Mathematical model of flow in the vitreous humor induced by saccadic eye rotations: effect of geometry". In: *Biomechanics and Modeling in Mechanobiology* 9.1 (2010), pp. 65–76.

Exercises

4.1 Derive the CSTR model of Section 4.1.1 on p. 80 without making the assumption of constant system volume.

4.2 How can we determine the steady state height of liquid in the preheater [see Section 4.1.3, p. 84] from Equation (4.8) on p. 86?

4.3 Extend the model of fluid flow between inclined parallel plates [see Section 4.2.2, p. 93] to the flow of two immiscible fluids with one flowing on top of the other.

4.4 Why cannot the microscopic equation of change for temperature be applied directly to model heat transfer in tapered fin [see Section 4.2.3, p. 96]?

4.5 In the heavy oil recovery process of Section 4.2.7 on p. 108, what is the mass of oil produced at a given time?

4.6 What happens at steady state in the heavy oil recovery process of Section 4.2.7?

4.7 For the hydrogel tablet dissolution process described in Section 4.2.8 on p. 112
  a. Obtain the equations of change for the mass fractions.
  b. Derive the expression for the amount of drug release at any time.

4.8 Explain the steady state of hydrogel dissolution process in Section 4.2.8.

4.9 Interpret the steady state of the reaction process in Section 4.2.11 on p. 122.

4.10 Derive the expression for the total flux in the porous layer around the solid reactant in Section 4.2.11.

4.11 What is the system in the model describing saccadic eye movement [see Section 4.3.1, p. 128]?

5

Model Transformation

In this chapter we introduce the tools to transform process models to more suitable forms. First we take up the transformation from Cartesian coordinate system to other orthogonal coordinate systems. Then we provide a general procedure to carry out the transformation between arbitrary coordinate systems. In the second part of the chapter, we present Laplace transformation, which is widely used to solve and analyze linear process models. Some important miscellaneous transformations are presented at the end.

5.1 Transformation between Orthogonal Coordinate Systems

Recall that we derived fundamental relations, and associated equations for distributed-parameter systems in Cartesian coordinate system. To obtain similar results in other orthogonal coordinate systems, especially the curvilinear ones, it is much easier to transform the equations in Cartesian coordinate system than to derive the results from scratch. To carry out this transformation, we need expressions for the following differential operators:

1. Gradient of scalar, e.g., ∇F

2. Divergence of a vector, e.g., $\nabla \cdot \mathbf{f}$

3. Curl of a vector, e.g., $\nabla \times \mathbf{f}$

4. Laplacian of a scalar, e.g., $\nabla^2 F$

5. Gradient of a vector, e.g., $\nabla \mathbf{f}$

6. Laplacian of a vector, e.g., $\nabla^2 \mathbf{f}$

7. Divergence of a tensor, e.g., $\nabla \cdot \boldsymbol{\tau}$

For an orthogonal coordinate system, these expressions are readily obtained from the scale factors relative to Cartesian coordinate system. We will first introduce the scale factors, and then derive important results leading to the expressions for the aforementioned operators.

5.1.1 Scale Factors

Consider a position vector in Cartesian coordinate system (x_1, x_2, x_3) given by

$$\mathbf{x} = x_1 \hat{\mathbf{x}}_1 + x_2 \hat{\mathbf{x}}_2 + x_3 \hat{\mathbf{x}}_3 \tag{5.1}$$

Process Modeling and Simulation for Chemical Engineers: Theory and Practice, First Edition. Simant Ranjan Upreti.
© 2017 John Wiley & Sons Ltd. Published 2017 by John Wiley & Sons Ltd.
Companion website: www.wiley.com/go/upreti/pms_for_chemical_engineers

where x_i and $\hat{x}_i$ are, respectively, the component of $\mathbf{x}$, and the unit vector along the x_i-direction. In an orthogonal coordinate system (y_1, y_2, y_3), we can represent $\mathbf{x}$ by the relation

$$\mathbf{x} = \mathbf{x}(y_1, y_2, y_3)$$

Using the chain rule of differentiation [see Section 8.7.3, p. 319], the differential change in $\mathbf{x}$ is given by

$$\mathrm{d}\mathbf{x} = \frac{\partial \mathbf{x}}{\partial y_1}\mathrm{d}y_1 + \frac{\partial \mathbf{x}}{\partial y_2}\mathrm{d}y_2 + \frac{\partial \mathbf{x}}{\partial y_3}\mathrm{d}y_3 \tag{5.2}$$

In the above equation, $\partial \mathbf{x}/\partial y_i$ [see Figure 5.1 below] is the tangent vector on the y_i-coordinate curve along which the differential change in $\mathbf{x}$ is $\mathrm{d}y_i$. Note that $\partial \mathbf{x}/\partial y_i$ is a vector along the y_i-direction.

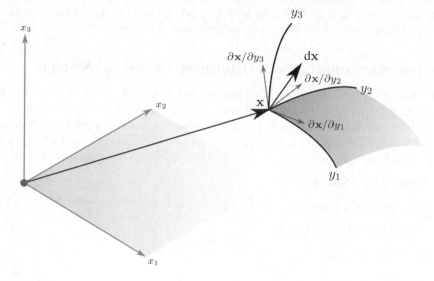

Figure 5.1   Tangent vectors to $\mathrm{d}\mathbf{x}$ in a system of orthogonal coordinates, (y_1, y_2, y_3)

Let $\hat{\mathbf{y}}_i$ denote the unit vector along $\partial \mathbf{x}/\partial y_i$, i.e., along the y_i-direction. Then

$$\hat{\mathbf{y}}_i = \frac{\partial \mathbf{x}}{\partial y_i} \bigg/ \underbrace{\left\| \frac{\partial \mathbf{x}}{\partial y_i} \right\|}_{h_i} = \frac{1}{h_i}\frac{\partial \mathbf{x}}{\partial y_i}; \qquad i = 1, 2, 3 \tag{5.3}$$

where h_i, which is the magnitude of the tangent vector, is called the metric coefficient, or **scale factor** along the y_i-direction.

Differentiation of Equation (5.1) on the previous page with respect to y_i yields

$$\frac{\partial \mathbf{x}}{\partial y_i} = \frac{\partial x_1}{\partial y_i}\hat{\mathbf{x}}_1 + \frac{\partial x_2}{\partial y_i}\hat{\mathbf{x}}_2 + \frac{\partial x_3}{\partial y_i}\hat{\mathbf{x}}_3; \qquad i = 1, 2, 3$$

using which the scale factors in Equation (5.3) on the previous page are given by

$$
h_i = \left\| \frac{\partial \mathbf{x}}{\partial y_i} \right\| = \sqrt{\frac{\partial \mathbf{x}}{\partial y_i} \cdot \frac{\partial \mathbf{x}}{\partial y_i}} = \sqrt{\left(\frac{\partial x_1}{\partial y_i}\right)^2 + \left(\frac{\partial x_2}{\partial y_i}\right)^2 + \left(\frac{\partial x_3}{\partial y_i}\right)^2}
$$

$$
i = 1, 2, 3 \tag{5.4}
$$

The above result stems from the fact that

$$
\hat{\mathbf{x}}_i \cdot \hat{\mathbf{x}}_j = \begin{cases} 0, & \text{if } i \neq j \\ 1, & \text{if } i = j \end{cases}
$$

since $\hat{\mathbf{x}}_1$, $\hat{\mathbf{x}}_2$ and $\hat{\mathbf{x}}_3$ are mutually perpendicular.

Example 5.1.1

Find the scale factors for cylindrical coordinate system.

Solution

The relations between Cartesian coordinates (x_1, x_2, x_3), and cylindrical coordinates $(r \equiv y_1,\ \theta \equiv y_2,\ z \equiv y_3)$ are [see Figure 5.2, next page]

$$
\begin{aligned}
x_1 &= r \cos\theta = y_1 \cos y_2 \\
x_2 &= r \sin\theta = y_1 \sin y_2 \\
x_3 &= z = y_3
\end{aligned}
$$

From Equation (5.4) above, the scale factors are

$$
h_1 = \sqrt{\left(\frac{\partial x_1}{\partial y_1}\right)^2 + \left(\frac{\partial x_2}{\partial y_1}\right)^2 + \underbrace{\left(\frac{\partial x_3}{\partial y_1}\right)^2}_{=\,0}} = \sqrt{\cos^2 y_2 + \sin^2 y_2} = 1
$$

$$
h_2 = \sqrt{\left(\frac{\partial x_1}{\partial y_2}\right)^2 + \left(\frac{\partial x_2}{\partial y_2}\right)^2 + \underbrace{\left(\frac{\partial x_3}{\partial y_2}\right)^2}_{=\,0}} = \sqrt{(-y_1 \sin y_2)^2 + (y_1 \cos y_2)^2} = \underbrace{y_1}_{r}
$$

$$
h_3 = \sqrt{\underbrace{\left(\frac{\partial x_1}{\partial y_3}\right)^2}_{=\,0} + \underbrace{\left(\frac{\partial x_2}{\partial y_3}\right)^2}_{=\,0} + \left(\frac{\partial x_3}{\partial y_3}\right)^2} = \sqrt{1^2} = 1
$$

❑

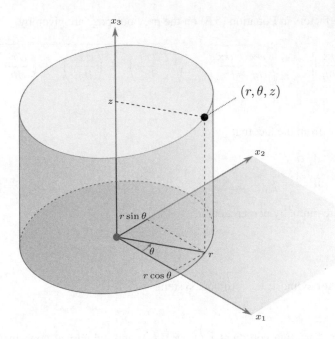

Figure 5.2   Relations between Cartesian and cylindrical coordinates

5.1.2   Differential Elements

From Equation (5.3) on p. 140, we have

$$\frac{\partial \mathbf{x}}{\partial y_i} = h_i \hat{\mathbf{y}}_i; \qquad i = 1, 2, 3$$

which when substituted in Equation (5.2) on p. 140 yields

$$\mathbf{dx} = \underbrace{h_1 dy_1}_{dl_1} \hat{\mathbf{y}}_1 + \underbrace{h_2 dy_2}_{dl_2} \hat{\mathbf{y}}_2 + \underbrace{h_3 dy_3}_{dl_3} \hat{\mathbf{y}}_3 = dl_1 \hat{\mathbf{y}}_1 + dl_2 \hat{\mathbf{y}}_2 + dl_3 \hat{\mathbf{y}}_3 \qquad (5.5)$$

In the above equation, $dl_i = h_i dy_i$ is the differential magnitude, or length along the y_i-direction.

Note that $\hat{\mathbf{y}}_1$, $\hat{\mathbf{y}}_2$ and $\hat{\mathbf{y}}_3$ are the basis vectors in the system of orthogonal coordinates (y_1, y_2, y_3). Consequently,

$$\hat{\mathbf{y}}_i \cdot \hat{\mathbf{y}}_j = \begin{cases} 0, & \text{if } i \neq j \\ 1, & \text{if } i = j \end{cases}$$

The length of the differential vector $\mathbf{dx}$ in the orthogonal coordinates (y_1, y_2, y_3) is given by

$$\|\mathbf{dx}\| = \sqrt{\mathbf{dx} \cdot \mathbf{dx}} = \sqrt{h_1^2 (dy_1)^2 + h_2^2 (dy_2)^2 + h_3^2 (dy_3)^2}$$

Differential Area and Volume

With the help of Equation (5.5) on the previous page, the differential areas in the (y_1, y_2, y_3) coordinates are given by

$$dA_1 = dl_2\, dl_3 = h_2 h_3\, dy_2\, dy_3 \tag{5.6}$$

$$dA_2 = dl_1\, dl_3 = h_1 h_3\, dy_1\, dy_3 \tag{5.7}$$

$$dA_3 = dl_1\, dl_2 = h_1 h_2\, dy_1\, dy_2 \tag{5.8}$$

In the above equations, dA_i is the differential area on the plane perpendicular to the y_i-direction. Thus, dA_1 is the differential area on the $y_2 y_3$-plane.

The differential volume is similarly given by

$$dV = dl_1\, dl_2\, dl_3 = h_1 h_2 h_3\, dy_1 dy_2\, dy_3$$

Example 5.1.2

Find the differential areas and volume in cylindrical coordinates.

Solution

Using the scale factors of cylindrical coordinates obtained in the last example, the differential areas and volume are

$$dA_1 = y_1\, dy_2\, dy_3 = r\, d\theta\, dz$$

$$dA_2 = dy_1\, dy_3 = dr\, dz$$

$$dA_3 = y_1\, dy_1\, dy_2 = r\, dr\, d\theta$$

$$dV = y_1\, dy_1\, dy_2\, dy_3 = r\, dr\, d\theta\, dz$$

❑

5.1.3   Vector Representation

Given an arbitrary vector $\mathbf{v}$ at a position $\mathbf{x}$, we can always vary the latter with a suitable parameter, say, t, such that $d\mathbf{x}/dt = \mathbf{v}$. Considering this fact, we divide Equation (5.5) on the previous page by dt to obtain

$$\underbrace{\frac{d\mathbf{x}}{dt}}_{\mathbf{v}} = h_1 \underbrace{\frac{dy_1}{dt}}_{v_1} \hat{\mathbf{y}}_1 + h_2 \underbrace{\frac{dy_2}{dt}}_{v_2} \hat{\mathbf{y}}_2 + h_3 \underbrace{\frac{dy_3}{dt}}_{v_3} \hat{\mathbf{y}}_3$$

$$\text{or} \quad \mathbf{v} = h_1 v_1 \hat{\mathbf{y}}_1 + h_2 v_2 \hat{\mathbf{y}}_2 + h_3 v_3 \hat{\mathbf{y}}_3$$

where v_is are the corresponding components of $\mathbf{v}$ in the (y_1, y_2, y_3) coordinates. The length of $\mathbf{v}$ is then given by

$$\|\mathbf{v}\| = \sqrt{\mathbf{v} \cdot \mathbf{v}} = \sqrt{h_1^2 v_1^2 + h_2^2 v_2^2 + h_3^2 v_3^2}$$

5.1.4 Derivatives of Unit Vectors

The derivative of a unit vector (say, $\hat{\mathbf{y}}_1$) with respect to a coordinate in a different direction (say, the y_2-direction) is zero in Cartesian coordinates. However, the derivative of a unit vector may be non-zero in other coordinate systems.

Cylindrical Coordinates

For these coordinates, Figure 5.3 below shows the unit vectors $\hat{\mathbf{r}}$, $\hat{\boldsymbol{\theta}}$ and $\hat{\mathbf{z}}$ at a point, along with the vector components on Cartesian axes. For example, the component of $\hat{\mathbf{r}}$ along the

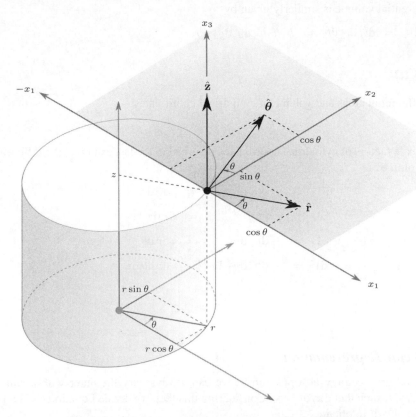

Figure 5.3 Unit vectors $\hat{\mathbf{r}}$, $\hat{\boldsymbol{\theta}}$ and $\hat{\mathbf{z}}$ in cylindrical coordinates, and their projections on Cartesian axes

x_1-direction is $\|\hat{\mathbf{r}}\| \cos\theta = \cos\theta$. Thus, in terms of these components,

$$\hat{\mathbf{r}} = \hat{\mathbf{x}}_1 \cos\theta + \hat{\mathbf{x}}_2 \sin\theta \tag{5.9}$$

$$\hat{\boldsymbol{\theta}} = -\hat{\mathbf{x}}_1 \sin\theta + \hat{\mathbf{x}}_2 \cos\theta \tag{5.10}$$

$$\hat{\mathbf{z}} = \hat{\mathbf{x}}_3$$

From the above relations,

$$\frac{\partial \hat{\mathbf{r}}}{\partial r} = \frac{\partial \hat{\mathbf{r}}}{\partial z} = \frac{\partial \hat{\boldsymbol{\theta}}}{\partial r} = \frac{\partial \hat{\boldsymbol{\theta}}}{\partial z} = \frac{\partial \hat{\mathbf{z}}}{\partial r} = \frac{\partial \hat{\mathbf{z}}}{\partial \theta} = \frac{\partial \hat{\mathbf{z}}}{\partial z} = \mathbf{0} \qquad (5.11)$$

$$\frac{\partial \hat{\mathbf{r}}}{\partial \theta} = -\hat{\mathbf{x}}_1 \sin \theta + \hat{\mathbf{x}}_2 \cos \theta = \hat{\boldsymbol{\theta}} \qquad (5.12)$$

$$\frac{\partial \hat{\boldsymbol{\theta}}}{\partial \theta} = -\hat{\mathbf{x}}_1 \cos \theta - \hat{\mathbf{x}}_2 \sin \theta = -\hat{\mathbf{r}} \qquad (5.13)$$

Spherical Coordinates

For these coordinates, Figure 5.4 below shows the unit vectors $\hat{\mathbf{r}}$, $\hat{\boldsymbol{\theta}}$ and $\hat{\boldsymbol{\phi}}$ at a point, along with the vector components on Cartesian axes. For example, the projection of $\hat{\mathbf{r}}$

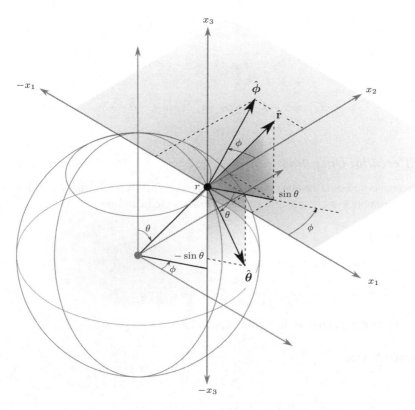

Figure 5.4   Unit vectors $\hat{\mathbf{r}}$, $\hat{\boldsymbol{\theta}}$ and $\hat{\boldsymbol{\phi}}$ in spherical coordinates, and their projections on Cartesian axes

on the $x_1 x_2$-plane is $\|\hat{\mathbf{r}}\| \sin \theta = \sin \theta$, which has components $\sin \theta \cos \phi$ and $\sin \theta \sin \phi$, respectively, along the x_1- and x_2-directions. Along the x_3-direction, the component of $\hat{\mathbf{r}}$

is $\|\hat{\mathbf{r}}\| \cos \theta = \cos \theta$. Resolving the components of $\hat{\boldsymbol{\theta}}$ and $\hat{\boldsymbol{\phi}}$ in the same manner, we obtain

$$
\begin{aligned}
\hat{\mathbf{r}} &= \hat{\mathbf{x}}_1 \sin \theta \cos \phi + \hat{\mathbf{x}}_2 \sin \theta \sin \phi + \hat{\mathbf{x}}_3 \cos \theta \\
\hat{\boldsymbol{\theta}} &= \hat{\mathbf{x}}_1 \cos \theta \cos \phi + \hat{\mathbf{x}}_2 \cos \theta \sin \phi - \hat{\mathbf{x}}_3 \sin \theta \\
\hat{\boldsymbol{\phi}} &= -\hat{\mathbf{x}}_1 \sin \phi + \hat{\mathbf{x}}_2 \cos \phi
\end{aligned}
$$

From the above relations,

$$
\frac{\partial \hat{\mathbf{r}}}{\partial r} = \frac{\partial \hat{\boldsymbol{\theta}}}{\partial r} = \frac{\partial \hat{\boldsymbol{\phi}}}{\partial r} = \frac{\partial \hat{\boldsymbol{\phi}}}{\partial \theta} = 0
$$

$$
\frac{\partial \hat{\mathbf{r}}}{\partial \theta} = \hat{\mathbf{x}}_1 \cos \theta \cos \phi + \hat{\mathbf{x}}_2 \cos \theta \sin \phi - \hat{\mathbf{x}}_3 \sin \theta = \hat{\boldsymbol{\theta}}
$$

$$
\frac{\partial \hat{\boldsymbol{\theta}}}{\partial \theta} = -\hat{\mathbf{x}}_1 \sin \theta \cos \phi - \hat{\mathbf{x}}_2 \sin \theta \sin \phi - \hat{\mathbf{x}}_3 \cos \theta = -\hat{\mathbf{r}}
$$

$$
\frac{\partial \hat{\mathbf{r}}}{\partial \phi} = (-\hat{\mathbf{x}}_1 \sin \phi + \hat{\mathbf{x}}_2 \cos \phi) \sin \theta = \hat{\boldsymbol{\phi}} \sin \theta
$$

$$
\frac{\partial \hat{\boldsymbol{\theta}}}{\partial \phi} = (-\hat{\mathbf{x}}_1 \sin \phi + \hat{\mathbf{x}}_2 \cos \phi) \cos \theta = \hat{\boldsymbol{\phi}} \cos \theta
$$

$$
\frac{\partial \hat{\boldsymbol{\phi}}}{\partial \phi} = -\hat{\mathbf{x}}_1 \cos \phi - \hat{\mathbf{x}}_2 \sin \phi = -\hat{\mathbf{r}} \sin \theta - \hat{\boldsymbol{\theta}} \cos \theta
$$

5.1.5   Differential Operators

For an orthogonal coordinate system (y_1, y_2, y_3) with scale factors h_1, h_2 and h_3 with respect to Cartesian coordinate system, the differential operators are as follows.

Gradient of a Scalar

The gradient of a scalar F is given by

$$
\nabla F = \frac{1}{h_1} \frac{\partial F}{\partial y_1} \hat{\mathbf{y}}_1 + \frac{1}{h_2} \frac{\partial F}{\partial y_2} \hat{\mathbf{y}}_2 + \frac{1}{h_3} \frac{\partial F}{\partial y_3} \hat{\mathbf{y}}_3 \tag{5.14}
$$

This expression for the gradient is derived in Appendix 5.A.1 on p. 180.

Divergence of a Vector

The divergence of a vector

$$
\mathbf{v} = v_1 \hat{\mathbf{y}}_1 + v_2 \hat{\mathbf{y}}_2 + v_3 \hat{\mathbf{y}}_3
$$

in the (y_1, y_2, y_3) coordinates is given by

$$
\nabla \cdot \mathbf{v} = \frac{1}{h_1 h_2 h_3} \left[\frac{\partial (h_2 h_3 v_1)}{\partial y_1} + \frac{\partial (h_1 h_3 v_2)}{\partial y_2} + \frac{\partial (h_1 h_2 v_3)}{\partial y_3} \right] \tag{5.15}
$$

The above expression for divergence is derived in Appendix 5.A.2 on p. 181.

Curl of a Vector

The curl of the vector **v** is given by

$$
\nabla \times \mathbf{v} = \frac{1}{h_1 h_2 h_3}
\begin{vmatrix}
h_1 \hat{\mathbf{y}}_1 & h_2 \hat{\mathbf{y}}_2 & h_3 \hat{\mathbf{y}}_3 \\[4pt]
\dfrac{\partial}{\partial y_1} & \dfrac{\partial}{\partial y_2} & \dfrac{\partial}{\partial y_3} \\[8pt]
h_1 v_1 & h_2 v_2 & h_3 v_3
\end{vmatrix}
\tag{5.16}
$$

This expression for the curl is derived in Appendix 5.A.4 on p. 184.

Laplacian of a Scalar

Using Equations (5.14) and (5.15) on the previous page, the Laplacian of a scalar F is

$$
\begin{aligned}
\nabla^2 F &= \nabla \cdot \nabla F \\[6pt]
&= \frac{1}{h_1 h_2 h_3} \left[\frac{\partial}{\partial y_1} \left(\frac{h_2 h_3}{h_1} \frac{\partial F}{\partial y_1} \right) + \frac{\partial}{\partial y_2} \left(\frac{h_1 h_3}{h_2} \frac{\partial F}{\partial y_2} \right) + \frac{\partial}{\partial y_3} \left(\frac{h_1 h_2}{h_3} \frac{\partial F}{\partial y_3} \right) \right]
\end{aligned}
\tag{5.17}
$$

Example 5.1.3

Find the expressions for the gradient of a scalar, divergence of a vector, curl of a vector, and Laplacian of a scalar in cylindrical coordinate system.

Solution

From Example 5.1.1 on p. 141, the scale factors for the cylindrical coordinate system are

$$
h_1 = 1, \qquad h_2 = y_1, \qquad \text{and} \qquad h_3 = 1.
$$

Gradient of a Scalar
From Equation (5.14) on the previous page, the gradient of a scalar F is given by

$$
\nabla F = \frac{\partial F}{\partial y_1} \hat{\mathbf{y}}_1 + \frac{1}{y_1} \frac{\partial F}{\partial y_2} \hat{\mathbf{y}}_2 + \frac{\partial F}{\partial y_3} \hat{\mathbf{y}}_3
$$

In the (r, θ, z) notation,

$$
\nabla F = \frac{\partial F}{\partial r} \hat{\mathbf{r}} + \frac{1}{r} \frac{\partial F}{\partial \theta} \hat{\boldsymbol{\theta}} + \frac{\partial F}{\partial z} \hat{\mathbf{z}}
\tag{5.18}
$$

Divergence of a Vector
From Equation (5.15) on the previous page, the divergence of a vector **v** is given by

$$
\nabla \cdot \mathbf{v} = \frac{1}{y_1} \left[\frac{\partial (y_1 v_1)}{\partial y_1} + \frac{\partial v_2}{\partial y_2} + y_1 \frac{\partial v_3}{\partial y_3} \right]
$$

In the (r, θ, z) notation where $\mathbf{v} = v_r \hat{\mathbf{r}} + v_\theta \hat{\boldsymbol{\theta}} + v_z \hat{\mathbf{z}}$,

$$\nabla \cdot \mathbf{v} = \frac{1}{r} \frac{\partial (r v_r)}{\partial r} + \frac{1}{r} \frac{\partial v_\theta}{\partial \theta} + \frac{\partial v_z}{\partial z} \tag{5.19}$$

Curl of a Vector

From Equation (5.16) on the previous page, the curl of $\mathbf{v}$ is given by

$$\nabla \times \mathbf{v} = \frac{1}{y_1} \begin{vmatrix} \hat{\mathbf{y}}_1 & y_1 \hat{\mathbf{y}}_2 & \hat{\mathbf{y}}_3 \\ \frac{\partial}{\partial y_1} & \frac{\partial}{\partial y_2} & \frac{\partial}{\partial y_3} \\ v_1 & y_1 v_2 & v_3 \end{vmatrix} = \frac{1}{r} \begin{vmatrix} \hat{\mathbf{r}} & r\hat{\boldsymbol{\theta}} & \hat{\mathbf{z}} \\ \frac{\partial}{\partial r} & \frac{\partial}{\partial \theta} & \frac{\partial}{\partial z} \\ v_r & r v_\theta & v_z \end{vmatrix} \tag{5.20}$$

$$= \frac{1}{y_1} \left[\frac{\partial v_3}{\partial y_2} - \frac{\partial}{\partial y_3} (y_1 v_2) \right] \hat{\mathbf{y}}_1 - \left[\frac{\partial v_3}{\partial y_1} - \frac{\partial v_1}{\partial y_3} \right] \hat{\mathbf{y}}_2 + \frac{1}{y_1} \left[\frac{\partial}{\partial y_1} (y_1 v_2) - \frac{\partial v_1}{\partial y_2} \right] \hat{\mathbf{y}}_3$$

In the (r, θ, z) notation,

$$\nabla \times \mathbf{v} = \frac{1}{r} \left[\frac{\partial v_z}{\partial \theta} - r \frac{\partial v_\theta}{\partial z} \right] \hat{\mathbf{r}} - \left[\frac{\partial v_z}{\partial r} - \frac{\partial v_r}{\partial z} \right] \hat{\boldsymbol{\theta}} + \frac{1}{r} \left[\frac{\partial (r v_\theta)}{\partial r} - \frac{\partial v_r}{\partial \theta} \right] \hat{\mathbf{z}}$$

considering that $\partial r / \partial z = 0$.

Laplacian of a Scalar

From Equation (5.17) on the previous page, the Laplacian of a scalar F is given by

$$\nabla^2 F = \frac{1}{y_1} \left[\frac{\partial}{\partial y_1} \left(y_1 \frac{\partial F}{\partial y_1} \right) + \frac{\partial}{\partial y_2} \left(\frac{1}{y_1} \frac{\partial F}{\partial y_2} \right) + \frac{\partial}{\partial y_3} \left(y_1 \frac{\partial F}{\partial y_3} \right) \right]$$

$$= \frac{1}{y_1} \frac{\partial}{\partial y_1} \left(y_1 \frac{\partial F}{\partial y_1} \right) + \frac{1}{y_1^2} \frac{\partial^2 F}{\partial y_2{}^2} + \frac{\partial^2 F}{\partial y_3{}^2}$$

In the (r, θ, z) notation,

$$\nabla^2 F = \frac{1}{r} \frac{\partial}{\partial r} \left(r \frac{\partial F}{\partial r} \right) + \frac{1}{r^2} \frac{\partial^2 F}{\partial \theta^2} + \frac{\partial^2 F}{\partial z^2}$$

❑

Gradient of a Vector

The gradient of a vector $\mathbf{v}$ is obtained from the long multiplication of the gradient operator ∇ [see Equation (5.14), p. 146], and $\mathbf{v}$ in a given orthogonal coordinate system, i.e.,

$$\nabla \mathbf{v} = \underbrace{\left(\frac{\hat{\mathbf{y}}_1}{h_1} \frac{\partial}{\partial y_1} + \frac{\hat{\mathbf{y}}_2}{h_2} \frac{\partial}{\partial y_2} + \frac{\hat{\mathbf{y}}_3}{h_3} \frac{\partial}{\partial y_3} \right)}_{\nabla} \underbrace{(\hat{\mathbf{y}}_1 v_1 + \hat{\mathbf{y}}_2 v_2 + \hat{\mathbf{y}}_3 v_3)}_{\mathbf{v}}$$

The resulting expression is then simplified by substituting the derivatives of unit vectors. These steps are illustrated in the next example.

Example 5.1.4

Find the gradient of a vector **v** in cylindrical coordinate system.

Solution

In terms of the (r, θ, z) notation for cylindrical coordinates,

$$\mathbf{v} = v_r \hat{\mathbf{r}} + v_\theta \hat{\boldsymbol{\theta}} + v_z \hat{\mathbf{z}} \tag{5.21}$$

and its gradient is given by

$$\nabla \mathbf{v} = \left(\hat{\mathbf{r}} \frac{\partial}{\partial r} + \frac{\hat{\boldsymbol{\theta}}}{r} \frac{\partial}{\partial \theta} + \hat{\mathbf{z}} \frac{\partial}{\partial z} \right) (\hat{\mathbf{r}} v_r + \hat{\boldsymbol{\theta}} v_\theta + \hat{\mathbf{z}} v_z)$$

$$= \hat{\mathbf{r}} \frac{\partial}{\partial r} (\hat{\mathbf{r}} v_r) + \hat{\mathbf{r}} \frac{\partial}{\partial r} (\hat{\boldsymbol{\theta}} v_\theta) + \hat{\mathbf{r}} \frac{\partial}{\partial r} (\hat{\mathbf{z}} v_z) + \frac{\hat{\boldsymbol{\theta}}}{r} \frac{\partial}{\partial \theta} (\hat{\mathbf{r}} v_r) + \frac{\hat{\boldsymbol{\theta}}}{r} \frac{\partial}{\partial \theta} (\hat{\boldsymbol{\theta}} v_\theta) + \frac{\hat{\boldsymbol{\theta}}}{r} \frac{\partial}{\partial \theta} (\hat{\mathbf{z}} v_z)$$

$$+ \ \hat{\mathbf{z}} \frac{\partial}{\partial z} (\hat{\mathbf{r}} v_r) + \hat{\mathbf{z}} \frac{\partial}{\partial z} (\hat{\boldsymbol{\theta}} v_\theta) + \hat{\mathbf{z}} \frac{\partial}{\partial z} (\hat{\mathbf{z}} v_z)$$

The right-hand side terms of the above equation simplify as follows:

$$\hat{\mathbf{r}} \frac{\partial}{\partial r} (\hat{\mathbf{r}} v_r) = \hat{\mathbf{r}} \underbrace{\frac{\partial \hat{\mathbf{r}}}{\partial r}}_{= 0} v_r + \hat{\mathbf{r}}\hat{\mathbf{r}} \frac{\partial v_r}{\partial r} = \frac{\partial v_r}{\partial r} \hat{\mathbf{r}}\hat{\mathbf{r}}$$

$$\hat{\mathbf{r}} \frac{\partial}{\partial r} (\hat{\boldsymbol{\theta}} v_\theta) = \hat{\mathbf{r}} \underbrace{\frac{\partial \hat{\boldsymbol{\theta}}}{\partial r}}_{= 0} v_\theta + \hat{\mathbf{r}}\hat{\boldsymbol{\theta}} \frac{\partial v_\theta}{\partial r} = \frac{\partial v_\theta}{\partial r} \hat{\mathbf{r}}\hat{\boldsymbol{\theta}}$$

$$\hat{\mathbf{r}} \frac{\partial}{\partial r} (\hat{\mathbf{z}} v_z) = \hat{\mathbf{r}} \underbrace{\frac{\partial \hat{\mathbf{z}}}{\partial r}}_{= 0} v_z + \hat{\mathbf{r}}\hat{\mathbf{z}} \frac{\partial v_z}{\partial r} = \frac{\partial v_z}{\partial r} \hat{\mathbf{r}}\hat{\mathbf{z}}$$

$$\frac{\hat{\boldsymbol{\theta}}}{r} \frac{\partial}{\partial \theta} (\hat{\mathbf{r}} v_r) = \frac{\hat{\boldsymbol{\theta}}}{r} \underbrace{\frac{\partial \hat{\mathbf{r}}}{\partial \theta}}_{= \hat{\boldsymbol{\theta}}} v_r + \hat{\boldsymbol{\theta}}\hat{\mathbf{r}} \frac{1}{r} \frac{\partial v_r}{\partial \theta} = \frac{v_r}{r} \hat{\boldsymbol{\theta}}\hat{\boldsymbol{\theta}} + \frac{1}{r} \frac{\partial v_r}{\partial \theta} \hat{\boldsymbol{\theta}}\hat{\mathbf{r}}$$

$$\frac{\hat{\boldsymbol{\theta}}}{r} \frac{\partial}{\partial \theta} (\hat{\boldsymbol{\theta}} v_\theta) = \frac{\hat{\boldsymbol{\theta}}}{r} \underbrace{\frac{\partial \hat{\boldsymbol{\theta}}}{\partial \theta}}_{= -\hat{\mathbf{r}}} v_\theta + \hat{\boldsymbol{\theta}}\hat{\boldsymbol{\theta}} \frac{1}{r} \frac{\partial v_\theta}{\partial \theta} = -\frac{v_\theta}{r} \hat{\boldsymbol{\theta}}\hat{\mathbf{r}} + \frac{1}{r} \frac{\partial v_\theta}{\partial \theta} \hat{\boldsymbol{\theta}}\hat{\boldsymbol{\theta}}$$

$$\frac{\hat{\theta}}{r}\frac{\partial}{\partial\theta}(\hat{z}v_z) = \frac{\hat{\theta}}{r}\frac{\partial\hat{z}}{\partial\theta}v_z + \hat{\theta}\hat{z}\frac{1}{r}\frac{\partial v_z}{\partial\theta} = \frac{1}{r}\frac{\partial v_z}{\partial\theta}\hat{\theta}\hat{z}$$
$$\underbrace{}_{=0}$$

$$\hat{z}\frac{\partial}{\partial z}(\hat{r}v_r) = \hat{z}\frac{\partial\hat{r}}{\partial z}v_r + \hat{z}\hat{r}\frac{\partial v_r}{\partial z} = \frac{\partial v_r}{\partial z}\hat{z}\hat{r}$$
$$\underbrace{}_{=0}$$

$$\hat{z}\frac{\partial}{\partial z}(\hat{\theta}v_\theta) = \hat{z}\frac{\partial\hat{\theta}}{\partial z}v_\theta + \hat{z}\hat{\theta}\frac{\partial v_\theta}{\partial z} = \frac{\partial v_\theta}{\partial z}\hat{z}\hat{\theta}$$
$$\underbrace{}_{=0}$$

$$\hat{z}\frac{\partial}{\partial z}(\hat{z}v_z) = \hat{z}\frac{\partial\hat{z}}{\partial z}v_z + \hat{z}\hat{z}\frac{\partial v_z}{\partial z} = \frac{\partial v_z}{\partial z}\hat{z}\hat{z}$$
$$\underbrace{}_{=0}$$

With the help of the above terms, the gradient is given by

$$\nabla\mathbf{v} = \frac{\partial v_r}{\partial r}\hat{r}\hat{r} + \frac{\partial v_\theta}{\partial r}\hat{r}\hat{\theta} + \frac{\partial v_z}{\partial r}\hat{r}\hat{z} + \left(\frac{1}{r}\frac{\partial v_r}{\partial\theta} - \frac{v_\theta}{r}\right)\hat{\theta}\hat{r} +$$

$$\left(\frac{v_r}{r} + \frac{1}{r}\frac{\partial v_\theta}{\partial\theta}\right)\hat{\theta}\hat{\theta} + \frac{1}{r}\frac{\partial v_z}{\partial\theta}\hat{\theta}\hat{z} + \frac{\partial v_r}{\partial z}\hat{z}\hat{r} + \frac{\partial v_\theta}{\partial z}\hat{z}\hat{\theta} + \frac{\partial v_z}{\partial z}\hat{z}\hat{z} \quad (5.22)$$

❑

Laplacian of a Vector

The Laplacian of a vector $\mathbf{f}$ is a vector. It can be determined using the above relations for the divergence and curl in the following identity [see Equation (8.3), p. 302]:

$$\nabla^2\mathbf{f} = \nabla(\nabla\cdot\mathbf{f}) - \nabla\times(\nabla\times\mathbf{f}) \quad (5.23)$$

Divergence of a Tensor

The divergence of a tensor is a vector. It is obtained from the long multiplication of the gradient operator [see Equation (5.14), p. 146], and the tensor. Thus,

$$\nabla\cdot\tau = \underbrace{\left(\frac{\hat{y}_1}{h_1}\frac{\partial}{\partial y_1} + \frac{\hat{y}_2}{h_2}\frac{\partial}{\partial y_2} + \frac{\hat{y}_3}{h_3}\frac{\partial}{\partial y_3}\right)}_{\nabla}\cdot\underbrace{\sum_{i=1}^{3}\sum_{j=1}^{3}\tau_{ij}\hat{y}_i\hat{y}_j}_{\tau}$$

where τ is a second order tensor [see Equation (8.7), p. 308]. The resulting expression is simplified by substituting the derivatives of unit vectors.

Example 5.1.5

Find the r-component of the divergence of a tensor τ in cylindrical coordinate system.

Solution

In terms of the (r, θ, z) notation for cylindrical coordinates, the last equation yields

$$\nabla \cdot \tau = \left(\hat{\mathbf{r}} \frac{\partial}{\partial r} + \frac{\hat{\boldsymbol{\theta}}}{r} \frac{\partial}{\partial \theta} + \hat{\mathbf{z}} \frac{\partial}{\partial z} \right) \cdot (\tau_{rr} \hat{\mathbf{r}} \hat{\mathbf{r}} + \tau_{r\theta} \hat{\mathbf{r}} \hat{\boldsymbol{\theta}} + \tau_{rz} \hat{\mathbf{r}} \hat{\mathbf{z}}$$

$$+ \ \tau_{\theta r} \hat{\boldsymbol{\theta}} \hat{\mathbf{r}} + \tau_{\theta\theta} \hat{\boldsymbol{\theta}} \hat{\boldsymbol{\theta}} + \tau_{\theta z} \hat{\boldsymbol{\theta}} \hat{\mathbf{z}} + \tau_{zr} \hat{\mathbf{z}} \hat{\mathbf{r}} + \tau_{z\theta} \hat{\mathbf{z}} \hat{\boldsymbol{\theta}} + \tau_{zz} \hat{\mathbf{z}} \hat{\mathbf{z}})$$

The right-hand side of the above equation results in 27 terms of which only the following contribute to the r-component of $\nabla \cdot \tau$:

$$\hat{\mathbf{r}} \frac{\partial}{\partial r} \cdot (\hat{\mathbf{r}} \hat{\mathbf{r}} \tau_{rr}) = \hat{\mathbf{r}} \cdot \underbrace{\frac{\partial}{\partial r} (\hat{\mathbf{r}} \hat{\mathbf{r}})}_{\substack{=0 \\ \text{since } \partial \hat{\mathbf{r}}/\partial r = 0}} \tau_{rr} + \underbrace{\hat{\mathbf{r}} \cdot \hat{\mathbf{r}} \hat{\mathbf{r}}}_{=1} \frac{\partial \tau_{rr}}{\partial r} = \frac{\partial \tau_{rr}}{\partial r} \hat{\mathbf{r}}$$

$$\frac{\hat{\boldsymbol{\theta}}}{r} \frac{\partial}{\partial \theta} \cdot (\hat{\mathbf{r}} \hat{\mathbf{r}} \tau_{rr}) = \frac{\hat{\boldsymbol{\theta}}}{r} \cdot \frac{\partial}{\partial \theta} (\hat{\mathbf{r}} \hat{\mathbf{r}}) \tau_{rr} + \frac{1}{r} \underbrace{\hat{\boldsymbol{\theta}} \cdot \hat{\mathbf{r}} \hat{\mathbf{r}}}_{=0} \frac{\partial \tau_{rr}}{\partial \theta}$$

$$= \frac{1}{r} \hat{\boldsymbol{\theta}} \cdot \underbrace{\frac{\partial \hat{\mathbf{r}}}{\partial \theta}}_{\substack{= \hat{\boldsymbol{\theta}} \\ =1}} \hat{\mathbf{r}} \tau_{rr} + \frac{1}{r} \underbrace{\hat{\boldsymbol{\theta}} \cdot \hat{\mathbf{r}}}_{=0} \frac{\partial \hat{\mathbf{r}}}{\partial \theta} \tau_{rr} = \frac{\tau_{rr}}{r} \hat{\mathbf{r}}$$

$$\frac{\hat{\boldsymbol{\theta}}}{r} \frac{\partial}{\partial \theta} \cdot (\hat{\boldsymbol{\theta}} \hat{\mathbf{r}} \tau_{\theta r}) = \frac{\hat{\boldsymbol{\theta}}}{r} \cdot \frac{\partial}{\partial \theta} (\hat{\boldsymbol{\theta}} \hat{\mathbf{r}}) \tau_{\theta r} + \frac{1}{r} \underbrace{\hat{\boldsymbol{\theta}} \cdot \hat{\boldsymbol{\theta}} \hat{\mathbf{r}}}_{=1} \frac{\partial \tau_{\theta r}}{\partial \theta}$$

$$= \frac{1}{r} \hat{\boldsymbol{\theta}} \cdot \underbrace{\frac{\partial \hat{\boldsymbol{\theta}}}{\partial \theta}}_{\substack{= -\hat{\mathbf{r}} \\ =0}} \hat{\mathbf{r}} \tau_{\theta r} + \frac{1}{r} \underbrace{\hat{\boldsymbol{\theta}} \cdot \hat{\boldsymbol{\theta}}}_{=1} \underbrace{\frac{\partial \hat{\mathbf{r}}}{\partial \theta}}_{= \hat{\boldsymbol{\theta}}} \tau_{\theta r} + \frac{\hat{\mathbf{r}}}{r} \frac{\partial \tau_{\theta r}}{\partial \theta} = \frac{\tau_{\theta r}}{r} \hat{\boldsymbol{\theta}} + \frac{1}{r} \frac{\partial \tau_{\theta r}}{\partial \theta} \hat{\mathbf{r}}$$

$$\frac{\hat{\boldsymbol{\theta}}}{r} \frac{\partial}{\partial \theta} \cdot (\hat{\boldsymbol{\theta}} \hat{\boldsymbol{\theta}} \tau_{\theta\theta}) = \frac{\hat{\boldsymbol{\theta}}}{r} \cdot \frac{\partial}{\partial \theta} (\hat{\boldsymbol{\theta}} \hat{\boldsymbol{\theta}}) \tau_{\theta\theta} + \frac{1}{r} \underbrace{\hat{\boldsymbol{\theta}} \cdot \hat{\boldsymbol{\theta}} \hat{\boldsymbol{\theta}}}_{=1} \frac{\partial \tau_{\theta\theta}}{\partial \theta}$$

$$= \frac{1}{r} \hat{\boldsymbol{\theta}} \cdot \underbrace{\frac{\partial \hat{\boldsymbol{\theta}}}{\partial \theta}}_{\substack{= -\hat{\mathbf{r}} \\ =0}} \hat{\boldsymbol{\theta}} \tau_{\theta\theta} + \frac{1}{r} \underbrace{\hat{\boldsymbol{\theta}} \cdot \hat{\boldsymbol{\theta}}}_{=1} \underbrace{\frac{\partial \hat{\boldsymbol{\theta}}}{\partial \theta}}_{= -\hat{\mathbf{r}}} \tau_{\theta\theta} + \frac{\hat{\boldsymbol{\theta}}}{r} \frac{\partial \tau_{\theta\theta}}{\partial \theta} = -\frac{\tau_{\theta\theta}}{r} \hat{\mathbf{r}} + \frac{1}{r} \frac{\partial \tau_{\theta\theta}}{\partial \theta} \hat{\boldsymbol{\theta}}$$

$$\hat{\mathbf{z}}\frac{\partial}{\partial z}\cdot(\hat{\mathbf{z}}\hat{\mathbf{r}}\tau_{zr}) = \hat{\mathbf{z}}\cdot\frac{\partial}{\partial z}(\hat{\mathbf{z}}\hat{\mathbf{r}})\tau_{zr} + \underbrace{\hat{\mathbf{z}}\cdot\hat{\mathbf{z}}}_{=1}\hat{\mathbf{r}}\frac{\partial\tau_{zr}}{\partial z}$$

$$= \hat{\mathbf{z}}\cdot\underbrace{\frac{\partial\hat{\mathbf{z}}}{\partial z}}_{=0}\hat{\mathbf{r}}\tau_{zr} + \hat{\mathbf{z}}\cdot\hat{\mathbf{z}}\underbrace{\frac{\partial\hat{\mathbf{r}}}{\partial z}}_{=0}\tau_{zr} + \hat{\mathbf{r}}\frac{\partial\tau_{zr}}{\partial z} = \frac{\partial\tau_{zr}}{\partial z}\hat{\mathbf{r}}$$

From the last five derivative expansions, we collect the terms with $\hat{\mathbf{r}}$ in common to obtain the r-component of $\nabla\cdot\boldsymbol{\tau}$, i.e.,

$$(\nabla\cdot\boldsymbol{\tau})_r = \frac{\partial\tau_{rr}}{\partial r} + \frac{\tau_{rr}}{r} + \frac{1}{r}\frac{\partial\tau_{\theta r}}{\partial\theta} - \frac{\tau_{\theta\theta}}{r} + \frac{\partial\tau_{zr}}{\partial z} \qquad (5.24)$$

❑

With the help of scale factors, derivatives of unit vectors, and differential operators derived above, we can transform the fundamental relations, and associated equations in Cartesian coordinate system to other orthogonal coordinate systems of interest.

Example 5.1.6

From the equation of change for velocity in Cartesian coordinate system [see p. 33],

$$\frac{\partial\mathbf{v}}{\partial t} = -\mathbf{v}\cdot\nabla\mathbf{v} - \frac{1}{\rho}(\nabla\cdot\boldsymbol{\tau} + \nabla P) + \mathbf{g} \qquad (2.25)$$

obtain the r-component of the equation in cylindrical coordinate system.

Solution

The r-component of the above equation is

$$\frac{\partial v_r}{\partial t} = -(\mathbf{v}\cdot\nabla\mathbf{v})_r - \frac{1}{\rho}\left[(\nabla\cdot\boldsymbol{\tau})_r + (\nabla P)_r\right] + g_r \qquad (5.25)$$

for which we need to obtain expressions for $(\mathbf{v}\cdot\nabla\mathbf{v})_r$ and $(\nabla P)_r$. Equation (5.24) above provides the expression for $(\nabla\cdot\boldsymbol{\tau})_r$.

Expression for $(\mathbf{v}\cdot\nabla\mathbf{v})_r$
From Equations (5.21) and (5.22) on pp. 149 and 150, respectively, we obtain

$$\mathbf{v}\cdot\nabla\mathbf{v} = (v_r\hat{\mathbf{r}} + v_\theta\hat{\boldsymbol{\theta}} + uv_z\hat{\mathbf{z}})\cdot\left[\frac{\partial v_r}{\partial r}\hat{\mathbf{r}}\hat{\mathbf{r}} + \frac{\partial v_\theta}{\partial r}\hat{\mathbf{r}}\hat{\boldsymbol{\theta}} + \frac{\partial v_z}{\partial r}\hat{\mathbf{r}}\hat{\mathbf{z}} + \left(\frac{1}{r}\frac{\partial v_r}{\partial\theta} - \frac{v_\theta}{r}\right)\hat{\boldsymbol{\theta}}\hat{\mathbf{r}}\right.$$

$$\left. + \left(\frac{v_r}{r} + \frac{1}{r}\frac{\partial v_\theta}{\partial\theta}\right)\hat{\boldsymbol{\theta}}\hat{\boldsymbol{\theta}} + \frac{1}{r}\frac{\partial v_z}{\partial\theta}\hat{\boldsymbol{\theta}}\hat{\mathbf{z}} + \frac{\partial v_r}{\partial z}\hat{\mathbf{z}}\hat{\mathbf{r}} + \frac{\partial v_\theta}{\partial z}\hat{\mathbf{z}}\hat{\boldsymbol{\theta}} + \frac{\partial v_z}{\partial z}\hat{\mathbf{z}}\hat{\mathbf{z}}\right]$$

The right-hand side of the above equation simplifies upon expansion, because of the orthogonality of the unit vectors. For example,

$$v_r\underbrace{\hat{\mathbf{r}}\cdot\hat{\mathbf{r}}}_{=1}\hat{\mathbf{r}}\frac{\partial v_r}{\partial r} = v_r\hat{\mathbf{r}}\frac{\partial v_r}{\partial r} \qquad \text{and} \qquad v_r\underbrace{\hat{\mathbf{r}}\cdot\hat{\boldsymbol{\theta}}}_{=0}\hat{\boldsymbol{\theta}}\frac{v_r}{r} = 0$$

Simplifying in this manner, we get

$$
\mathbf{v}\cdot\nabla\mathbf{v} = \left(v_r\frac{\partial v_r}{\partial r} + \frac{v_\theta}{r}\frac{\partial v_r}{\partial\theta} - \frac{v_\theta^2}{r} + v_z\frac{\partial v_r}{\partial z}\right)\hat{\mathbf{r}} + (\ldots)\hat{\boldsymbol{\theta}} + (\ldots)\hat{\mathbf{z}}
$$

The r-component of the equation is

$$
(\mathbf{v}\cdot\nabla\mathbf{v})_r = v_r\frac{\partial v_r}{\partial r} + \frac{v_\theta}{r}\frac{\partial v_r}{\partial\theta} - \frac{v_\theta^2}{r} + v_z\frac{\partial v_r}{\partial z} \tag{5.26}
$$

Expression for $(\nabla P)_r$

From Equation (5.14) on p. 146,

$$
\nabla P = \frac{\partial P}{\partial r}\hat{\mathbf{r}} + \frac{1}{r}\frac{\partial P}{\partial\theta}\hat{\boldsymbol{\theta}} + \frac{\partial P}{\partial z}\hat{\mathbf{z}}
$$

from which

$$
(\nabla P)_r = \frac{\partial P}{\partial r} \tag{5.27}
$$

Substituting the above expressions for $(\mathbf{v}\cdot\nabla\mathbf{v})_r$, $(\nabla\cdot\hat{\tau})_r$ and $(\nabla P)_r$ in Equation (5.25) on the previous page, we finally obtain the desired equation, i.e.,

$$
\begin{aligned}
\frac{\partial v_r}{\partial t} = {}& -v_r\frac{\partial v_r}{\partial r} - \frac{v_\theta}{r}\frac{\partial v_r}{\partial\theta} - v_z\frac{\partial v_r}{\partial z} + \frac{v_\theta^2}{r} \\
& -\frac{1}{\rho}\left(\frac{\partial\tau_{rr}}{\partial r} + \frac{1}{r}\frac{\partial\tau_{r\theta}}{\partial\theta} + \frac{\partial\tau_{zr}}{\partial z} + \frac{\tau_{rr}-\tau_{\theta\theta}}{r} - \frac{\partial P}{\partial r}\right) + g_r
\end{aligned}
$$

Note that τ_{rr}/r and $-\tau_{\theta\theta}/r$ in the above equation arise because of non-zero $\partial\hat{\mathbf{r}}/\partial\theta$ and $\partial\hat{\boldsymbol{\theta}}/\partial\theta$ [see Equations (5.12) and (5.13), p. 145].

❑

Example 5.1.7

Obtain the r-component of the Navier–Stokes equation [see Equation (3.1), p. 62] for an incompressible fluid of constant viscosity cylindrical coordinates.

Solution

From Equation (3.1), the r-component of the equation of the Navier–Stokes equation is given by

$$
\frac{\partial v_r}{\partial t} = -(\mathbf{v}\cdot\nabla\mathbf{v})_r + \frac{1}{\rho}\left[\mu(\nabla^2\mathbf{v})_r - (\nabla P)_r\right] + g_r \tag{5.28}
$$

The expressions for $(\mathbf{v}\cdot\nabla\mathbf{v})_r$ and $(\nabla P)_r$ were derived in the last example. We now derive $(\nabla^2\mathbf{v})_r$, which from Equation (5.23) on p. 150 is given by

$$
(\nabla^2\mathbf{v})_r = [\nabla(\nabla\cdot\mathbf{v})]_r - [\nabla\times(\nabla\times\mathbf{v})]_r
$$

Expression for $[\nabla(\nabla \cdot \mathbf{v})]_r$

From Equation (5.18) on p. 147,

$$\nabla(\nabla \cdot \mathbf{v}) \;=\; \frac{\partial(\nabla \cdot \mathbf{v})}{\partial r}\hat{\mathbf{r}} + \frac{1}{r}\frac{\partial(\nabla \cdot \mathbf{v})}{\partial \theta}\hat{\boldsymbol{\theta}} + \frac{\partial(\nabla \cdot \mathbf{v})}{\partial z}\hat{\mathbf{z}}$$

The r-component of the above equation is

$$[\nabla(\nabla \cdot \mathbf{v})]_r \;=\; \frac{\partial}{\partial r}(\nabla \cdot \mathbf{v}) \;=\; \frac{\partial}{\partial r}\underbrace{\left(\frac{v_r}{r} + \frac{\partial v_r}{\partial r} + \frac{1}{r}\frac{\partial v_\theta}{\partial \theta} + \frac{\partial v_z}{\partial z}\right)}_{\text{from Equation (5.19) on p. 148}}$$

$$= \;-\frac{v_r}{r^2} + \frac{1}{r}\frac{\partial v_r}{\partial r} + \frac{\partial^2 v_r}{\partial r^2} - \frac{1}{r^2}\frac{\partial v_\theta}{\partial \theta} + \frac{1}{r}\underbrace{\frac{\partial^2 v_\theta}{\partial r \partial \theta}}_{=\,0} + \underbrace{\frac{\partial^2 v_z}{\partial r \partial z}}_{=\,0}$$

$$= \;-\frac{v_r}{r^2} + \frac{1}{r}\frac{\partial v_r}{\partial r} + \frac{\partial^2 v_r}{\partial r^2} - \frac{1}{r^2}\frac{\partial v_\theta}{\partial \theta}$$

Expression for $[\nabla \times (\nabla \times \mathbf{v})]_r$

Using the vector $(\nabla \times \mathbf{v})$ instead of $\mathbf{v}$ in Equation (5.20) on p. 148, we get

$$\nabla \times (\nabla \times \mathbf{v}) \;=\; \underbrace{\frac{1}{r}\left[\frac{\partial(\nabla \times \mathbf{v})_z}{\partial \theta} - r\frac{\partial(\nabla \times \mathbf{v})_\theta}{\partial z}\right]}_{[\nabla \times (\nabla \times \mathbf{v})]_r}\hat{\mathbf{r}} \;-\; [\cdots]\hat{\boldsymbol{\theta}} \;+\; [\cdots]\hat{\mathbf{z}}$$

We are interested in $[\nabla \times (\nabla \times \mathbf{v})]_r$, which depends on the following derivatives:

$$\frac{\partial(\nabla \times \mathbf{v})_z}{\partial \theta} \;=\; \frac{\partial}{\partial \theta}\left\{\frac{1}{r}\left[\frac{\partial(rv_\theta)}{\partial r} - \frac{\partial v_r}{\partial \theta}\right]\right\} \;=\; \frac{\partial}{\partial \theta}\left[\frac{v_\theta}{r} + \frac{\partial v_\theta}{\partial r} - \frac{1}{r}\frac{\partial v_r}{\partial \theta}\right]$$

$$= \;\frac{1}{r}\frac{\partial v_\theta}{\partial \theta} + \underbrace{\frac{\partial^2 v_\theta}{\partial \theta \partial r}}_{=\,0} - \frac{1}{r}\frac{\partial^2 v_r}{\partial \theta^2} \qquad \text{and}$$

$$\frac{\partial(\nabla \times \mathbf{v})_\theta}{\partial z} \;=\; -\frac{\partial}{\partial z}\left(\frac{\partial v_z}{\partial r} - \frac{\partial v_r}{\partial z}\right) \;=\; -\underbrace{\frac{\partial^2 v_z}{\partial z \partial r}}_{=\,0} + \frac{\partial^2 v_r}{\partial z^2} \;=\; \frac{\partial^2 v_r}{\partial z^2}$$

Utilizing these derivatives,

$$[\nabla \times (\nabla \times \mathbf{v})]_r \;=\; \frac{1}{r}\frac{\partial(\nabla \times \mathbf{v})_z}{\partial \theta} - \frac{\partial(\nabla \times \mathbf{v})_\theta}{\partial z} \;=\; \frac{1}{r^2}\frac{\partial v_\theta}{\partial \theta} - \frac{1}{r^2}\frac{\partial^2 v_r}{\partial \theta^2} - \frac{\partial^2 v_r}{\partial z^2}$$

With the help of the above expressions for $[\nabla(\nabla \cdot \mathbf{v})]_r$ and $[\nabla \times (\nabla \times \mathbf{v})]_r$, we get

$$(\nabla^2 \mathbf{v})_r \;=\; \underbrace{-\frac{v_r}{r^2} + \frac{1}{r}\frac{\partial v_r}{\partial r} + \frac{\partial^2 v_r}{\partial r^2} - \frac{1}{r^2}\frac{\partial v_\theta}{\partial \theta}}_{[\nabla(\nabla \cdot \mathbf{v})]_r} - \underbrace{\left(\frac{1}{r^2}\frac{\partial v_\theta}{\partial \theta} - \frac{1}{r^2}\frac{\partial^2 v_r}{\partial \theta^2} - \frac{\partial^2 v_r}{\partial z^2}\right)}_{[\nabla \times (\nabla \times \mathbf{v})]_r}$$

Substituting in Equation (5.28) on p. 153, the expressions for (i) $(\nabla^2 \mathbf{v})_r$ above, (ii) $(\mathbf{v} \cdot \nabla \mathbf{v})_r$ [see Equation (5.26), p. 153], and (iii) $(\nabla P)_r$ [see Equation (5.27), p. 153], we finally obtain

$$\frac{\partial v_r}{\partial t} = -v_r \frac{\partial v_r}{\partial r} - \frac{v_\theta}{r} \frac{\partial v_r}{\partial \theta} - v_z \frac{\partial v_r}{\partial z} + \frac{\mu}{\rho} \left(\frac{\partial^2 v_r}{\partial r^2} + \frac{1}{r} \frac{\partial v_r}{\partial r} + \frac{1}{r^2} \frac{\partial^2 v_r}{\partial \theta^2} \right.$$

$$\left. + \frac{\partial^2 v_r}{\partial z^2} - \frac{2}{r^2} \frac{\partial v_\theta}{\partial \theta} - \frac{v_r}{r^2} \right) + \frac{v_\theta^2}{r} - \frac{1}{\rho} \frac{\partial P}{\partial r} + g_r$$

❑

5.2 Transformation between Arbitrary Coordinate Systems

In an arbitrary coordinate system, the axes are not necessarily orthogonal, i.e., perpendicular to each other. In the absence of orthogonal axes, the results of Section 5.1 on p. 139 are not applicable.

A general approach to transform equations between arbitrary coordinate systems is to transform the velocity, and spatial derivatives based on the chain rule of differentiation. We present this approach to transform equations from one arbitrary coordinate system (y_1, y_2, y_2) to another arbitrary coordinate system (z_1, z_2, z_3).

5.2.1 Transformation of Velocity

The starting point is the set of relations

$$y_i = y_i(z_1, z_2, z_3) \qquad i = 1, 2, 3$$

between the two coordinate systems. From the chain rule of differentiation [Section 8.7.3, p. 319], the velocity component along the y_1-direction is given by

$$u_1 = \frac{\mathrm{d}y_1}{\mathrm{d}t} = \frac{\partial y_1}{\partial z_1}\bigg|_{\substack{z_2 \\ z_3}} \cdot \underbrace{\frac{\mathrm{d}z_1}{\mathrm{d}t}}_{v_1} + \frac{\partial y_1}{\partial z_2}\bigg|_{\substack{z_1 \\ z_3}} \cdot \underbrace{\frac{\mathrm{d}z_2}{\mathrm{d}t}}_{v_2} + \frac{\partial y_1}{\partial z_3}\bigg|_{\substack{z_1 \\ z_2}} \cdot \underbrace{\frac{\mathrm{d}z_3}{\mathrm{d}t}}_{v_3}$$

In general, the velocity components along the y_1-, y_2- and y_3-directions are given by

$$u_i = \sum_{j=1}^{3} \frac{\partial y_i}{\partial z_j}\bigg|_{\tilde{\mathbf{z}}_j} v_j ; \qquad i = 1, 2, 3$$

where $\tilde{\mathbf{z}}_j$ is the vector $\begin{bmatrix} z_1 & z_2 & z_3 \end{bmatrix}^\top$ without the element z_j. The above equations in matrix notation are given by

$$\mathbf{u} = \frac{\partial \mathbf{y}}{\partial \mathbf{z}} \mathbf{v} \qquad\qquad (5.29)$$

where (i) $\mathbf{u}$ and $\mathbf{v}$ are the velocity vectors, respectively, in the (y_1, y_2, y_2) and (z_1, z_2, z_3) coordinate systems, and (ii) $\partial \mathbf{y}/\partial \mathbf{z}$ is the Jacobian of $\mathbf{y}$ with respect to $\mathbf{z}$. With the Jacobian expressed solely in terms of $\mathbf{z}$, we can use the last equation to transform $\mathbf{u}$ to the (z_1, z_2, z_3) coordinate system.

The equation to transform $\mathbf{v}$ to the (y_1, y_2, y_2) coordinate system is similarly given by

$$\mathbf{v} = \frac{\partial \mathbf{z}}{\partial \mathbf{y}} \mathbf{u} \tag{5.30}$$

where $\partial \mathbf{z}/\partial \mathbf{y}$ is the Jacobian of $\mathbf{z}$ with respect to $\mathbf{y}$, and expressed solely in terms of $\mathbf{y}$.

5.2.2 Transformation of Spatial Derivatives

Let a quantity q be represented in the two arbitrary coordinate systems as

$$q = A(y_1, y_2, y_3) = B(z_1, z_2, z_3)$$

Then from the chain rule of differentiation [Section 8.7.3, p. 319], the partial derivative of q with respect to y_1 is

$$\left.\frac{\partial A}{\partial y_1}\right|_{\substack{y_2 \\ y_3}} = \left.\frac{\partial B}{\partial z_1}\right|_{\substack{z_2 \\ z_3}} \cdot \left.\frac{\partial z_1}{\partial y_1}\right|_{\substack{y_2 \\ y_3}} + \left.\frac{\partial B}{\partial z_2}\right|_{\substack{z_1 \\ z_3}} \cdot \left.\frac{\partial z_2}{\partial y_1}\right|_{\substack{y_2 \\ y_3}} + \left.\frac{\partial B}{\partial z_3}\right|_{\substack{z_1 \\ z_2}} \cdot \left.\frac{\partial z_3}{\partial y_1}\right|_{\substack{y_2 \\ y_3}}$$

In general, we can write

$$\underbrace{\left.\frac{\partial A}{\partial y_i}\right|_{\tilde{\mathbf{y}}_i}}_{\frac{\partial q}{\partial y_i}} = \sum_{j=1}^{3} \left.\frac{\partial z_j}{\partial y_i}\right|_{\tilde{\mathbf{y}}_i} \underbrace{\left.\frac{\partial B}{\partial z_j}\right|_{\tilde{\mathbf{z}}_j}}_{\frac{\partial q}{\partial z_i}} \; ; \qquad i = 1, 2, 3$$

where $\tilde{\mathbf{y}}_i$ is the vector $\begin{bmatrix} y_1 & y_2 & y_3 \end{bmatrix}^\top$ without the element y_i. The above set of equations can be written in matrix notation as

$$q_{\mathbf{y}} = \frac{\partial \mathbf{z}}{\partial \mathbf{y}}^\top q_{\mathbf{z}} \tag{5.31}$$

where $q_{\mathbf{y}}$ and $q_{\mathbf{z}}$ are the vectors of partial derivatives of q with respect to $\mathbf{y}$ and $\mathbf{z}$, and $\partial \mathbf{z}/\partial \mathbf{y}$ is the Jacobian of $\mathbf{z}$ with respect to $\mathbf{y}$. With the Jacobian expressed solely in terms of $\mathbf{z}$, we can use the above equation to transform $q_{\mathbf{y}}$ to the (z_1, z_2, z_3) coordinate system.

The equation to transform $q_{\mathbf{z}}$ to the (y_1, y_2, y_2) coordinate system is similarly given by

$$q_{\mathbf{z}} = \frac{\partial \mathbf{y}}{\partial \mathbf{z}}^\top q_{\mathbf{y}}$$

with $\partial \mathbf{y}/\partial \mathbf{z}$ expressed solely in terms of $\mathbf{y}$.

5.2.3 Correctness of Transformation Matrices

From Equation (5.29) on the previous page, and Equation (5.30) above, note that

$$\mathbf{u} = \frac{\partial \mathbf{y}}{\partial \mathbf{z}} \mathbf{v} = \frac{\partial \mathbf{y}}{\partial \mathbf{z}} \frac{\partial \mathbf{z}}{\partial \mathbf{y}} \mathbf{u} \quad \Rightarrow \quad \frac{\partial \mathbf{y}}{\partial \mathbf{z}} \frac{\partial \mathbf{z}}{\partial \mathbf{y}} = \mathbf{I} \quad \text{(identity matrix)}$$

This result should be used to verify the correctness of the Jacobian matrices involved in the transformation of velocity, and spatial derivatives.

Example 5.2.1

Obtain the equation of continuity for a polymer melt flowing through the helical channel of an extruder shown in Figure 5.5 below. The helical coordinates (z_1, z_2, z_3) are related to the

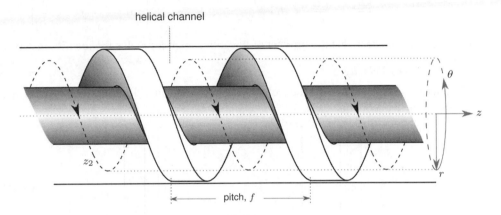

Figure 5.5   Flow of polymer melt in the helical channel of an extruder

cylindrical coordinates (r, θ, z) through the following equations:

$$z_1 = \frac{r}{R}, \qquad z_2 = -\theta a, \qquad z_3 = \frac{z}{R} + f\theta \qquad (5.32)$$

where $a = \sqrt{(r/R)^2 + f^2}$, and R and f are constants.[1]

Solution

For convenience, we will denote r, θ and z by y_1, y_2 and y_3, respectively. The above equations can be manipulated to get the inverse coordinate transformation, i.e., y_is expressed solely in terms of z_is as

$$y_1 = z_1 R, \qquad y_2 = -\frac{z_2}{a}, \qquad y_3 = \left(\frac{f z_2}{a} + z_3 \right) R \qquad (5.33)$$

where $a = \sqrt{z_1^2 + f^2}$.

The equation of continuity in cylindrical coordinate system [see Equation (2.66), p. 54] can be written as

$$\frac{\partial \rho}{\partial t} = -\frac{1}{y_1} \left[\frac{\partial}{\partial y_1} (\overbrace{\rho y_1 u_1}^{q_1}) + \frac{\partial}{\partial y_2} (\overbrace{\rho u_2}^{q_2}) \right] - \frac{\partial}{\partial y_3} (\overbrace{\rho u_3}^{q_3}) \qquad (5.34)$$

Based on Equations (5.32) and (5.33) on the previous page, the Jacobians needed to transform Equation (5.34) to helical coordinate system (z_1, z_2, z_3) are given by

$$
\frac{\partial \mathbf{y}}{\partial \mathbf{z}} =
\begin{bmatrix}
R & 0 & 0 \\
\dfrac{z_1 z_2}{a^3} & -\dfrac{1}{a} & 0 \\
-\dfrac{z_1 z_2 f R}{a^3} & \dfrac{f R}{a} & R
\end{bmatrix}
\quad \text{and} \quad
\frac{\partial \mathbf{z}}{\partial \mathbf{y}} =
\begin{bmatrix}
\dfrac{1}{R} & 0 & 0 \\
\dfrac{z_1 z_2}{a^2 R} & -a & 0 \\
0 & f & \dfrac{1}{R}
\end{bmatrix}
$$

Readers may like to check that the product of the above matrices yields the identity matrix.

Transformation of Velocity

From Equation (5.29) on p. 155, the velocity components in cylindrical coordinates are expressible in helical coordinates as

$$
\begin{bmatrix}
u_1 \\
u_2 \\
u_3
\end{bmatrix}
=
\underbrace{
\begin{bmatrix}
R & 0 & 0 \\
\dfrac{z_1 z_2}{a^3} & -\dfrac{1}{a} & 0 \\
-\dfrac{z_1 z_2 f R}{a^3} & \dfrac{f R}{a} & R
\end{bmatrix}
}_{\partial \mathbf{y}/\partial \mathbf{z}}
\begin{bmatrix}
v_1 \\
v_2 \\
v_3
\end{bmatrix}
\tag{5.35}
$$

Transformation of Spatial Derivatives

From Equation (5.31) on p. 156, the spatial derivatives in cylindrical coordinates are expressible in helical coordinates as

$$
\underbrace{
\begin{bmatrix}
\dfrac{\partial q}{\partial y_1} \\[2mm]
\dfrac{\partial q}{\partial y_2} \\[2mm]
\dfrac{\partial q}{\partial y_3}
\end{bmatrix}
}_{q_\mathbf{y}}
=
\underbrace{
\begin{bmatrix}
\dfrac{1}{R} & -\dfrac{z_1 z_2}{a^2 R} & 0 \\
0 & -a & f \\
0 & 0 & \dfrac{1}{R}
\end{bmatrix}
}_{\partial \mathbf{z}/\partial \mathbf{y}^\top}
\underbrace{
\begin{bmatrix}
\dfrac{\partial q}{\partial z_1} \\[2mm]
\dfrac{\partial q}{\partial z_2} \\[2mm]
\dfrac{\partial q}{\partial z_3}
\end{bmatrix}
}_{q_\mathbf{z}}
\tag{5.36}
$$

Transformation of Equation of Continuity

To transform the equation of continuity [Equation (5.34), previous page] to helical coordinate system, we need to express the following in that system – the y_is, u_is and the derivatives with respect to y_is. This is done with the help of Equation (5.33) on the previous page, and Equations (5.35) and (5.36) as explained next.

In Equation (5.34), we carry out the following steps:

1. We replace

    a. the coordinates y_is with z_is, using the inverse coordinate transformation, and

    b. the velocities u_is with v_is, using the velocity transformation.

2. We introduce temporary variables (q_is) for the arguments of spatial derivatives.

3. We repeat Step 1 for each q_i.

4. We express $\partial q_i / \partial y_j$s in terms of $\partial q_i / \partial z_j$s, using the derivative transformation, and expand derivatives as needed.

The incorporation of expressions of Step 4 into the equation of Step 2 transforms the equation of continuity to helical coordinate system (z_1, z_2, z_3).

The aforementioned steps are carried out below.

Steps 1 and 2
We use Equation (5.33) to introduce helical coordinates in Equation (5.34) on p. 157. In terms of q_is, the equation is

$$\frac{\partial \rho}{\partial t} = -\frac{1}{z_1 R}\left(\frac{\partial q_1}{\partial y_1} + \frac{\partial q_2}{\partial y_2}\right) - \frac{\partial q_3}{\partial y_3} \qquad (5.37)$$

Step 3
We expand the q_is, and repeat Step 1. Velocity transformation [see Equation (5.35), previous page] is applicable this time. The following expressions are obtained as a result:

$$q_1 = \rho \cdot z_1 R \cdot R v_1 = R^2 \rho z_1 v_1$$

$$q_2 = \rho u_2 = \rho\left(\frac{z_1 z_2 v_1}{a^3} - \frac{v_2}{a}\right)$$

$$q_3 = \rho u_3 = \rho\left(-\frac{z_1 z_2 f R}{a^3} v_1 + \frac{f R}{a} v_2 + R v_3\right)$$

Step 4
Using Equation (5.36) on the previous page for derivative transformation, and the expressions for q_is in Step 3, we get

$$\frac{\partial q_1}{\partial y_1} = \frac{1}{R}\frac{\partial q_1}{\partial z_1} - \frac{z_1 z_2}{a^2 R}\frac{\partial q_1}{\partial z_2} \qquad \text{where}$$

$$\frac{\partial q_1}{\partial z_1} = R^2\left[\rho\frac{\partial}{\partial z_1}(z_1 v_1) + z_1 v_1\frac{\partial \rho}{\partial z_1}\right] = R^2\left[\rho\left(z_1\frac{\partial v_1}{\partial z_1} + v_1\right) + z_1 v_1\frac{\partial \rho}{\partial z_1}\right]$$

$$\frac{\partial q_1}{\partial z_2} = R^2\left[\rho\frac{\partial}{\partial z_2}(z_1 v_1) + z_1 v_1\frac{\partial \rho}{\partial z_2}\right] = R^2\left[\rho z_1\frac{\partial v_1}{\partial z_2} + z_1 v_1\frac{\partial \rho}{\partial z_2}\right]$$

$$\frac{\partial q_2}{\partial y_2} = -a\frac{\partial q_2}{\partial z_2} + f\frac{\partial q_2}{\partial z_3} \qquad \text{where}$$

$$\frac{\partial q_2}{\partial z_2} = \rho\frac{\partial}{\partial z_2}\left(\underbrace{\frac{z_1 z_2}{a^3}}_{\equiv b}v_1 - \frac{v_2}{a}\right) + \left(\overbrace{bv_1 - \frac{v_2}{a}}^{q_2/\rho}\right)\frac{\partial \rho}{\partial z_2}$$

$$= \rho\left[\frac{z_1}{a^3}\left(z_2\frac{\partial v_1}{\partial z_2} + v_1\right) - \frac{1}{a}\frac{\partial v_2}{\partial z_2}\right] + \frac{q_2}{\rho}\frac{\partial \rho}{\partial z_2}$$

$$\frac{\partial q_2}{\partial z_3} = \rho\frac{\partial}{\partial z_3}\left(bv_1 - \frac{v_2}{a}\right) + \frac{q_2}{\rho}\frac{\partial \rho}{\partial z_3} = \rho\left(b\frac{\partial v_1}{\partial z_3} - \frac{1}{a}\frac{\partial v_2}{\partial z_3}\right) + \frac{q_2}{\rho}\frac{\partial \rho}{\partial z_3}$$

$$\frac{\partial q_3}{\partial y_3} = \frac{1}{R}\frac{\partial q_3}{\partial z_3} \qquad \text{where}$$

$$\frac{\partial q_3}{\partial z_3} = \rho\frac{\partial}{\partial z_3}\left(-bfRv_1 + \frac{fR}{a}v_2 + Rv_3\right) + \left(\overbrace{-bfRv_1 + \frac{fR}{a}v_2 + Rv_3}^{q_3/\rho}\right)\frac{\partial \rho}{\partial z_3}$$

$$= \rho\left(-bfR\frac{\partial v_1}{\partial z_3} + \frac{fR}{a}\frac{\partial v_2}{\partial z_3} + R\frac{\partial v_3}{\partial z_3}\right) + \frac{q_3}{\rho}\frac{\partial \rho}{\partial z_3}$$

The above set of equations, derived in Steps 1–4, represents the equation of continuity in helical coordinates. By gathering expressions for $\partial q_i/\partial y_j$s from this set, and substituting them in Equation (5.37) on the previous page, we can obtain the helical continuity equation. Doing so, however, would result in a messy form of the equation, and be prone to errors. Since a sophisticated differential equation such as this one will have to be integrated numerically, a better strategy is to order the equations obtained in Steps 1–4 such that when each equation is evaluated (for the left-hand term), the right-hand side terms are always known. Thus, equations ordered in this way can be sequentially evaluated beginning with the first equation.

Ordered Set of Transformed Equations

Arranged in the aforementioned order of evaluation, the set of equations representing the helical continuity equation is as follows:

$$\frac{\partial q_1}{\partial z_1} = R^2\left[\rho\left(z_1\frac{\partial v_1}{\partial z_1} + v_1\right) + z_1 v_1\frac{\partial \rho}{\partial z_1}\right]$$

$$\frac{\partial q_1}{\partial z_2} = R^2\left[\rho z_1\frac{\partial v_1}{\partial z_2} + z_1 v_1\frac{\partial \rho}{\partial z_2}\right]$$

$$\frac{\partial q_1}{\partial y_1} = \frac{1}{R}\frac{\partial q_1}{\partial z_1} - \frac{z_1 z_2}{a^2 R}\frac{\partial q_1}{\partial z_2}$$

$$b = \frac{z_1 z_2}{a^3}$$

$$q_2 = \rho \left(b v_1 - \frac{v_2}{a} \right)$$

$$\frac{\partial q_2}{\partial z_2} = \rho \left[\frac{z_1}{a^3} \left(z_2 \frac{\partial v_1}{\partial z_2} + v_1 \right) - \frac{1}{a} \frac{\partial v_2}{\partial z_2} \right] + \frac{q_2}{\rho} \frac{\partial \rho}{\partial z_2}$$

$$\frac{\partial q_2}{\partial z_3} = \rho \left(b \frac{\partial v_1}{\partial z_3} - \frac{1}{a} \frac{\partial v_2}{\partial z_3} \right) + \frac{q_2}{\rho} \frac{\partial \rho}{\partial z_3}$$

$$\frac{\partial q_2}{\partial y_2} = -a \frac{\partial q_2}{\partial z_2} + f \frac{\partial q_2}{\partial z_3}$$

$$q_3 = \rho \left(-bfR v_1 + \frac{fR}{a} v_2 + R v_3 \right)$$

$$\frac{\partial q_3}{\partial z_3} = \rho \left(-bfR \frac{\partial v_1}{\partial z_3} + \frac{fR}{a} \frac{\partial v_2}{\partial z_3} + R \frac{\partial v_3}{\partial z_3} \right) + \frac{q_3}{\rho} \frac{\partial \rho}{\partial z_3}$$

$$\frac{\partial q_3}{\partial y_3} = \frac{1}{R} \frac{\partial q_3}{\partial z_3}$$

$$\frac{\partial \rho}{\partial t} = -\frac{1}{z_1 R} \left(\frac{\partial q_1}{\partial y_1} + \frac{\partial q_2}{\partial y_2} \right) - \frac{\partial q_3}{\partial y_3} \tag{5.37}$$

In the ordered set of equations above, Equation (5.37) is evaluated in the end. Its evaluation yields the left-hand side derivative needed for the numerical integration of the continuity equation in helical coordinates.

❏

5.3 Laplace Transformation

It is an integral transformation defined as

$$F(s) = \underbrace{\int_0^\infty f(t) e^{-st} \, \mathrm{d}t}_{\equiv \mathcal{L}[f(t)]} \tag{5.38}$$

where (i) $F(s)$ is called the **Laplace transform** of the function $f(t)$ in the t-domain, (ii) s is the variable of the Laplace transform domain, and the inverse of time, and (iii) $\mathcal{L}$ is the Laplace transform operator.

The motivation behind Laplace transformation is to have a simpler solution of equations in the s-domain than that in the original t-domain. Once a solution is obtained in the s-domain, the **inverse Laplace transform** provides the actual solution in the t-domain. This objective is indeed achievable for a large class of linear differential equations used in process control.

Laplace transformation of linear differential equations results in algebraic equations which are easier to solve. Moreover, the algebraic equations are made of *transfer functions*, which help characterize the solution, i.e., the response of the system modeled by the differential equations. The inverse Laplace transformation of the solution of algebraic equations yields the actual solution.

5.3.1   Examples

We will derive the Laplace transforms of some simple functions using the definition given by Equation (5.38) on the previous page.

Unit Step Function

This function is defined as

$$u(t) = \begin{cases} 0, & t < 0 \\ 1, & t \geq 0 \end{cases}$$

The Laplace transform of $u(t)$ is given by

$$\mathcal{L}[u(t)] = \int_0^\infty u(t)e^{-st}\, dt = \left[\frac{-e^{-st}}{s}\right]_0^\infty = \frac{1}{s}$$

Pulse Function

This function is defined as

$$f(t) = \begin{cases} 0, & t < 0, \quad t > T \\ H, & 0 \leq t \leq T \end{cases}$$

The Laplace transform of $f(t)$ is given by

$$\mathcal{L}[f(t)] = \int_0^\infty f(t)e^{-st}\, dt = \int_0^T He^{-st} dt = \left[\frac{-He^{-st}}{s}\right]_0^T = \frac{H}{s}\left[1 - e^{-sT}\right]$$

Unit Pulse Function

Also known as the Dirac delta function, this function is defined as

$$\delta(t) = \lim_{T \to 0} \begin{cases} 0, & t < 0, \quad t > T \\ \dfrac{1}{T}, & 0 \leq t \leq T \end{cases}$$

The Laplace transform of $\delta(t)$ is given by

$$\mathcal{L}[\delta(t)] = \int_0^\infty \delta(t)e^{-st}\, dt = \lim_{T \to 0}\int_0^T \frac{1}{T}e^{-st} dt = \lim_{T \to 0}\left[\frac{-e^{-st}}{sT}\right]_0^T$$

$$= \lim_{T \to 0}\frac{1 - e^{-sT}}{sT} = 1$$

Sine Wave

From Euler's formulas [see p. 327], a sine wave of unit amplitude, and frequency ω is given by

$$\sin(\omega t) = \frac{e^{i\omega t} - e^{-i\omega t}}{2i}$$

The Laplace transform of $\sin(\omega t)$ is given by

$$\mathcal{L}[\sin(\omega t)] = \int_0^\infty \frac{e^{i\omega t} - e^{-i\omega t}}{2i} e^{-st} \, dt \quad - \quad \frac{1}{2i}\left[-\frac{e^{-(s-i\omega)t}}{s - i\omega} + \frac{e^{-(s+i\omega)t}}{s + i\omega} \right]_0^\infty$$

$$= \frac{1}{2i}\left[-\frac{0-1}{s - i\omega} + \frac{0-1}{s + i\omega} \right] = \frac{\omega}{s^2 + \omega^2}$$

An Exponential Function

Given an exponential function, $t^n e^{-at}$, its Laplace transform is obtained using its definition followed by integration by parts [see p. 327] as follows.

$$\mathcal{L}[t^n e^{-at}] = \int_0^\infty t^n e^{-at} e^{-st} \, dt = \left[t^n \int e^{-(s+a)t} \, dt \right]_0^\infty - \int_0^\infty \left[n t^{n-1} \int e^{-(s+a)t} \, dt \right] dt$$

$$= -\left[\frac{t^n e^{-(s+a)t}}{s + a} \right]_0^\infty + \frac{n}{s + a} \int_0^\infty t^{n-1} e^{-(s+a)t} \, dt$$

$$= -\left[\frac{t^n e^{-(s+a)t}}{s + a} \right]_0^\infty + \frac{n}{s + a} \int_0^\infty t^{n-1} e^{-at} e^{-st} \, dt$$

Applying L'Hôpital's rule [see p. 326] successively to the first term on the right-hand side of the last equation, we get

$$\left[\frac{t^n e^{-(s+a)t}}{s + a} \right]_\infty^\infty = \left[\frac{n t^{n-1}}{(s + a)^2 e^{(s+a)t}} \right]_\infty^\infty = \left[\frac{n(n - 1) t^{n-2}}{(s + a)^3 e^{(s+a)t}} \right]_\infty^\infty = \cdots$$

$$= \left[\frac{n!}{(s + a)^{n+1} e^{(s+a)t}} \right]_\infty^\infty = 0$$

Therefore,

$$\mathcal{L}[t^n e^{-at}] = \left(\frac{n}{s + a} \right) \mathcal{L}[t^{n-1} e^{-at}] = \left(\frac{n}{s + a} \right)\left(\frac{n - 1}{s + a} \right) \mathcal{L}[t^{n-2} e^{-at}] = \cdots$$

$$= \frac{n!}{(s + a)^n} \mathcal{L}[e^{-at}] = \frac{n!}{(s + a)^{n+1}}$$

Table 5.1 on the next page lists Laplace transforms of some common functions.

Table 5.1 Laplace transforms of some common functions

$f(t)$	$F(s)$		$f(t)$	$F(s)$
$\delta(t)$	1		$u(t)$	$\dfrac{1}{s}$
t^n	$\dfrac{n!}{s^{n+1}}$	$(n = 1, 2, \dots)$	$t^n e^{-at}$	$\dfrac{n!}{(s+a)^{n+1}}$
$\sin(\omega t)$	$\dfrac{\omega}{s^2 + \omega^2}$		$\cos(\omega t)$	$\dfrac{s}{s^2 + \omega^2}$
$e^{-at}\sin(\omega t)$	$\dfrac{\omega}{(s+a)^2 + \omega^2}$		$e^{-at}\cos(\omega t)$	$\dfrac{s+a}{(s+a)^2 + \omega^2}$
$\sinh(\omega t)$	$\dfrac{\omega}{s^2 - \omega^2}$		$\cosh(\omega t)$	$\dfrac{s}{s^2 - \omega^2}$

5.3.2   *Properties of Laplace Transforms*

Properties of Laplace transforms are as follows.

Linearity

Given two constants, a and b,

$$\mathcal{L}[af(t) + bg(t)] \;=\; aF(s) + bG(s)$$

This property can be proved by using the definition of the Laplace transform.

Real Differentiation Theorem

According to this theorem, the Laplace transform of the derivative of a function is an algebraic expression given by

$$\mathcal{L}\left[\frac{\mathrm{d}f(t)}{\mathrm{d}t}\right] \;=\; sF(s) - f(0)$$

Proof

Using the definition of the Laplace transform, and integration by parts, we get

$$\mathcal{L}\left[\frac{\mathrm{d}f(t)}{\mathrm{d}t}\right] = \int_0^\infty \frac{\mathrm{d}f(t)}{\mathrm{d}t} e^{-st}\,\mathrm{d}t = \left[e^{-st}f(t)\right]_0^\infty - \int_0^\infty -se^{-st}f(t)\,\mathrm{d}t$$

$$= \left[0 - f(0)\right] + s\int_0^\infty e^{-st}f(t)\,\mathrm{d}t = sF(s) - f(0)$$

Similarly, the Laplace transform of the second derivative of a function is given by

$$\mathcal{L}\left[\frac{\mathrm{d}^2 f(t)}{\mathrm{d}t^2}\right] = \mathcal{L}\left[\frac{\mathrm{d}}{\mathrm{d}t}\left(\frac{\mathrm{d}f}{\mathrm{d}t}\right)\right] = s\mathcal{L}\left[\frac{\mathrm{d}f}{\mathrm{d}t}\right] - \left[\frac{\mathrm{d}f}{\mathrm{d}t}\right]_{t=0}$$

$$= s\left[sF(s) - f(0)\right] - \left[\frac{\mathrm{d}f}{\mathrm{d}t}\right]_{t=0} = s^2 F(s) - sf(0) - \left[\frac{\mathrm{d}f}{\mathrm{d}t}\right]_{t=0}$$

In general, the Laplace transform of the nth derivative of a function is given by

$$\mathcal{L}\left[\frac{\mathrm{d}^n f(t)}{\mathrm{d}t^n}\right] = s^n F(s) - s^{n-1}f(0) - s^{n-2}\left[\frac{\mathrm{d}f}{\mathrm{d}t}\right]_{t=0} - s^{n-3}\left[\frac{\mathrm{d}^2 f}{\mathrm{d}t^2}\right]_{t=0}$$

$$- \cdots - \left[\frac{\mathrm{d}^{n-1}f}{\mathrm{d}t^{n-1}}\right]_{t=0} \tag{5.39}$$

Zero-Valued, Initial Steady State Condition

If at $t = 0$ the function is zero as well as at steady state then

$$\mathcal{L}\left[\frac{\mathrm{d}^n f(t)}{\mathrm{d}t^n}\right] = s^n F(s)$$

The above transform does not involve any initial value of the function or its derivatives with respect to t.

Real Integration Theorem

According to this theorem, the Laplace transform of the integral of a function is an algebraic expression given by

$$\mathcal{L}\left[\int_0^t f(t)\mathrm{d}t\right] = \frac{1}{s}F(s)$$

Proof

Using the definition of the Laplace transform, and integration by parts, we get

$$\mathcal{L}\left[\int_0^t f(t)\,dt\right] = \int_0^\infty \left[\int_0^t f(t)\,dt \times e^{-st}\right] dt$$

$$= \left[\int_0^t f(t)\,dt \int e^{-st}\,dt\right]_0^\infty - \int_0^\infty \left[f(t)\int e^{-st}\,dt\right] dt$$

$$= -\left[\frac{e^{-st}}{s}\int_0^t f(t)\,dt\right]_0^\infty + \frac{1}{s}\int_0^\infty f(t)e^{-st}\,dt$$

$$= \underset{\lim t\to 0}{-0} + \frac{1}{s}\int_0^t f(t)\,dt + \frac{1}{s}\mathcal{L}[f(t)] = \frac{F(s)}{s}$$

Similarly, the Laplace transform of the double integral of the function is given by

$$\mathcal{L}\left[\int_0^t\int_0^t f(t)\,dt\,dt\right] = \int_0^\infty \left[\int_0^t\int_0^t f(t)\,dt\,dt \times e^{-st}\right] dt$$

$$= \left[\int_0^t\int_0^t f(t)\,dt\,dt \int e^{-st}\,dt\right]_0^\infty - \int_0^\infty \left[\int_0^t f(t)\,dt \int e^{-st}\,dt\right] dt$$

$$= -\left[\frac{e^{-st}}{s}\int_0^t\int_0^t f(t)\,dt\,dt\right]_0^\infty + \frac{1}{s}\int_0^\infty \left[\int_0^t f(t)\,dt\right]e^{-st}\,dt$$

$$= 0 + \frac{1}{s}\mathcal{L}\left[\int_0^t f(t)\,dt\right] = \frac{F(s)}{s^2}$$

In general, the Laplace transform of the nth integral of the function is given by

$$\mathcal{L}\left[\int_0^t\int_0^t\cdots\int_0^t f(t)\cdots dt^n\right] = \frac{F(s)}{s^n}$$

Real Translation Theorem

According to this theorem, the Laplace transform of a function with a time delay t_d is given by

$$\mathcal{L}[f(t - t_d)] = e^{-st_d}F(s) \tag{5.40}$$

Proof

Applying the definition of the Laplace transform, and introducing $\tau \equiv t - t_d$, we get

$$\mathcal{L}[f(t - t_d)] = \int_0^\infty f(t - t_d)e^{-st}\, dt = \int_{-t_d}^\infty f(\tau)e^{-s(t_d + \tau)}\, d\tau$$

Since $f(\tau) = 0$ for $\tau < 0$ (i.e., $t < t_d$), and t_d is a constant, the above equation simplifies to

$$\mathcal{L}[f(t - t_d)] = e^{-st_d} \int_0^\infty f(\tau)e^{-s\tau}\, d\tau = e^{-st_d} F(s)$$

Initial Value Theorem

This theorem relates the initial value of a function to an s-domain expression as

$$\lim_{t \to 0} f(t) = \lim_{s \to \infty} sF(s)$$

Proof

$$\lim_{s \to \infty} sF(s) = \lim_{s \to \infty} s \int_0^\infty f(t)e^{-st}\, dt = \lim_{s \to \infty} \left\{ \left[sf(t)\frac{e^{-st}}{-s} \right]_0^\infty - s\int_0^\infty \frac{\mathrm{d}f}{\mathrm{d}t}\frac{e^{-st}}{-s}\, dt \right\}$$

$$= \lim_{t \to 0} f(t) + \underbrace{\int_0^\infty \frac{\mathrm{d}f}{\mathrm{d}t}\left(\lim_{s \to \infty} e^{-st} \right) dt}_{= 0} = \lim_{t \to 0} f(t)$$

Final Value Theorem

This theorem relates the final value of a function to an s-domain expression as

$$\lim_{t \to \infty} f(t) = \lim_{s \to 0} sF(s)$$

Proof

$$\lim_{s \to 0} sF(s) = \lim_{s \to 0} s \int_0^\infty f(t)e^{-st}\, dt = \lim_{s \to 0} \left\{ \left[sf(t)\frac{e^{-st}}{-s} \right]_0^\infty - s\int_0^\infty \frac{\mathrm{d}f}{\mathrm{d}t}\frac{e^{-st}}{-s}\, dt \right\}$$

$$= f(0) + \int_0^\infty \frac{\mathrm{d}f}{\mathrm{d}t}\left(\lim_{s \to 0} e^{-st} \right) dt = f(0) + \int_0^\infty \frac{\mathrm{d}f}{\mathrm{d}t}\, dt = f(0) + \left[f(t) \right]_0^\infty$$

$$= \lim_{t \to \infty} f(t)$$

5.3.3 Solution of Linear Differential Equations

Linear differential equations are made of additive terms that are linear functions of the dependent variables, and their derivatives with respect to the independent variables. Such an equation gets converted to an algebraic equation upon taking the Laplace transform. This algebraic equation yields the s-domain solution. With the help of the inverse Laplace transform, this solution is converted to the t-domain solution.

Example 5.3.1

Obtain the Laplace transform of the following differential equation:

$$a\frac{d^2y}{dt^2} + b\frac{dy}{dt} + cy = dx, \qquad y(0) = \left[\frac{dy}{dt}\right]_{t=0} = 0$$

where a, b, c and d are constants.

Solution

Taking Laplace transform on both sides of the above differential equation, we get

$$\mathcal{L}\left[a\frac{d^2y}{dt^2} + b\frac{dy}{dt} + cy\right] = \mathcal{L}[dx]$$

Because of the linearity property of Laplace transforms, the above equation becomes

$$a\mathcal{L}\left[\frac{d^2y}{dt^2}\right] + b\mathcal{L}\left[\frac{dy}{dt}\right] + c\mathcal{L}[y] = d\mathcal{L}[x]$$

Using the real differentiation theorem [Equation (5.39), p. 165], we get

$$a\left\{s^2Y(s) - sy(0) - \left[\frac{dy}{dt}\right]_{t=0}\right\} + b[sY(s) - y(0)] + cY(s) = dX(s)$$

where $X(s)$ and $Y(s)$ denote, respectively, the Laplace transforms of $x(t)$ and $y(t)$. Note that the above equation is algebraic. We apply the initial conditions, and rearrange the result to obtain the s-domain solution, i.e.,

$$Y(s) = \left(\frac{d}{as^2 + bs + c}\right) \times X(s) \tag{5.41}$$

$$\underbrace{}_{\text{output variable}} \qquad \underbrace{\phantom{\frac{d}{as^2+bs+c}}}_{\text{transfer function}} \qquad \underbrace{}_{\text{input variable}}$$

In the above equation, Y and X are the s-domain counterparts of y (system output), and x (input function) in the t-domain. The coefficient of X is called **transfer function**.

❑

In general, the Laplace transform of a linear differential equation results in an explicit s-domain solution. It requires to be transferred to the original t-domain of interest. Hence, in

Example 5.3.1 on the previous page, if x is the unit step function $u(t)$, which signifies a unit value for $t \geq 0$, then $X(s) = 1/s$ and

$$Y(s) = \frac{d}{as^2 + bs + c} \times \frac{1}{s}$$

Taking the inverse Laplace transform of the above equation, we get

$$y(t) = \mathcal{L}^{-1}[Y(s)] = \mathcal{L}^{-1}\left[\frac{d}{as^2 + bs + c} \times \frac{1}{s}\right]$$

where y is the desired solution of the differential equation in the t-domain. The evaluation of the inverse Laplace transform requires its argument, i.e., $Y(s)$, expressed in terms of partial fractions. Their inverse Laplace transforms are readily obtained from previous results [see Table 5.1, p. 164].

Solution Procedure

The following steps summarize the procedure to solve a linear differential equation using Laplace transforms:

1. Take the Laplace transform of the differential equation.

2. Get the solution in the s-domain.

3. Express the solution in terms of partial fractions.

4. Take the inverse Laplace transform ($\mathcal{L}^{-1}$) of the partial fractions to obtain the final solution in the t-domain.

Obtaining Partial Fractions and Inverse Laplace Transforms

The s-domain solution $Y(s)$ is of the form $A(s)/B(s)$, where $A(s)$ and $B(s)$ are polynomials of s. Based on the roots ($r_1, r_2, \ldots, r_n$) of $B(s)$, $Y(s)$ is expressed as the sum of partial fractions comprising factors $(s - r_1), (s - r_2), \ldots, (s - r_n)$ in their denominators. The procedure to obtain partial fractions, and the final solution depends on whether or not the roots repeat.

Non-Repeating Roots

When the roots of $B(s)$ are distinct, $Y(s)$ is expressible as

$$Y(s) \equiv \frac{A_1}{s - r_1} + \frac{A_2}{s - r_2} + \cdots \frac{A_n}{s - r_n} \qquad (5.42)$$

where the coefficients A_is are given by

$$A_i = \lim_{s \to r_i} (s - r_i)Y(s), \quad i = 1, \ldots, n \qquad (5.43)$$

Utilizing Table 5.1 on p. 164, the inverse Laplace transform of Equation (5.42) above yields

$$y(t) = A_1 e^{r_1 t} + A_2 e^{r_2 t} + \cdots A_n e^{r_n t}$$

which is the desired t-domain solution of the differential equation.

Repeating Roots

If m out of the n roots of $B(s)$ repeat (i.e., have the same value) then

$$Y(s) \equiv \sum_{i=1}^{n-m} \frac{A_i}{s - r_i} + \frac{\tilde{A}_1}{(s - r)^m} + \frac{\tilde{A}_2}{(s - r)^{m-1}} + \cdots + \frac{\tilde{A}_m}{s - r} \qquad (5.44)$$

where r is the value of the repeating roots, and the coefficients $\tilde{A}_i$s are given by

$$\tilde{A}_1 = \left. (s - r)^m Y(s) \right|_{\lim s \to r} \quad \text{and} \quad \tilde{A}_i = \left. \frac{1}{(i-1)!} \frac{d^{(i-1)}}{ds^{(i-1)}} \left[(s-r)^m Y(s) \right] \right|_{\lim s \to r}$$

$$i = 2, 3, \ldots, m \qquad (5.45)$$

Utilizing Table 5.1 on p. 164, the inverse Laplace transform of Equation (5.44) above yields

$$y(t) = \sum_{i=1}^{n-m} A_i e^{r_i t} + \left[\frac{\tilde{A}_1 t^{m-1}}{(m-1)!} + \frac{\tilde{A}_2 t^{m-2}}{(m-2)!} + \cdots + \tilde{A}_m \right] e^{rt} \qquad (5.46)$$

Example 5.3.2

Using Laplace transforms, solve the differential equation of Example 5.3.1 on p. 168 for $a = 1$, $b = 3$, $c = 2$, $d = 1$ and $x = 2$.

Solution

For $x = 2$, $X(s) = 2/s$. Using Equation (5.41) on p. 168, and substituting for $X(s)$, we get

$$Y(s) = \frac{2d}{s\underbrace{(as^2 + bs + c)}_{B(s)}} = \frac{2}{s(s^2 + 3s + 2)}$$

where the roots of $B(s)$ are $r_1 = 0$, $r_2 = -1$ and $r_3 = -2$, which are all distinct. Using Equation (5.42) on the previous page, we get

$$Y(s) = \frac{2}{s(s+1)(s+2)} \equiv \frac{A_1}{s} + \frac{A_2}{s+1} + \frac{A_3}{s+2}$$

where, from Equation (5.43),

$$A_1 = \left. \frac{2}{(s+1)(s+2)} \right|_{\lim s \to 0} = 1$$

$$A_2 = \left. \frac{2}{s(s+2)} \right|_{\lim s \to -1} = -2$$

$$A_3 = \left. \frac{2}{s(s+1)} \right|_{\lim s \to -2} = 1$$

The solution $y(t)$ is given by the inverse Laplace transform of $Y(s)$. With the help of Table 5.1 on p. 164, we obtain

$$y(t) = \mathcal{L}^{-1}[Y(s)] = u(t) - 2e^{-t} + 2e^{-2t}$$

❑

Example 5.3.3

Using Laplace transforms, solve the differential equation of Example 5.3.1 on p. 168 for $a = 1$, $b = 4$, $c = 4$ and $d = 1$.

Solution

Here the roots of $B(s)$ are $r_1 = 0$ and two repeating roots $r_2 = r_3 = -2$. From Equation (5.44) on the previous page with $m = 2$,

$$Y(s) = \frac{2}{s(s+2)^2} \equiv \frac{A_1}{s} + \frac{\tilde{A}_1}{(s+2)^2} + \frac{\tilde{A}_2}{s+2}$$

Using Equation (5.43) on p. 169 for the non-repeating root r_1,

$$A_1 = \left. \frac{2}{(s+2)^2} \right|_{\lim s \to 0} = \frac{1}{2}$$

Using Equation (5.45) on the previous page for the repeating roots r_1 and r_2,

$$\tilde{A}_1 = \left. \frac{2}{s} \right|_{\lim s \to -2} = -1 \quad \text{and}$$

$$\tilde{A}_2 = \left. \frac{1}{1!} \frac{d}{ds}\left(\frac{2}{s}\right) \right|_{\lim s \to -2} = \left. -\frac{2}{s^2} \right|_{\lim s \to -2} = -\frac{1}{2}$$

The solution $y(t)$ is given by the inverse Laplace transform of $Y(s)$. With the help of Table 5.1, we obtain

$$y(t) = \mathcal{L}^{-1}[Y(s)] = \frac{u(t)}{2} - \left(t + \frac{1}{2}\right)e^{-2t}$$

❑

Example 5.3.4

Using Laplace transforms, solve the differential equation of Example 5.3.1 on p. 168 for $a = 1$, $b = -2$, $c = 5$ and $d = 20$.

Solution

Here the roots of $B(s)$ are $r_1 = 0$, $r_2 = -1 + 2i$ and $r_3 = -1 - 2i$, which are all distinct. The last two roots are complex. From Equation (5.42) on p. 169,

$$Y(s) = \frac{20}{s(s+1-2i)(s+1+2i)} \equiv \frac{A_1}{s} + \frac{A_2}{s+1-2i} + \frac{A_3}{s+1+2i}$$

Using Equation (5.43) on p. 169,

$$A_1 = \frac{20}{(s+1-2i)(s+1+2i)} \bigg|_{\lim s \to 0} = 4$$

$$A_2 = \frac{20}{s(s+1+2i)} \bigg|_{\lim s \to -1+2i} = -2+i$$

$$A_3 = \frac{20}{s(s+1-2i)} \bigg|_{\lim s \to -1-2i} = -2-i$$

Thus, the solution is given by the inverse Laplace transform of $Y(s)$, i.e.,

$$y(t) = \mathcal{L}^{-1}[Y(s)] = 4u(t) + (-2+i)e^{(-1+2i)t} + (-2-i)e^{(-1-2i)t}$$

The above solution has exponential function with complex powers, which is related to sinusoidal functions through Euler's formulas [see p. 327]. To express the solution in terms of a sine wave, we rearrange the solution, and use the formulas as follows:

$$y(t) = 4u(t) - 2e^{-t}e^{2it} + ie^{-t}e^{2it} - 2e^{-t}e^{-2it} - ie^{-t}e^{-2it}$$

$$= 4u(t) - 2e^{-t}\underbrace{(e^{2it} + e^{-2it})}_{2\cos(2t)} + ie^{-t}\underbrace{(e^{2it} - e^{-2it})}_{2i\sin(2t)}$$

$$= 4u(t) - 2e^{-t}\left[2\cos(2t) + \sin(2t)\right]$$

In the last equation, dividing and multiplying the last two terms by $\sqrt{2^2+1^2}$, we get

$$y(t) = 4u(t) - 2e^{-t}\left[\underbrace{\frac{2}{\sqrt{2^2+1^2}}}_{\sin\theta}\cos(2t) + \underbrace{\frac{1}{\sqrt{2^2+1^2}}}_{\cos\theta}\sin(2t)\right]\sqrt{2^2+1^2}$$

where $\theta = \tan^{-1}(2/1) = 1.1072$ rad or $63.44°$. Utilizing the trigonometric identity

$$\sin(p+q) = \sin p \cos q + \cos p \sin q$$

we finally obtain

$$y(t) = 4u(t) - 2\sqrt{5}e^{-t}\sin(2t+\theta)$$

Thus, if t is in s, the solution has a sinusoidal term with angular frequency $\omega = 2\text{ s}^{-1}$, and time lag $\theta = 63.44°$.

❑

Transformation to Deviation Variables

So far, we have solved linear differential equations with zero-valued, initial steady state conditions, which simplified the involved Laplace transforms. For linear differential equations with non-zero initial conditions, the same simplification can be achieved by transforming the dependent variables to deviation variables.

Given a dependent variable $z(t)$ of a differential equation, a **deviation variable** is defined as the difference of $z(t)$ from its initial value, i.e.,

$$\check{z}(t) \equiv z(t) - z(0) \tag{5.47}$$

From the above definition,

$$\check{z}(0) = 0 \quad \text{and} \quad \frac{d^k \check{z}}{dt^k} = \frac{d^k z}{dt^k}; \qquad k = 1, 2, \ldots$$

Thus, the initial value of a deviation variable is zero. Moreover, its derivative is the same as that of the dependent variable. We utilize these results in a general linear differential equation

$$a_n \frac{d^n y(t)}{dt^n} + a_{n-1} \frac{d^{n-1} y(t)}{dt^{n-1}} + \cdots + a_0 y(t) =$$

$$b_m \frac{d^m x(t)}{dt^m} + b_{m-1} \frac{d^{m-1} x(t)}{dt^{m-1}} + \cdots + b_0 x(t) + c \tag{5.48}$$

where $n \geq m$ for linear differential equations describing real-world processes.

Let $\check{x}(t)$ and $\check{y}(t)$ be the deviation variables, respectively, for $x(t)$ and $y(t)$. Then in the above differential equation, the substitution of (i) $\check{x}(t) + x(0)$ for x, (ii) $\check{y}(t) + y(0)$ for y, (iii) the derivatives of $\check{x}$ and $\check{y}$ for those of x and y, respectively, and (iv) the result obtained from the differential equation at the initial steady state, i.e.,

$$a_0 y(0) = b_0 x(0) + c$$

yields

$$a_n \frac{d^n \check{y}(t)}{dt^n} + a_{n-1} \frac{d^{n-1} \check{y}(t)}{dt^{n-1}} + \cdots + a_0 \check{y}(t) =$$

$$b_m \frac{d^m \check{x}(t)}{dt^m} + b_{m-1} \frac{d^{m-1} \check{x}(t)}{dt^{m-1}} + \cdots + b_0 \check{x}(t)$$

Note that the initial condition of the above differential equation is $\check{y}(0) = \check{x}(0) = 0$. The Laplace transform of the equation yields

$$\check{Y}(s) = \left(\frac{b_m s^m + b_{m-1} s^{m-1} + \cdots + b_0}{a_n s^n + a_{n-1} s^{n-1} + \cdots + a_0} \right) \check{X}(s) \equiv \frac{A(s)}{B(s)}$$

From the above equation, $\check{y}(t)$ is obtainable as usual from Equations (5.42)–(5.46) on p. 169–170. Then from the definition of $\check{y}(t)$, the solution of Equation (5.48) above is given by

$$y(t) = \check{y}(t) + y(0)$$

Handling Time Delays

Process models can involve functions with time delays. A simple example is liquid flow rate at a tank inlet. Any change in the flow rate appears after a certain time duration (delay) required by the liquid to travel the distance between the pump (which varies the flow rate), and the tank. If the distance is large then the delay might be large enough to be taken into account.

As illustrated in Figure 5.6 below, a function y with a time delay t_d is expressible as $y(t - t_d)$ by shifting the t-axis such that initial function value (y_i) is at the zero shifted time, i.e., $(t - t_d)$.

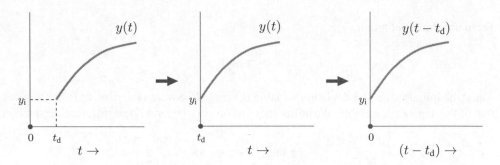

Figure 5.6 Representation of a function $y(t)$ with a time lag t_d as $y(t - t_d)$

Laplace transformation of a function with time delay involves the application of the real translation theorem [Equation (5.40), p. 166], as shown in the following example.

Example 5.3.5

Find the Laplace transform of the function

$$y(t) \;=\; u(t-5)\left[1 - \exp\left(-\frac{t-5}{7}\right)\right]$$

Solution

The function has a delay of 5 units of time. Applying the Laplace transformation, we get

$$Y(s) \;=\; \mathcal{L}\left[\underbrace{u(t-5)\left(1 - e^{-\frac{t-5}{7}}\right)}_{\equiv\, z(t-5)}\right] \;=\; \mathcal{L}[z(t-5)]$$

Then according to the real translation theorem

$$Y(s) \;=\; e^{-5s}\mathcal{L}[z(t)] \;=\; e^{-5s}\mathcal{L}\left[u(t)\left(1 - e^{-\frac{t}{7}}\right)\right] \;=\; e^{-5s}\left\{\mathcal{L}[u(t)] - \mathcal{L}\left[u(t)e^{-\frac{t}{7}}\right]\right\}$$

$$=\; e^{-5s}\left\{\frac{1}{s} - \frac{1}{s+\frac{1}{7}}\right\} \;=\; \frac{e^{-5s}}{s(7s+1)}$$

❏

Solving Linear Differential Equations with Time Delays

When using Laplace transforms to solve such an equation, the real translation theorem is applied as shown in the last example. This results in a multiplicative exponential delay term, which is retained until the inverse Laplace transformation is carried out. The real translation theorem is applied again to get the solution. The following example illustrates this procedure.

Example 5.3.6

Using Laplace transforms, solve the following differential equation:

$$\frac{dy(t)}{dt} + 3y(t) = u(t-2), \quad y(0) = 0$$

Solution

Applying the Laplace transformation to the above differential equation, we get

$$sY(s) + 3Y(s) = e^{-2s}\frac{1}{s}$$

Rearranging the above equation yields

$$Y(s) = e^{-2s}\left[\frac{1}{s(s+3)}\right] = e^{-2s}\left[\frac{A_1}{s} + \frac{A_2}{s+3}\right]$$

In the above equation, we have set the delay term e^{-2s} aside, and expressed the remaining terms in the square brackets as partial fractions. Using Equation (5.43) on p. 169, we obtain

$$A_1 = \underset{\lim s \to 0}{\frac{1}{s+3}} = \frac{1}{3} \quad \text{and} \quad A_2 = \underset{\lim s \to -3}{\frac{1}{s}} = -\frac{1}{3}$$

Thus, we have

$$Y(s) = e^{-2s}\underbrace{\left[\frac{1}{3s} - \frac{1}{3(s+3)}\right]}_{\equiv Y_1(s)} = e^{-2s}Y_1(s) \tag{5.49}$$

Now the inverse Laplace transform of $Y_1(s)$ as indicated above is

$$y_1(t) = \frac{1}{3}u(t) - \frac{1}{3}e^{-3t} = \frac{u(t)}{3}\left[1 - e^{-3t}\right]$$

Since $Y(s) = e^{-2s}Y_1(s)$ from Equation (5.49) above, the inverse Laplace transform of $Y(s)$ is the solution

$$y(t) = \underbrace{\mathcal{L}^{-1}\left[e^{-2s}Y_1(s)\right]}_{\substack{= y_1(t-2) \\ \text{from real translation} \\ \text{theorem}}} = \frac{u(t-2)}{3}\left[1 - e^{-3(t-2)}\right]$$

Thus, because of the delayed input function $u(t-2)$, the solution $y(t)$ is delayed accordingly.

❏

Simultaneous Differential Equations

Laplace transformation can be used to obtain the solution of simultaneous linear differential equations. Essentially, the transformed equations are linear and algebraic. They are manipulated to obtain the expression in terms of partial fractions for each dependent variable in the Laplace domain. Then the inverse Laplace transforms of the expressions yield the overall solution in the time domain. The following example illustrates this procedure.

Example 5.3.7

A CSTR carries out a reversible reaction

$$A \underset{k_1}{\overset{k_0}{\rightleftharpoons}} B$$

where k_0 and k_1 are the forward and backward reaction rate coefficients. The concentrations of species A and B, respectively, c_0 and c_1 in the reactor are given by

$$\frac{dc_0}{dt} = f(\tilde{c}_0 - c_0) - k_0 c_0 + k_1 c_1, \qquad c_0(0) = \bar{c}_0$$

$$\frac{dc_1}{dt} = f(\tilde{c}_1 - c_1) - k_1 c_1 + k_0 c_0, \qquad c_1(0) = \bar{c}_1$$

where f is the volumetric flow rate per unit volume of the reaction mixture, and $\tilde{c}_0$ and $\tilde{c}_1$ are, respectively, the feed concentrations of A and B. Using Laplace transforms, find $c_0(t)$ for the parameters given in Table 5.2 below.

Table 5.2 Values of parameters for the CSTR of Example 5.3.7

parameter	value	parameter	value
f	$0.2\,\text{s}^{-1}$	$\tilde{c}_0$	$1\,\text{kmol}\,\text{m}^{-3}$
$\tilde{c}_1$	$0\,\text{kmol}\,\text{m}^{-3}$	$\bar{c}_0, \bar{c}_1$	$0\,\text{kmol}\,\text{m}^{-3}$
k_0	$200\,\text{s}^{-1}$	k_1	$10\,\text{s}^{-1}$

Solution

The above differential equations are linear, and therefore can be solved using Laplace transformation. Taking Laplace transforms of these equations, we get

$$s\underbrace{C_0}_{\mathcal{L}[c_0]} - \bar{c}_0 = \frac{f\tilde{c}_0}{s} - (f+k_0)C_0 + k_1 C_1 \Rightarrow C_0 = \underbrace{\frac{\bar{c}_0 + \frac{f\tilde{c}_0}{s} + k_1 C_1}{s+f+k_0}}_{\equiv a_0}$$

$$s\underbrace{C_1}_{\mathcal{L}[c_1]} - \bar{c}_1 = \frac{f\tilde{c}_1}{s} - (f+k_1)C_1 + k_0 C_0 \Rightarrow C_1 = \underbrace{\frac{\bar{c}_1 + \frac{f\tilde{c}_1}{s} + k_0 C_0}{s+f+k_1}}_{\equiv a_1}$$

Substituting the expression for C_1 into the one for C_0, we get after some rearrangement

$$C_0 = \frac{\bar{c}_0(s+a_1)}{(s+a_0)(s+a_1)-k_0k_1} + \frac{f\tilde{c}_0(s+a_1)}{s[(s+a_0)(s+a_1)-k_0k_1]}$$

$$+ \frac{k_1\bar{c}_1}{(s+a_0)(s+a_1)-k_0k_1} + \frac{fk_1\tilde{c}_1}{s[(s+a_0)(s+a_1)-k_0k_1]} \qquad (5.50)$$

For the purpose of illustration, let us assume that the values of a_0, a_1, k_0 and k_1 are such that the roots of the quadratic, $(s+a_0)(s+a_1)-k_0k_1$, in the above equation, are non-zero and distinct. In terms of these roots, namely,

$$s_{0,1} = \frac{-(a_0+a_1)\pm\sqrt{(a_0+a_1)^2-4(a_0a_1-k_0k_1)}}{2} \qquad (5.51)$$

we can express the right-hand side of Equation (5.50) above as the sum of partial fractions using the formula for distinct roots. For example, the last term of that equation can be expressed as

$$\frac{fk_1\tilde{c}_1}{s(s-s_0)(s-s_1)} \equiv \frac{A}{s} + \frac{B}{s-s_0} + \frac{C}{s-s_1}$$

where

$$A = \lim_{s\to 0}\left[\frac{fk_1\tilde{c}_1}{(s-s_0)(s-s_1)}\right] = \frac{fk_1\tilde{c}_1}{s_0s_1}$$

$$B = \lim_{s\to s_0}\left[\frac{fk_1\tilde{c}_1}{s(s-s_1)}\right] = \frac{fk_1\tilde{c}_1}{s_0(s_0-s_1)}$$

$$C = \lim_{s\to s_1}\left[\frac{fk_1\tilde{c}_1}{s(s-s_0)}\right] = \frac{fk_1\tilde{c}_1}{s_1(s_1-s_0)}$$

Note that the inverse Laplace transform of the last term of Equation (5.50) is

$$A + Be^{s_0t} + Ce^{s_1t}$$

Expressing the remaining right-hand side terms of Equation (5.50) as partial fractions, and taking the inverse Laplace of the resulting equation, we obtain

$$c_0(t) = \frac{f[\tilde{c}_0(s_0+a_1)+k_1\tilde{c}_1]}{s_0s_1} + \left\{\frac{\bar{c}_0(s_0+a_1)+k_1\bar{c}_1}{s_0-s_1} + \frac{f[\tilde{c}_0(s_0+a_1)+k_1\tilde{c}_1]}{s_0(s_0-s_1)}\right\}e^{s_0t}$$

$$+ \left\{\frac{\bar{c}_0(s_0+a_1)+k_1\bar{c}_1}{s_1-s_0} + \frac{f[\tilde{c}_0(s_0+a_1)+k_1\tilde{c}_1]}{s_1(s_1-s_0)}\right\}e^{s_1t}$$

For the parameters given in Table 5.2 on the previous page, $s_0 = -0.2$ and $s_1 = -210.2$ from Equation (5.51) above. Note that these roots are non-zero and distinct as assumed. Using these roots and parameters, we finally obtain

$$c_0(t) = 4.7574 \times 10^{-2} - 4.7619 \times 10^{-2}e^{-0.2t} - 4.5308 \times 10^{-5}e^{-210.2t}$$

which is the desired expression for the concentration of species A in the reactor. The concentration of the other species can be found in the similar manner.

❑

5.4 Miscellaneous Transformations

In addition to the major transformations presented above, there are some useful ones, which help with the handling of equations in process models. These transformations are as follows.

5.4.1 Higher Order Derivatives

Consider the nth order ordinary differential equation of the form

$$\frac{dy_0}{dt} = f\left(\frac{d^n y_0}{dt^n}, \frac{d^{n-1} y_0}{dt^{n-1}}, \dots, \frac{d^3 y_0}{dt^3}, \frac{d^2 y_0}{dt^2}, y_0\right)$$

in which the right-hand side is a function of the dependent variable, and its derivatives of order higher than one. With the introduction of

$$y_1 \equiv \frac{dy_0}{dt}$$

$$y_2 \equiv \frac{dy_1}{dt} = \frac{d^2 y_0}{dt^2}$$

$$y_3 \equiv \frac{dy_2}{dt} = \frac{d^2 y_1}{dt^2} = \frac{d^3 y_0}{dt^3}$$

$$\vdots \quad \vdots \qquad \vdots$$

$$y_n \equiv \frac{dy_{n-1}}{dt} = \frac{d^2 y_{n-1}}{dt^2} = \cdots = \frac{d^n y_0}{dt^n}$$

the original differential equation gets transformed into the following set of $(n+1)$ differential equations:

$$\frac{dy_0}{dt} = f(y_n, y_{n-1}, \dots, y_3, y_2, y_0)$$

$$\frac{dy_1}{dt} = y_2$$

$$\frac{dy_2}{dt} = y_3$$

$$\vdots \quad \vdots \quad \vdots$$

$$\frac{dy_n}{dt} = y_{n-1}$$

When the initial values of y_0 and its derivatives are known, the above set of equations can be integrated with standard solvers such as Runge–Kutta methods.

5.4.2 Scaling

Consider the set of n ordinary differential equations

$$\frac{dy_i}{dt} = f_i(y_0, y_1, \dots, y_{n-1}); \qquad i = 0, 1, \dots, n-1 \tag{5.52}$$

with $y_i(0)$ specified as $\bar{y}_i$ for each ith variable. For better handling of errors in y_is during numerical integration with step-size control, we should normalize or scale each dependent variable y_i, based on its maximum expected value.

Let $\hat{y}_i$ be the maximum absolute value of y_i over the integration interval. The maximum value may be known a priori, or could be an estimate. Then based on such values, we define scaled variables as

$$\tilde{y}_i \;\equiv\; \frac{y_i}{\hat{y}_i}; \qquad i = 0, 1, \ldots, n-1$$

Differentiating both sides of the above equations with respect to t, and using Equation (5.52) on the previous page, we obtain the set of scaled differential equations

$$\frac{\mathrm{d}\tilde{y}_i}{\mathrm{d}t} \;=\; \frac{f_i}{\hat{y}_i}; \qquad i = 0, 1, \ldots, n-1 \tag{5.53}$$

where $\tilde{y}_i(0) = \bar{y}_i/\hat{y}_i$ is the initial condition of the ith, scaled dependent variable.

5.4.3 Change of Independent Variable

Sometimes it is desired to solve differential equations until a specified value of a dependent variable is attained. One way to do that is to transform the equations such that the dependent variable becomes the independent variable.

For example, to render y_0 as the independent variable in the set of equations in Equation (5.52), we invert the first equation of the set, and use the chain rule of differentiation [Section 8.7.3, p. 319] on the left-hand sides of the remaining equations to obtain

$$\frac{\mathrm{d}t}{\mathrm{d}y_0} \;=\; \frac{1}{f_0} \quad \text{and} \quad \frac{\mathrm{d}y_i}{\mathrm{d}y_0} \;=\; \frac{f_i}{f_0}; \qquad i = 1, 2, \ldots, n-1$$

The above differential equations can be integrated from $\bar{y}_0$ to a final value of y_0 with $t = 0$, and $y_i = \bar{y}_i$ for $i = 1, 2, \ldots, (n-1)$ as the initial conditions at $y_0 = \bar{y}_0$.

5.4.4 Semi-Infinite Domain

Sometimes we need to solve differential equations in a semi-infinite domain, i.e., with conditions specified at the infinite value of the independent variable. Consider for example, the set of following ordinary differential equations:

$$\frac{\mathrm{d}y_0}{\mathrm{d}r} \;=\; f_0(y_0, y_1); \qquad y_0(r = \infty) \;=\; \hat{y}_0$$

$$\frac{\mathrm{d}y_1}{\mathrm{d}r} \;=\; f_1(y_0, y_1); \qquad y_1(r = 0) \;=\; \bar{y}_1$$

To handle the semi-infinite r-domain $[0, \infty]$, we utilize the hyperbolic function

$$\tanh r \;=\; \frac{\sinh r}{\cosh r} \;=\; \frac{e^r - e^{-r}}{e^r + e^{-r}}$$

This function has a special property – it is 0 at $r = 0$, and 1 at $r = \infty$. Thus, with

$$s \;\equiv\; \tanh r$$

as the new independent variable, the domain of integration is $[0, 1]$ for the last set of differential equations. Using the chain rule of differentiation [Section 8.7.3, p. 319],

$$\frac{dy_i}{dr} = \frac{dy_i}{ds}\frac{ds}{dr} = \frac{dy_i}{ds}\underbrace{(1 - \tanh^2 r)}_{1-s^2}; \qquad i = 0, 1$$

With the help of the above result, the differential equations get transformed to

$$\frac{dy_0}{ds} = \frac{f_0}{1 - s^2}; \qquad y_0(s = 1) = \hat{y}_0$$

$$\frac{dy_1}{ds} = \frac{f_1}{1 - s^2}; \qquad y_1(s = 0) = \bar{y}_1$$

over the finite s-domain of $[0, 1]$.

5.4.5   Non-Autonomous to Autonomous Differential Equation

An autonomous differential equation does not carry the independent variable explicitly, e.g.,

$$\frac{dy_0}{dt} = y_0 - u$$

On the other hand, the following is a non-autonomous differential equation:

$$\frac{dy_1}{dt} = y_1 t - ut^2$$

Any non-autonomous differential equation can be transformed into a set of autonomous differential equations by prescribing the independent variable as an additional dependent variable. For example, with the introduction of the new state variable

$$y_2 \equiv t$$

the non-autonomous differential equation above is transformed into the following set of autonomous differential equations:

$$\frac{dy_1}{dt} = y_1 y_2 - u y_2^2 \qquad \text{and} \qquad \frac{dy_2}{dt} = 1$$

5.A   Differential Operators in an Orthogonal Coordinate System

In this appendix, we derive the expressions for important differential operators in an orthogonal coordinate system (y_1, y_2, y_3).

5.A.1   Gradient of a Scalar

The differential change in a scalar function $F(y_1, y_2, y_3)$ is given by

$$dF = \frac{\partial F}{\partial y_1}dy_1 + \frac{\partial F}{\partial y_2}dy_2 + \frac{\partial F}{\partial y_3}dy_3$$

The last equation can be written as

$$dF = \frac{h_1 dy_1}{h_1} \frac{\partial F}{\partial y_1} + \frac{h_2 dy_2}{h_2} \frac{\partial F}{\partial y_2} + \frac{h_3 dy_3}{h_3} \frac{\partial F}{\partial y_3} \qquad (5.54)$$

where h_is are the scale factors of the orthogonal coordinate system. Now, taking the dot product of Equation (5.5) on p. 142 with $\hat{\mathbf{y}}_1$, $\hat{\mathbf{y}}_2$ and $\hat{\mathbf{y}}_3$ yields respectively,

$$h_1 dy_1 = \hat{\mathbf{y}}_1 \cdot d\mathbf{x}, \qquad h_2 dy_2 = \hat{\mathbf{y}}_2 \cdot d\mathbf{x}, \qquad \text{and} \qquad h_3 dy_3 = \hat{\mathbf{y}}_3 \cdot d\mathbf{x}.$$

Upon substituting the last three relations in Equation (5.54) above, we get

$$dF = \left(\hat{\mathbf{y}}_1 \frac{1}{h_1} \frac{\partial F}{\partial y_1} + \hat{\mathbf{y}}_2 \frac{1}{h_2} \frac{\partial F}{\partial y_2} + \hat{\mathbf{y}}_3 \frac{1}{h_3} \frac{\partial F}{\partial y_3} \right) \cdot d\mathbf{x} \qquad (5.55)$$

But in Cartesian coordinate system (x_1, x_2, x_3),

$$
\begin{aligned}
dF &= \frac{\partial F}{\partial x_1} dx_1 + \frac{\partial F}{\partial x_1} dx_2 + \frac{\partial F}{\partial x_3} dx_3 \\
&= \underbrace{\left(\hat{\mathbf{x}}_1 \frac{\partial F}{\partial x_1} + \hat{\mathbf{x}}_2 \frac{\partial F}{\partial x_2} + \hat{\mathbf{x}}_3 \frac{\partial F}{\partial x_3} \right)}_{\nabla F} \cdot \underbrace{(dx_1 \hat{\mathbf{x}}_1 + dx_2 \hat{\mathbf{x}}_2 + dx_3 \hat{\mathbf{x}}_3)}_{d\mathbf{x}} = \nabla F \cdot d\mathbf{x}
\end{aligned}
$$

where the vector ∇F is called the gradient of F. Comparing the last equation with Equation (5.55) above, we obtain

$$\nabla F = \underbrace{\hat{\mathbf{y}}_1 \frac{1}{h_1} \frac{\partial F}{\partial y_1}}_{(\nabla F)_1} + \underbrace{\hat{\mathbf{y}}_2 \frac{1}{h_2} \frac{\partial F}{\partial y_2}}_{(\nabla F)_2} + \underbrace{\hat{\mathbf{y}}_3 \frac{1}{h_3} \frac{\partial F}{\partial y_3}}_{(\nabla F)_3} \qquad (5.56)$$

with $(\nabla F)_i$ as the component of ∇F along the y_i-direction. Note that ∇ is the gradient operator given by

$$\nabla = \hat{\mathbf{y}}_1 \frac{1}{h_1} \frac{\partial}{\partial y_1} + \hat{\mathbf{y}}_2 \frac{1}{h_2} \frac{\partial}{\partial y_2} + \hat{\mathbf{y}}_3 \frac{1}{h_3} \frac{\partial}{\partial y_3}$$

5.A.2   Divergence of a Vector

The divergence of a vector $\mathbf{f}$ is given by $\nabla \cdot \mathbf{f}$. In the orthogonal coordinate system (y_1, y_2, y_3),

$$\nabla \cdot \mathbf{f} = \overbrace{\left(\hat{\mathbf{y}}_1 \frac{1}{h_1} \frac{\partial}{\partial y_1} + \hat{\mathbf{y}}_2 \frac{1}{h_2} \frac{\partial}{\partial y_2} + \hat{\mathbf{y}}_3 \frac{1}{h_3} \frac{\partial}{\partial y_3} \right)}^{\nabla} \cdot \overbrace{(f_1 \hat{\mathbf{y}}_1 + f_2 \hat{\mathbf{y}}_2 + f_3 \hat{\mathbf{y}}_3)}^{\mathbf{f}}$$

Expanding the right-hand side of the last equation, we get

$$
\nabla \cdot \mathbf{f} = \hat{\mathbf{y}}_1 \frac{1}{h_1} \cdot \overbrace{\frac{\partial(f_1\hat{\mathbf{y}}_1)}{\partial y_1}}^{\equiv t_{11}} + \hat{\mathbf{y}}_2 \frac{1}{h_2} \cdot \overbrace{\frac{\partial(f_1\hat{\mathbf{y}}_1)}{\partial y_2}}^{\equiv t_{12}} + \hat{\mathbf{y}}_3 \frac{1}{h_3} \cdot \overbrace{\frac{\partial(f_1\hat{\mathbf{y}}_1)}{\partial y_3}}^{\equiv t_{13}}
$$

$$
{\equiv\, t_1}
$$

$$
+ \hat{\mathbf{y}}_1 \frac{1}{h_1} \cdot \frac{\partial(f_2\hat{\mathbf{y}}_2)}{\partial y_1} + \hat{\mathbf{y}}_2 \frac{1}{h_2} \cdot \frac{\partial(f_2\hat{\mathbf{y}}_2)}{\partial y_2} + \hat{\mathbf{y}}_3 \frac{1}{h_3} \cdot \frac{\partial(f_2\hat{\mathbf{y}}_2)}{\partial y_3}
$$

$$
{\equiv\, t_2}
$$

$$
+ \hat{\mathbf{y}}_1 \frac{1}{h_1} \cdot \frac{\partial(f_3\hat{\mathbf{y}}_3)}{\partial y_1} + \hat{\mathbf{y}}_2 \frac{1}{h_2} \cdot \frac{\partial(f_3\hat{\mathbf{y}}_3)}{\partial y_2} + \hat{\mathbf{y}}_3 \frac{1}{h_3} \cdot \frac{\partial(f_3\hat{\mathbf{y}}_3)}{\partial y_3} \qquad (5.57)
$$

$$
{\equiv\, t_3}
$$

We simplify the above equation in the following three steps:

Step 1

Since the magnitude of the unit vector $\hat{\mathbf{y}}_1$ is unity,

$$
\underbrace{\hat{\mathbf{y}}_1 \cdot \hat{\mathbf{y}}_1}_{\|\hat{\mathbf{y}}_1\|^2} = 1
$$

Differentiating both sides of the above equation with respect to y_j, we get

$$
\frac{\partial}{\partial y_j}(\hat{\mathbf{y}}_1 \cdot \hat{\mathbf{y}}_1) = 0
$$

Expanding the left-hand side of the above equation results in

$$
\hat{\mathbf{y}}_1 \cdot \frac{\partial \hat{\mathbf{y}}_1}{\partial y_j} + \hat{\mathbf{y}}_1 \cdot \frac{\partial \hat{\mathbf{y}}_1}{\partial y_j} = 0 \qquad \Rightarrow \qquad \hat{\mathbf{y}}_1 \cdot \frac{\partial \hat{\mathbf{y}}_1}{\partial y_j} = 0 \qquad (5.58)
$$

Step 2

Next, we consider

$$
\frac{\partial^2 \mathbf{x}}{\partial y_2 \partial y_1} = \frac{\partial}{\partial y_2}\left(\frac{\partial \mathbf{x}}{\partial y_1}\right) = \frac{\partial}{\partial y_2}(h_1\hat{\mathbf{y}}_1) = \underbrace{h_1\frac{\partial \hat{\mathbf{y}}_1}{\partial y_2} + \hat{\mathbf{y}}_1\frac{\partial h_1}{\partial y_2}}_{\mathbf{s}_1}
$$

$$
= h_1\hat{\mathbf{y}}_1
$$

[from Equation (5.3), p. 140]

Similarly,

$$
\frac{\partial^2 \mathbf{x}}{\partial y_1 \partial y_2} = \frac{\partial}{\partial y_1}\left(\frac{\partial \mathbf{x}}{\partial y_2}\right) = \underbrace{h_2\frac{\partial \hat{\mathbf{y}}_2}{\partial y_1} + \hat{\mathbf{y}}_2\frac{\partial h_2}{\partial y_1}}_{\mathbf{s}_2}
$$

$$
= h_2\hat{\mathbf{y}}_2
$$

[from Equation (5.3)]

Since the last two mixed partial derivatives are the same, the vectors denoted as s_1 and s_2 are the same as well. Equating their dot products with $\hat{\mathbf{y}}_1$, we get

$$\mathbf{s}_1 \cdot \hat{\mathbf{y}}_1 = \mathbf{s}_2 \cdot \hat{\mathbf{y}}_1 \quad \Rightarrow \quad h_1 \underbrace{\frac{\partial \hat{\mathbf{y}}_1}{\partial y_2} \cdot \hat{\mathbf{y}}_1}_{=0} + \underbrace{\hat{\mathbf{y}}_1 \cdot \hat{\mathbf{y}}_1}_{=1} \frac{\partial h_1}{\partial y_2} = h_2 \underbrace{\frac{\partial \hat{\mathbf{y}}_2}{\partial y_1} \cdot \hat{\mathbf{y}}_1}_{} + \underbrace{\hat{\mathbf{y}}_1 \cdot \hat{\mathbf{y}}_2}_{=0} \frac{\partial h_2}{\partial y_1}$$

[from Equation (5.58), previous page]

The above result can be written as

$$\hat{\mathbf{y}}_1 \cdot \frac{\partial \hat{\mathbf{y}}_2}{\partial y_1} = \frac{1}{h_2} \frac{\partial h_1}{\partial y_2} \tag{5.59}$$

For $i, j \in \{1, 2, 3\}$, Equations (5.58) and (5.59) generalize, respectively, to

$$\hat{\mathbf{y}}_i \cdot \frac{\partial \hat{\mathbf{y}}_i}{\partial y_j} = 0 \qquad \text{(including } i = j\text{)}, \qquad \text{and} \tag{5.60}$$

$$\hat{\mathbf{y}}_i \cdot \frac{\partial \hat{\mathbf{y}}_j}{\partial y_i} = \frac{1}{h_j} \frac{\partial h_i}{\partial y_j}, \qquad i \neq j \tag{5.61}$$

Step 3
We now resolve the terms of Equation (5.57) on the previous page as follows.

$$t_{11} = \frac{\hat{\mathbf{y}}_1}{h_1} \cdot \left(f_1 \frac{\partial \hat{\mathbf{y}}_1}{\partial y_1} + \hat{\mathbf{y}}_1 \frac{\partial f_1}{\partial y_1} \right) = \frac{f_1}{h_1} \underbrace{\hat{\mathbf{y}}_1 \cdot \frac{\partial \hat{\mathbf{y}}_1}{\partial y_1}}_{=0} + \frac{1}{h_1} \frac{\partial f_1}{\partial y_1} \underbrace{\hat{\mathbf{y}}_1 \cdot \hat{\mathbf{y}}_1}_{=1}$$

[from Equation (5.60) above]

$$= \frac{1}{h_1} \frac{\partial f_1}{\partial y_1}$$

$$t_{12} = \frac{\hat{\mathbf{y}}_2}{h_2} \cdot \left(f_1 \frac{\partial \hat{\mathbf{y}}_1}{\partial y_2} + \hat{\mathbf{y}}_1 \frac{\partial f_1}{\partial y_2} \right) = \frac{f_1}{h_2} \hat{\mathbf{y}}_2 \cdot \frac{\partial \hat{\mathbf{y}}_1}{\partial y_2} + \frac{1}{h_2} \frac{\partial f_1}{\partial y_2} \underbrace{\hat{\mathbf{y}}_2 \cdot \hat{\mathbf{y}}_1}_{=0}$$

With the help of Equation (5.61) above,

$$t_{12} = \frac{f_1}{h_1 h_2} \frac{\partial h_2}{\partial y_1}$$

Similarly, we obtain

$$t_{13} = \frac{f_1}{h_1 h_3} \frac{\partial h_3}{\partial y_1}$$

Thus,

$$t_1 = t_{11} + t_{12} + t_{13} = \frac{1}{h_1} \frac{\partial f_1}{\partial y_1} + \frac{f_1}{h_1 h_2} \frac{\partial h_2}{\partial y_1} + \frac{f_1}{h_1 h_3} \frac{\partial h_3}{\partial y_1}$$

which can be expressed as

$$t_1 = \frac{1}{h_1 h_2 h_3} \frac{\partial}{\partial y_1} (f_1 h_2 h_3)$$

In the same manner as above, the terms t_2 and t_3 in Equation (5.57) on p. 182 are obtained as

$$t_2 = \frac{1}{h_1 h_2 h_3}\frac{\partial}{\partial y_2}(f_2 h_1 h_3) \quad \text{and} \quad t_3 = \frac{1}{h_1 h_2 h_3}\frac{\partial}{\partial y_3}(f_3 h_1 h_2)$$

Substituting the t_is in Equation (5.57) on p. 182, we finally obtain

$$\nabla \cdot \mathbf{f} = \frac{1}{h_1 h_2 h_3}\left[\frac{\partial}{\partial y_1}(f_1 h_2 h_3) + \frac{\partial}{\partial y_2}(f_2 h_1 h_3) + \frac{\partial}{\partial y_3}(f_3 h_1 h_2)\right] \qquad (5.62)$$

5.A.3 Laplacian of a Scalar

The Laplacian of a scalar function F is given by

$$\nabla^2 F = \nabla \cdot \underbrace{\nabla F}_{\equiv \mathbf{g}} = \nabla \cdot \mathbf{g}$$

with $\mathbf{g} = \begin{bmatrix} g_1 & g_2 & g_3 \end{bmatrix}^\top$. Using Equation (5.62) above,

$$\nabla^2 F = \nabla \cdot \mathbf{g} = \frac{1}{h_1 h_2 h_3}\left[\frac{\partial}{\partial y_1}(g_1 h_2 h_3) + \frac{\partial}{\partial y_2}(g_2 h_1 h_3) + \frac{\partial}{\partial y_3}(g_3 h_1 h_2)\right]$$

In the above equation, we substitute from Equation (5.56) on p. 181,

$$g_i = \frac{1}{h_i}\frac{\partial F}{\partial y_i}; \qquad i = 1, 2, 3$$

and obtain

$$\nabla^2 F = \frac{1}{h_1 h_2 h_3}\left[\frac{\partial}{\partial y_1}\left(\frac{h_2 h_3}{h_1}\frac{\partial F}{\partial y_1}\right) + \frac{\partial}{\partial y_2}\left(\frac{h_1 h_3}{h_2}\frac{\partial F}{\partial y_2}\right) + \frac{\partial}{\partial y_3}\left(\frac{h_1 h_2}{h_3}\frac{\partial F}{\partial y_3}\right)\right]$$

5.A.4 Curl of a Vector

The curl of a vector $\mathbf{f} = \begin{bmatrix} f_1 & f_2 & f_3 \end{bmatrix}^\top$ in the orthogonal coordinate system (y_1, y_2, y_3) is given by

$$\nabla \times \mathbf{f} = \nabla \times f_1 \hat{\mathbf{y}}_1 + \nabla \times f_2 \hat{\mathbf{y}}_2 + \nabla \times f_3 \hat{\mathbf{y}}_3$$

Using the identity

$$\nabla \times a\mathbf{b} = a\nabla \times \mathbf{b} + \nabla a \times \mathbf{b} \qquad (5.63)$$

where a is a scalar, and $\mathbf{b}$ is a vector,

$$\nabla \times f_1 \hat{\mathbf{y}}_1 = \nabla \times \left[(f_1 h_1)\left(\frac{\hat{\mathbf{y}}_1}{h_1}\right)\right] = \underbrace{(f_1 h_1)\nabla \times \left(\frac{\hat{\mathbf{y}}_1}{h_1}\right)}_{\equiv \mathbf{p}} + \underbrace{\nabla(f_1 h_1) \times \left(\frac{\hat{\mathbf{y}}_1}{h_1}\right)}_{\equiv \mathbf{q}}$$

$$(5.64)$$

In the last equation, the term denoted $\mathbf{p}$ vanishes as follows:

$$\mathbf{p} = (f_1 h_1)\nabla \times \underbrace{\left(\frac{\hat{\mathbf{y}}_1}{h_1}\right)}_{=\nabla y_1} = (f_1 h_1)\underbrace{\nabla \times \nabla y_1}_{=0} = \mathbf{0}$$

[from Equation (5.56) on p. 181
with y_1 as F]

In the last step, we have used the following result that the curl of a gradient is zero:
From the definition of the cross product [see Equation (8.2), p. 301],

$$\nabla \times \nabla f = \begin{vmatrix} \hat{\mathbf{y}}_1 & \hat{\mathbf{y}}_2 & \hat{\mathbf{y}}_3 \\ \dfrac{1}{h_1}\dfrac{\partial}{\partial y_1} & \dfrac{1}{h_2}\dfrac{\partial}{\partial y_2} & \dfrac{1}{h_3}\dfrac{\partial}{\partial y_3} \\ \dfrac{1}{h_1}\dfrac{\partial f}{\partial y_1} & \dfrac{1}{h_2}\dfrac{\partial f}{\partial y_2} & \dfrac{1}{h_3}\dfrac{\partial f}{\partial y_3} \end{vmatrix}$$

$$= \left[\frac{1}{h_2 h_3}\left(\frac{\partial^2 f}{\partial y_2 \partial y_3} - \frac{\partial^2 f}{\partial y_3 \partial y_2}\right)\right]\hat{\mathbf{y}}_1 - \left[\frac{1}{h_1 h_3}\left(\frac{\partial^2 f}{\partial y_1 \partial y_3} - \frac{\partial^2 f}{\partial y_3 \partial y_1}\right)\right]\hat{\mathbf{y}}_2$$

$$+ \left[\frac{1}{h_1 h_2}\left(\frac{\partial^2 f}{\partial y_1 \partial y_2} - \frac{\partial^2 f}{\partial y_2 \partial y_1}\right)\right]\hat{\mathbf{y}}_3 = \mathbf{0}$$

due to the equality of mixed partial derivatives.

Using Equation (5.56) on p. 181, and the definition of the cross product of two vectors, we get the last term of Equation (5.64) on the previous page as

$$\mathbf{q} = \begin{vmatrix} \hat{\mathbf{y}}_1 & \hat{\mathbf{y}}_2 & \hat{\mathbf{y}}_3 \\ \dfrac{1}{h_1}\dfrac{\partial(f_1 h_1)}{\partial y_1} & \dfrac{1}{h_2}\dfrac{\partial(f_1 h_1)}{\partial y_2} & \dfrac{1}{h_3}\dfrac{\partial(f_1 h_1)}{\partial y_3} \\ \dfrac{1}{h_1} & 0 & 0 \end{vmatrix}$$

$$= \frac{1}{h_1 h_3}\frac{\partial(f_1 h_1)}{\partial y_3}\hat{\mathbf{y}}_2 - \frac{1}{h_1 h_2}\frac{\partial(f_1 h_1)}{\partial y_2}\hat{\mathbf{y}}_3$$

which is equal to $\nabla \times f_1 \hat{\mathbf{y}}_1$. Thus,

$$\nabla \times f_1 \hat{\mathbf{y}}_1 = \frac{1}{h_1 h_2 h_3}\left[\frac{\partial(f_1 h_1)}{\partial y_3}h_2 \hat{\mathbf{y}}_2 - \frac{\partial(f_1 h_1)}{\partial y_2}h_3 \hat{\mathbf{y}}_3\right]$$

Similarly, we obtain

$$\nabla \times f_2 \hat{\mathbf{y}}_2 = \frac{1}{h_1 h_2 h_3}\left[\frac{\partial(f_2 h_2)}{\partial y_1}h_3 \hat{\mathbf{y}}_3 - \frac{\partial(f_2 h_2)}{\partial y_3}h_1 \hat{\mathbf{y}}_1\right] \quad \text{and}$$

$$\nabla \times f_3 \hat{\mathbf{y}}_3 \;=\; \frac{1}{h_1 h_2 h_3}\left[\frac{\partial(f_3 h_3)}{\partial y_2}h_1\hat{\mathbf{y}}_1 - \frac{\partial(f_3 h_3)}{\partial y_1}h_2\hat{\mathbf{y}}_2\right]$$

Combining the last three equations, we finally obtain

$$\nabla \times \mathbf{f} \;=\; \frac{1}{h_1 h_2 h_3}\begin{vmatrix} h_1\hat{\mathbf{y}}_1 & h_2\hat{\mathbf{y}}_2 & h_3\hat{\mathbf{y}}_3 \\[6pt] \dfrac{\partial}{\partial y_1} & \dfrac{\partial}{\partial y_2} & \dfrac{\partial}{\partial y_3} \\[6pt] h_1 f_1 & h_2 f_2 & h_3 f_3 \end{vmatrix}$$

References

[1] J. Nebrensky, J.F.T. Pittman, and J.M. Smith. "Flow and Heat Transfer in Screw Extruders: I. A Variational Analysis Applied in Helical Coordinates". In: *Polymer Engineering and Science* 13.3 (1973), pp. 209–215.

Bibliography

[1] P. Moon and D.E. Spencer. *Field Theory Handbook Including Coordinate Systems, Differential Equations and Their Solutions.* 2nd edition. Berlin: Springer-Verlag, 1971.

[2] L.E. Malvern. *Introduction to the Mechanics of a Continuous Medium.* Appendix II. New Jersey: Prentice-Hall, Inc., 1977.

[3] R.B. Bird, W.E. Stewart, and E.N. Lightfoot. *Transport Phenomena.* 2nd edition. Appendix A. New York: John Wiley & Sons, Inc., 2007.

[4] C.A. Smith and A.B. Corripio. *Principles and Practice of Automatic Process Control.* 3rd edition. Chapters 2–4. New Jersey: John Wiley & Sons, Inc., 2006.

Exercises

5.1 Find the scale factors of spherical coordinate system.

5.2 Find the expressions for the gradient of a scalar, divergence of a vector, curl of a vector, and Laplacian of a scalar in spherical coordinate system.

5.3 Derive the equations to transform a position-independent vector from cylindrical coordinate system to the helical coordinate system of Example 5.2.1 on p. 157.

5.4 Obtain the helical equation of motion for the polymer melt in Example 5.2.1. Assume that the melt is a Newtonian fluid of constant density and viscosity.

5.5 Find the concentration of species B as a function of time in Example 5.3.7 on p. 176.

5.6 Prove the following properties of Laplace transforms:

$$\mathcal{L}[tf(t)] = -\frac{d}{ds}F(s) \qquad \text{(complex differentiation theorem)}$$

$$\mathcal{L}[e^{at}f(t)] = F(s-a) \qquad \text{(complex translation theorem)}$$

where $F(s)$ is the Laplace transform of $f(t)$.

5.7 Use Laplace transforms to solve the following differential equation:

$$\frac{dy}{dt} + ay = u(t-d) + u(t-2d) + \cdots, \qquad y(0) = 0$$

5.8 Obtain the Jacobian of scaled differential equations [see Equation (5.53), p. 179] in terms of the Jacobian of the original differential equations.

5.9 Explain the rationale behind the logarithmic transformation of functions.

5.10 Propose a transformation to solve a set of differential equations with some dependent variables of very small orders of magnitude.

5.11 Prove the identity given by Equation (5.63) on p. 184.

6

Model Simplification and Approximation

In this chapter, we present methods to simplify process models as well as to approximate them. The motivation behind these methods is to render complicated process models more amenable to solution, or to develop simpler models based on experimental data.

For model simplification, we present the methods of scaling and ordering analysis, and model linearization. The methods of dimensional analysis, and model fitting are presented for model approximation.

6.1 Model Simplification

As a general rule, a process model can be simplified by increasing, or moderating assumptions. Doing that makes one or more terms unnecessary in the model. The elimination of those terms therefore results in a simpler model.

We can simplify a process model by assuming the following:

1. Steady state condition, which eliminates the time derivatives in the model.

   For example, the differential equations of the isothermal CSTR model [Section 4.1.1, p. 80] get simplified to algebraic equations under the steady state assumption.

2. Insignificant spatial variations in a property, and using its space-averaged, or lumped value in order to free the model from the corresponding spatial derivatives.

   For example, in the model of tapered fin [Section 4.2.3, p. 96], assuming insignificant variations in T with respect to y, and using the lumped temperature eliminates from Equation (4.25) on p. 98 the terms having partial derivatives with respect to y. This results in a simpler model, which consists of an ordinary differential equation.

3. Negligible changes in a property, and treating it as a constant.

   For example, in the model of solvent induced heavy oil recovery [see Section 4.2.7, p. 108], if diffusivity is additionally assumed to be constant then the term involving the derivative of diffusivity drops out from the solvent mass balance [Equation (4.41), p. 110].

4. Linearity of certain terms in the model.

Process Modeling and Simulation for Chemical Engineers: Theory and Practice, First Edition. Simant Ranjan Upreti.
© 2017 John Wiley & Sons Ltd. Published 2017 by John Wiley & Sons Ltd.
Companion website: www.wiley.com/go/upreti/pms_for_chemical_engineers

For example, we can convert the non-linear differential equations to linear differential equations, which are easier to handle and solve. Linear equations are in fact extensively used in process control.

However, model simplification is fruitful only when the added, simplifying assumptions are justified. The criterion for a valid assumption is that the ensuing model predictions should be satisfactorily close to reality. This is likely to happen when the terms that get eliminated due to model simplification are sufficiently weak, i.e., their contributions to the model are not significant enough.

6.1.1   Scaling and Ordering Analysis

This analysis helps in the identification of weak terms in process models. Given a process model, the variables are scaled with respect to their maximum expected values, and introduced in the model. The additive terms of the model are compared with each other to reveal the weak terms for elimination from the model.

Consider the following process model of an isothermal plug flow reactor with axial mixing:

$$\frac{\partial c}{\partial t} = D\frac{\partial^2 c}{\partial z^2} - v\frac{\partial c}{\partial z} - kc^2 \tag{6.1}$$

$$t = 0: \quad c = 0 \ \forall \ 0 \le z \le L \tag{6.2}$$

$$z = \begin{cases} 0: & vc_0 = vc - D\dfrac{\partial c}{\partial z} \\[2mm] L: & \dfrac{\partial c}{\partial z} = 0 \end{cases} \right\} \ \forall \ 0 \le t \le t_f \tag{6.3}$$

In the above equations, c is reactant concentration, t is time, D is reactant diffusivity, z is axial variable, v is fluid velocity, k is reaction rate coefficient, L is reactor length, c_0 is c at the reactor inlet, and t_f is final time.

We begin the analysis by defining the following scaled variables:

$$\hat{c} \equiv \frac{c}{c_0}, \quad \hat{t} \equiv \frac{t}{t_f}, \quad \hat{z} \equiv \frac{z}{L} \tag{6.4}$$

Observe that the original variables are normalized with respect to their maximum possible values so that the resulting scaled variables have the maximum value of unity. With the help of these definitions, the derivatives of c are given by

$$\frac{\partial c}{\partial t} = \frac{\partial}{\partial t}(c_0\hat{c}) = c_0\frac{\partial \hat{c}}{\partial \hat{t}}\underbrace{\frac{\partial \hat{t}}{\partial t}}_{1/t_f} = \frac{c_0}{t_f}\frac{\partial \hat{c}}{\partial \hat{t}} \tag{6.5}$$

$$\underbrace{\frac{\partial c}{\partial z}}_{\equiv f} = \frac{\partial}{\partial z}(c_0\hat{c}) = c_0\frac{\partial \hat{c}}{\partial \hat{z}}\underbrace{\frac{\partial \hat{z}}{\partial z}}_{1/L} = \frac{c_0}{L}\frac{\partial \hat{c}}{\partial \hat{z}} \tag{6.6}$$

$$\frac{\partial^2 c}{\partial z^2} = \frac{\partial f}{\partial z} = \frac{\partial f}{\partial \hat{z}}\frac{\partial \hat{z}}{\partial z} = \left(\frac{c_0}{L}\frac{\partial^2 \hat{c}}{\partial \hat{z}^2}\right)\frac{1}{L} = \frac{c_0}{L^2}\frac{\partial^2 \hat{c}}{\partial \hat{z}^2} \tag{6.7}$$

For the comparison of terms of the process model later on, we assume that the magnitude of $\partial \hat{c}/\partial \hat{t}$ is close to the ratio of the overall change in $\hat{c}$ to that in $\hat{t}$, i.e., 1/1 or unity. Similarly, the values of $\partial \hat{c}/\partial \hat{z}$ and, by extension, $\partial^2 \hat{c}/\partial \hat{z}^2$ are close to unity. Thus, the order is unity for the derivatives involving only the scaled variables.

Substituting the expressions for t, z, c, and the derivatives of c from Equations (6.4)–(6.7) on the previous page into Equations (6.1)–(6.3), we obtain the following scaled process model of the reactor:

$$\frac{1}{t_f}\frac{\partial \hat{c}}{\partial \hat{t}} = \frac{D}{L^2}\frac{\partial^2 \hat{c}}{\partial \hat{z}^2} - \frac{v}{L}\frac{\partial \hat{c}}{\partial \hat{z}} - kc_0\hat{c}^2 \tag{6.8}$$

$$\hat{t} = 0: \quad \hat{c} = 0 \ \forall \ 0 \le \hat{z} \le 1$$

$$\hat{z} = \begin{cases} 0: & v = v\hat{c} - \dfrac{D}{L}\dfrac{\partial \hat{c}}{\partial \hat{z}} \\[2mm] 1: & \dfrac{\partial \hat{c}}{\partial \hat{z}} = 0 \end{cases} \quad \forall \ 0 \le \hat{t} \le 1 \tag{6.9}$$

Equivalent to the original model given by Equations (6.1)–(6.3), the scaled model offers the advantage of model simplification depending on the relative magnitude of terms in the scaled equations. To illustrate this, we use the parameter values shown in Table 6.1 below.

Table 6.1   Parameter values for the plug flow reactor

parameter	value	parameter	value
t_f	10^3 s	L	70 m
c_0	0.2 kmol m^{-3}	k	$5.2 \times 10^{-3} \text{ m}^3 \text{ kmol}^{-1} \text{ s}^{-1}$
D	10^{-8} m s^{-2}	v	$6.4 \times 10^{-2} \text{ m s}^{-1}$

Substituting the parameter values of Table 6.1 in Equation (6.8) above, we get

$$\underbrace{10^{-3}}_{\frac{1}{t_f}} \cdot \left[\frac{\partial \hat{c}}{\partial \hat{t}}\right] = \underbrace{2 \times 10^{-12}}_{\frac{D}{L^2}} \cdot \left[\frac{\partial^2 \hat{c}}{\partial \hat{z}^2}\right] - \underbrace{9.1 \times 10^{-4}}_{\frac{v}{L}} \cdot \left[\frac{\partial \hat{c}}{\partial \hat{z}}\right] - \underbrace{10^{-3}}_{kc_0} \cdot [\hat{c}^2]$$

where the bracketed terms are of order unity. Upon rearranging the above equation as

$$\frac{\partial \hat{c}}{\partial \hat{t}} = \underbrace{2 \times 10^{-9} \cdot \left[\frac{\partial^2 \hat{c}}{\partial \hat{z}^2}\right]}_{\text{smallest term}} - 9.1 \times 10^{-1} \cdot \left[\frac{\partial \hat{c}}{\partial \hat{z}}\right] - [\hat{c}^2] \tag{6.10}$$

we observe that the second term of the above equation is the smallest of all the terms. In fact, that term is significantly smaller – by at least eight orders of magnitude – than the rest of the terms in the equation. Thus, the relative contribution of the second term being negligible, it can be ignored and eliminated from the equation.

Substituting the parameter values of Table 6.1 on the previous page in the first boundary condition (for $\hat{z} = 0$ and $0 \leq \hat{t} \leq 1$) in Equation (6.9), we get

$$\underbrace{6.4 \times 10^{-2}}_{v} = \underbrace{6.4 \times 10^{-2}}_{v} \cdot [\hat{c}] - \underbrace{1.4 \times 10^{-10}}_{\frac{D}{L}} \cdot \left[\frac{\partial \hat{c}}{\partial \hat{z}}\right]$$

which can be rearranged as

$$1 = [\hat{c}] - 2.2 \times 10^{-9} \cdot \left[\frac{\partial \hat{c}}{\partial \hat{z}}\right]$$

In the above equation, the last term is significantly smaller – by nine orders of magnitude – than the rest of the terms, and can therefore be eliminated.

With the aforementioned eliminations, the scaled model of the reactor simplifies to

$$\frac{1}{t_f}\frac{\partial \hat{c}}{\partial \hat{t}} = -\frac{v}{L}\frac{\partial \hat{c}}{\partial \hat{z}} - kc_0\hat{c}^2 \tag{6.11}$$

$$\hat{t} = 0: \quad \hat{c} = 0 \;\; \forall \;\; 0 \leq \hat{z} \leq 1$$

$$\hat{z} = \begin{cases} 0: & \hat{c} = 1 \\ 1: & \dfrac{\partial \hat{c}}{\partial \hat{z}} = 0 \end{cases} \;\; \forall \;\; 0 \leq \hat{t} \leq 1$$

Remarks

It must be noted that the above model simplification rests on the assumption that the involved derivatives are sufficiently close to unity. If, for example, $\partial^2\hat{c}/\partial\hat{z}^2$ at some $\hat{z} > 0$ is very high, say, of the order 10^9, then the second term of Equation (6.10) on the previous page would not be negligible but comparable to other terms. The elimination of this term would then render the model predictions erroneous.

In general, before accepting a simplified model, its predictions should be carefully analyzed, and checked against experimental data. Seemingly negligible terms of a model can become significant during the process if there are (i) coupled equations that influence dependent variables strongly, and (ii) combinations of variables and parameters that can change by several orders of magnitude.

Further Simplification

If the reactor has to run continuously for many months (as is normal in chemical plants) then Equation (6.11) above could be further simplified after a certain time duration. For example, when $t_f = 10^6$ s (i.e., 11.6 days), that equation becomes

$$10^{-3} \cdot \left[\frac{\partial \hat{c}}{\partial \hat{t}}\right] = -9.1 \times 10^{-1} \cdot \left[\frac{\partial \hat{c}}{\partial \hat{z}}\right] - [\hat{c}^2]$$

in which the first term is at least two orders of magnitude smaller than the remaining terms, and could be ignored. In other words, the reactor may be assumed to be under steady state after 11.6 days from the startup, and describable by

$$\frac{\partial \hat{c}}{\partial \hat{z}} = -\frac{kLc_0}{v}\hat{c}^2$$

$$\hat{z} = \begin{cases} 0: & \hat{c} = 1 \\ 1: & \dfrac{\partial \hat{c}}{\partial \hat{z}} = 0 \end{cases}$$

Note that the above set of equations constitutes a steady state model, which describes the reaction process when the contribution of the time derivative term in Equation (6.11) on the previous page becomes relatively insignificant.

6.1.2 Linearization

Linearization is the approximation of a non-linear function by a linear function passing through a reference point. In its vicinity, the handling of the linear function, and its evaluation are simpler compared to the original non-linear function. Appendix 6.A on p. 220 provides a formal description of a linear function.

Consider for example a non-linear function $f(x)$ shown in Figure 6.1 below. The linearized approximation, or linear function at a reference point x is a tangent line of slope $f'(x)$. From

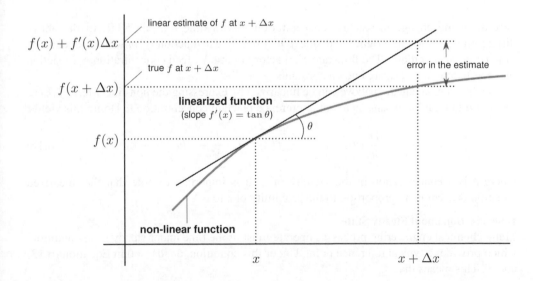

Figure 6.1 Linearization of a non-linear function

this approximation, the function value at a nearby location $(x + \Delta x)$ is $[f(x) + f'(x)\Delta x]$. Note that this estimate involves some error relative to the true function value. As can be seen from the figure, the smaller the Δx the smaller the error in the estimate.

Linearization of process models is readily carried out with the help of the first order Taylor expansion.

Linearization using First Order Taylor Expansion

A function that has first order derivatives with respect to its variables can be linearized using the first order Taylor expansion [see Section 8.8.2, p. 323]. The expansion of a function $f(\mathbf{x})$ at the reference point $\bar{\mathbf{x}}$ is given by

$$f(\mathbf{x}) = f(\bar{\mathbf{x}}) + \sum_{i=1}^{n} f_{x_i}(\bar{\mathbf{x}})\,(x_i - \bar{x}_i) \tag{6.12}$$

where $\mathbf{x} = \begin{bmatrix} x_1 & x_2 & \cdots & x_n \end{bmatrix}^{\top}$, and $f_{x_i}(\bar{\mathbf{x}})$ is $\partial f / \partial x_i$ evaluated at $\mathbf{x} = \bar{\mathbf{x}}$. Observe that the above equation provides f as a linear function of $\mathbf{x}$. This f is an approximation or estimate of the actual function value at an $\mathbf{x}$ in the vicinity of $\bar{\mathbf{x}}$. The closer the $\mathbf{x}$ to $\bar{\mathbf{x}}$ the better is the function estimate.

The first order Taylor expansion is widely used to linearize differential equations.

Linearization of Differential Equations

Consider the ordinary differential equation

$$\frac{dc}{dt} = \underbrace{\frac{F}{V}(c_{\text{in}} - c) - kc^2}_{\equiv f(c)}, \qquad c(0) = c_0 \tag{6.13}$$

describing the change in reactant concentration c with time t in a CSTR. In the above differential equation, the right-hand side f is a non-linear function of c, which depends on time t, and is c_0 initially. The flow rate F, reactor volume V, feed concentration c_{in}, reaction rate coefficient k, and c_0 are known constants.

The differential equation above can be linearized by expressing the non-linear term $f(c)$ as the first order Taylor expansion about a reference reactant concentration $\bar{c}$. Doing this yields

$$\frac{dc}{dt} = f(\bar{c}) + \frac{df}{dc}\bigg|_{\bar{c}}(c - \bar{c}) = f(\bar{c}) + \left(-\frac{F}{V} - 2k\bar{c}\right)(c - \bar{c}) \tag{6.14}$$

where c is a concentration in the vicinity of $\bar{c}$. It is important to note that the linearized equation is accurate in proportion to the proximity of c to $\bar{c}$.

Linearization about Steady State

Many chemical engineering processes operate most of the time under steady state condition, which provides a typical reference point. Under this condition, $dc/dt = 0$ in Equation (6.13) above. This means that

$$f(\bar{c}) = 0$$

where $\bar{c}$ is c at the steady state, serving as the reference concentration at $t = 0$. Using the above result in Equation (6.14) above, we obtain the linearized equation in the vicinity of the steady state, i.e.,

$$\frac{dc}{dt} = -\left(\frac{F}{V} + 2k\bar{c}\right)(c - \bar{c}), \qquad c(0) = \bar{c} \tag{6.15}$$

Use of Deviation Variables

It is convenient to express Equation (6.15) on the previous page in terms of the deviation variable [see Equation (5.47), p. 173] defined as

$$\check{c} \equiv c - \bar{c}$$

From the above definition,

$$c = \check{c} + \bar{c}, \qquad \check{c}(0) = 0, \qquad \text{and} \qquad \frac{dc}{dt} = \frac{d\check{c}}{dt}$$

Using the above expressions of c, $c(0)$ and dc/dt in Equation (6.15) on the previous page, we obtain

$$\frac{d\check{c}}{dt} = -\left(\frac{F}{V} + 2k\bar{c}\right)\check{c}, \qquad \check{c}(0) = 0$$

which is the linearization of Equation (6.13) on the previous page in terms of the deviation variable. This procedure is generalized for a set of ordinary differential equations as follows.

Generalization

Consider a process model given by the following set of ordinary differential equations:

$$\frac{d\mathbf{y}}{dt} = \mathbf{f}(\mathbf{y}, \mathbf{p}) \equiv \mathbf{f}(\mathbf{z}), \qquad \mathbf{y}(0) = \bar{\mathbf{y}} \tag{6.16}$$

where $\mathbf{y}$, $\mathbf{f}$ and $\mathbf{p}$ are the vectors of the dependent variables, right-hand sides of the above equations, and parameters given, respectively, by

$$\mathbf{y} = \begin{bmatrix} y_1 & y_2 & \cdots & y_n \end{bmatrix}^\mathsf{T}, \qquad \begin{bmatrix} f_1 & f_2 & \cdots & f_n \end{bmatrix}^\mathsf{T} \qquad \text{and}$$

$$\mathbf{p} = \begin{bmatrix} p_1 & p_2 & \cdots & p_m \end{bmatrix}^\mathsf{T}$$

The vector $\mathbf{z}$ is defined as

$$\mathbf{z} = \begin{bmatrix} \mathbf{y} \\ \mathbf{p} \end{bmatrix}$$

Around a reference point $\bar{\mathbf{z}}$, the linearization of Equation (6.16) above is then given by

$$\frac{d\mathbf{y}}{dt} = \mathbf{f}(\bar{\mathbf{z}}) + \mathbf{J}(\bar{\mathbf{z}})(\mathbf{z} - \bar{\mathbf{z}}), \qquad \mathbf{y}(0) = \bar{\mathbf{y}} \tag{6.17}$$

where $\mathbf{J}$ is the matrix of partial derivatives of $\mathbf{f}$ with respect to $\mathbf{z}$, i.e., the Jacobian

$$\mathbf{J} = \begin{bmatrix} \dfrac{\partial f_1}{\partial z_1} & \dfrac{\partial f_1}{\partial z_2} & \cdots & \dfrac{\partial f_1}{\partial z_l} \\[2ex] \dfrac{\partial f_2}{\partial z_1} & \dfrac{\partial f_2}{\partial z_2} & \cdots & \dfrac{\partial f_2}{\partial z_l} \\[2ex] \vdots & \vdots & \ddots & \vdots \\[2ex] \dfrac{\partial f_n}{\partial z_1} & \dfrac{\partial f_n}{\partial z_2} & \cdots & \dfrac{\partial f_n}{\partial z_l} \end{bmatrix}, \qquad l = n + m$$

Note that Equation (6.17) on the previous page involves $\mathbf{J}(\bar{\mathbf{z}})$, which is $\mathbf{J}$ evaluated at the reference point $\bar{\mathbf{z}}$.

For linearization around the steady state, $\mathbf{f}(\bar{\mathbf{z}}) = \mathbf{0}$. The linearized equation is then

$$\frac{d\mathbf{y}}{dt} = \mathbf{J}(\bar{\mathbf{z}})(\mathbf{z} - \bar{\mathbf{z}}), \qquad \mathbf{y}(0) = \bar{\mathbf{y}}$$

In terms of deviation variables,

$$\check{\mathbf{y}} \equiv \mathbf{y} - \bar{\mathbf{y}}, \quad \check{\mathbf{p}} \equiv \mathbf{p} - \bar{\mathbf{p}}, \quad \text{i.e.,} \quad \check{\mathbf{z}} \equiv \mathbf{z} - \bar{\mathbf{z}}$$

the linearized equation is given by

$$\frac{d\check{\mathbf{y}}}{dt} = \mathbf{J}(\bar{\mathbf{z}})\check{\mathbf{z}}, \qquad \check{\mathbf{y}}(0) = \mathbf{0}$$

Example 6.1.1

Linearize the CSTR model given by Equations (4.1)–(4.5) on p. 82. Express the linearized model using deviation variables.

Solution

The CSTR model comprises the differential equations in which the vectors of dependent variables, and right-hand side functions are, respectively,

$$\mathbf{y} = \begin{bmatrix} c_A \\ c_B \\ c_C \\ c_D \end{bmatrix} \quad \text{and} \quad \mathbf{f} = \begin{bmatrix} \dfrac{F_{Af}c_{Af} - Fc_A}{V} - k_1 c_A c_B \\ \dfrac{F_{Bf}c_{Bf} - Fc_B}{V} - k_1 c_A c_B - k_2 c_B c_C \\ -\dfrac{Fc_C}{V} + k_1 c_A c_B - k_2 c_B c_C \\ -\dfrac{Fc_D}{V} + k_1 c_B c_C \end{bmatrix}$$

Let the vector of the other variables (except t) be $\mathbf{p} = \begin{bmatrix} c_{Af} & c_{Bf} \end{bmatrix}^T$. Then the augmented vector $\mathbf{z}$ and the matrix of partial derivatives $\mathbf{J}$ are, respectively,

$$\mathbf{z} = \begin{bmatrix} \mathbf{y} \\ \mathbf{p} \end{bmatrix} \quad \text{and} \quad \mathbf{J} = \begin{bmatrix} \dfrac{\partial f_1}{\partial c_A} & \dfrac{\partial f_1}{\partial c_B} & \dfrac{\partial f_1}{\partial c_C} & \dfrac{\partial f_1}{\partial c_D} & \dfrac{\partial f_1}{\partial c_{Af}} & \dfrac{\partial f_1}{\partial c_{Bf}} \\[2mm] \dfrac{\partial f_2}{\partial c_A} & \dfrac{\partial f_2}{\partial c_B} & \dfrac{\partial f_2}{\partial c_C} & \dfrac{\partial f_2}{\partial c_D} & \dfrac{\partial f_2}{\partial c_{Af}} & \dfrac{\partial f_2}{\partial c_{Bf}} \\[2mm] \dfrac{\partial f_3}{\partial c_A} & \dfrac{\partial f_3}{\partial c_B} & \dfrac{\partial f_3}{\partial c_C} & \dfrac{\partial f_3}{\partial c_D} & \dfrac{\partial f_3}{\partial c_{Af}} & \dfrac{\partial f_3}{\partial c_{Bf}} \\[2mm] \dfrac{\partial f_4}{\partial c_A} & \dfrac{\partial f_4}{\partial c_B} & \dfrac{\partial f_4}{\partial c_C} & \dfrac{\partial f_4}{\partial c_D} & \dfrac{\partial f_4}{\partial c_{Af}} & \dfrac{\partial f_4}{\partial c_{Bf}} \end{bmatrix}$$

Expanding the terms of $\mathbf{J}$, and evaluating them at the reference point $\bar{z}$, we get

$$\mathbf{J}(\bar{z}) = \begin{bmatrix} -\dfrac{F}{V} - k_1\bar{c}_{\text{B}} & -k_1\bar{c}_{\text{A}} & 0 & 0 & \dfrac{F_{\text{Af}}}{V} & 0 \\[2ex] -k_1\bar{c}_{\text{B}} & -\dfrac{F}{V} - k_1\bar{c}_{\text{A}} - k_2\bar{c}_{\text{C}} & -k_2\bar{c}_{\text{B}} & 0 & 0 & \dfrac{F_{\text{Bf}}}{V} \\[2ex] k_1\bar{c}_{\text{B}} & k_1\bar{c}_{\text{A}} - k_2\bar{c}_{\text{C}} & -\dfrac{F}{V} - k_2\bar{c}_{\text{B}} & 0 & 0 & 0 \\[2ex] 0 & k_1\bar{c}_{\text{C}} & k_1\bar{c}_{\text{B}} & -\dfrac{F}{V} & 0 & 0 \end{bmatrix}$$

Substituting for $\mathbf{y}$, $\mathbf{f}(\bar{z})$, $\mathbf{J}(\bar{z})$ and $\bar{z}$ in Equation (6.17) on p. 195, we get the linearized CSTR model, i.e.,

$$\frac{dc_{\text{A}}}{dt} = \underbrace{\frac{F_{\text{Af}}\bar{c}_{\text{Af}} - F\bar{c}_{\text{A}}}{V} - k_1\bar{c}_{\text{A}}\bar{c}_{\text{B}}}_{f_1(\bar{z})} - \left(\frac{F}{V} + k_1\bar{c}_{\text{B}}\right)(c_{\text{A}} - \bar{c}_{\text{A}}) - k_1\bar{c}_{\text{A}}(c_{\text{B}} - \bar{c}_{\text{B}})$$

$$+ \frac{F_{\text{Af}}}{V}(c_{\text{Af}} - \bar{c}_{\text{Af}})$$

$$\frac{dc_{\text{B}}}{dt} = \underbrace{\frac{F_{\text{Bf}}\bar{c}_{\text{Bf}} - F\bar{c}_{\text{B}}}{V} - k_1\bar{c}_{\text{A}}\bar{c}_{\text{B}} - k_2\bar{c}_{\text{B}}\bar{c}_{\text{C}}}_{f_2(\bar{z})} - k_1\bar{c}_{\text{B}}(c_{\text{A}} - \bar{c}_{\text{A}})$$

$$- \left(\frac{F}{V} + k_1\bar{c}_{\text{A}} + k_2\bar{c}_{\text{C}}\right)(c_{\text{B}} - \bar{c}_{\text{B}}) - k_2\bar{c}_{\text{B}}(c_{\text{C}} - \bar{c}_{\text{C}}) + \frac{F_{\text{Bf}}}{V}(c_{\text{Bf}} - \bar{c}_{\text{Bf}})$$

$$\frac{dc_{\text{C}}}{dt} = \underbrace{-\frac{F\bar{c}_{\text{C}}}{V} + k_1\bar{c}_{\text{A}}\bar{c}_{\text{B}} - k_2\bar{c}_{\text{B}}\bar{c}_{\text{C}}}_{f_3(\bar{z})} + k_1\bar{c}_{\text{B}}(c_{\text{A}} - \bar{c}_{\text{A}}) + (k_1\bar{c}_{\text{A}} - k_2\bar{c}_{\text{C}})(c_{\text{B}} - \bar{c}_{\text{B}})$$

$$- \left(\frac{F}{V} + k_2\bar{c}_{\text{B}}\right)(c_{\text{C}} - \bar{c}_{\text{C}})$$

$$\frac{dc_{\text{D}}}{dt} = \underbrace{-\frac{F\bar{c}_{\text{D}}}{V} + k_1\bar{c}_{\text{B}}\bar{c}_{\text{C}}}_{f_4(\bar{z})} + k_1\bar{c}_{\text{C}}(c_{\text{B}} - \bar{c}_{\text{B}}) + k_1\bar{c}_{\text{B}}(c_{\text{C}} - \bar{c}_{\text{C}}) - \frac{F}{V}(c_{\text{D}} - \bar{c}_{\text{D}})$$

If $\bar{z}$ corresponds to the steady state then the elements of $f(\bar{z})$ as indicated above vanish. Then, in terms of deviation variables, the linearized CSTR model is given by

$$\frac{d\check{c}_A}{dt} = -\left(\frac{F}{V} + k_1 \bar{c}_B\right)\check{c}_A - k_1 \bar{c}_A \check{c}_B + \frac{F_{Af}}{V}\check{c}_{Af}$$

$$\frac{d\check{c}_B}{dt} = -k_1 \bar{c}_B \check{c}_A - \left(\frac{F}{V} + k_1 \bar{c}_A + k_2 \bar{c}_C\right)\check{c}_B - k_2 \bar{c}_B \check{c}_C + \frac{F_{Bf}}{V}\check{c}_{Bf}$$

$$\frac{d\check{c}_C}{dt} = k_1 \bar{c}_B \check{c}_A + (k_1 \bar{c}_A - k_2 \bar{c}_C)\check{c}_B - \left(\frac{F}{V} + k_2 \bar{c}_B\right)\check{c}_C$$

$$\frac{d\check{c}_D}{dt} = k_1 \bar{c}_C \check{c}_B + k_1 \bar{c}_B \check{c}_C - \frac{F}{V}\check{c}_D$$

with zero initial conditions.

❏

Linearization of Partial Differential Equations

Partial differential equations are linearized in the same manner as above using the first order Taylor expansion.

Consider Equation (4.16) on p. 92, which after expanding, and dropping the subscript 's', can be written as

$$\frac{\partial \omega}{\partial t} = D\left[\underbrace{\frac{1}{1-\omega}\frac{\partial^2 \omega}{\partial y^2}}_{\omega_{yy}} + \left(\underbrace{\frac{1}{1-\omega}\frac{\partial \omega}{\partial y}}_{\omega_y}\right)^2\right] \equiv f(\underbrace{\omega, \omega_y, \omega_{yy}, D}_{x}) \qquad (6.18)$$

where f is the right-hand side as a function, which is dependent on the variable vector, $x = \begin{bmatrix} \omega & \omega_y & \omega_{yy} & D \end{bmatrix}^{\mathsf{T}}$. Applying the first order Taylor expansion to f, we get

$$\frac{\partial \omega}{\partial t} = f(\bar{x}) + \left[\frac{\partial f}{\partial \omega}\right]_{\bar{x}}(\omega - \bar{\omega}) + \left[\frac{\partial f}{\partial \omega_y}\right]_{\bar{x}}(\omega_y - \bar{\omega}_y) + \left[\frac{\partial f}{\partial \omega_{yy}}\right]_{\bar{x}}(\omega_{yy} - \bar{\omega}_{yy})$$

$$+ \left[\frac{\partial f}{\partial D}\right]_{\bar{x}}(D - \bar{D}) \qquad (6.19)$$

where the right-hand side is the first order Taylor expansion of $f(x)$ in the neighborhood of some reference value of x given by $\bar{x} = \begin{bmatrix} \bar{\omega} & \bar{\omega}_y & \bar{\omega}_{yy} & \bar{D} \end{bmatrix}^{\mathsf{T}}$. Note that $\bar{\omega}_y$ and $\bar{\omega}_{yy}$ are the partial derivatives evaluated at $\bar{x}$, i.e.,

$$\bar{\omega}_y = \left[\frac{\partial \omega}{\partial y}\right]_{\bar{x}} \quad \text{and} \quad \bar{\omega}_{yy} = \left[\frac{\partial^2 \omega}{\partial y^2}\right]_{\bar{x}}$$

From Equation (6.18) above,

$$f(\bar{x}) = \bar{D}\left[\frac{\bar{\omega}_{yy}}{1-\bar{\omega}} + \left(\frac{\bar{\omega}_y}{1-\bar{\omega}}\right)^2\right]$$

and the partial derivatives of f are as follows:

$$\frac{\partial f}{\partial \omega} = D\left[\frac{\omega_{yy}}{(1-\omega)^2} + \frac{2\omega_y^2}{(1-\omega)^3}\right] \qquad \frac{\partial f}{\partial \omega_y} = \frac{2\omega_y D}{(1-\omega)^2}$$

$$\frac{\partial f}{\partial \omega_{yy}} = \frac{D}{1-\omega} \qquad\qquad \frac{\partial f}{\partial D} = \frac{\omega_{yy}}{1-\omega} + \left(\frac{\omega_y}{1-\omega}\right)^2$$

Expressing the above derivatives at $\bar{\mathbf{x}}$, and substituting them into Equation (6.19) on the previous page along with $f(\bar{\mathbf{x}})$ yields the linearization of Equation (4.16) on p. 92, i.e.,

$$\frac{\partial \omega}{\partial t} = \overbrace{\bar{D}\left[\frac{\bar{\omega}_{yy}}{1-\bar{\omega}} + \left(\frac{\bar{\omega}_y}{1-\bar{\omega}}\right)^2\right]}^{f(\bar{\mathbf{x}})} + \bar{D}\left[\frac{\bar{\omega}_{yy}}{(1-\bar{\omega})^2} + \frac{2\bar{\omega}_y^2}{(1-\bar{\omega})^3}\right](\omega - \bar{\omega})$$

$$+ \frac{2\bar{\omega}_y \bar{D}}{(1-\bar{\omega})^2}(\omega_y - \bar{\omega}_y) + \frac{\bar{D}}{1-\bar{\omega}}(\omega_{yy} - \bar{\omega}_{yy})$$

$$+ \underbrace{\left[\frac{\bar{\omega}_{yy}}{1-\bar{\omega}} + \left(\frac{\bar{\omega}_y}{1-\bar{\omega}}\right)^2\right]}_{f(\bar{\mathbf{x}})/\bar{D}}(D - \bar{D}) \qquad\qquad (6.20)$$

Linearization around Steady State

Equation (6.20) above becomes the steady state linearization with the following provisions: With $\bar{\mathbf{x}} = \bar{\mathbf{x}}(y)$ as the steady state reference, $\partial \omega / \partial t = f(\bar{\mathbf{x}}) = 0$. It means that the first and second terms of Equation (6.20) vanish. Furthermore, assuming $\bar{D}$ to be non-zero, the last term of the equation vanishes as well, and $f(\bar{\mathbf{x}}) = 0$ yields

$$\bar{\omega}_{yy} = -\frac{\bar{\omega}_y^2}{1-\bar{\omega}}$$

The solution of the above differential equation for the specified boundary conditions provides $\bar{\omega}(y)$, $\bar{\omega}_y(y)$ and $\bar{\omega}_{yy}(y)$, which are needed in the steady state linearization.

Linearization of Functions for Error Analysis

The linearized form of a function can be used to get an estimate of the error in the function value due to uncertainties in the variable values. Based on the first order Taylor expansion of a function $f(\mathbf{x})$ at a reference point $\bar{\mathbf{x}}$ [see Equation (6.12) on p. 194], we can write

$$\underbrace{f(\mathbf{x}) - f(\bar{\mathbf{x}})}_{\Delta f} \approx \underbrace{\sum_{i=1}^{n}\left[\frac{\partial f}{\partial x_i}\right]_{\bar{\mathbf{x}}}}_{f'(\bar{\mathbf{x}})} \underbrace{(x_i - \bar{x}_i)}_{\Delta x_i}$$

where Δx_i is the uncertainty in the ith variable from the reference value $\bar{x}_i$, and Δf is the associated change or error in the function value. Note that the right-hand side obeys the operator inequality [see Section 8.5.3, p. 305],

$$\sum_{i=1}^{n} f'(\bar{\mathbf{x}})\Delta x_i \leq \sum_{i=1}^{n}|f'(\bar{\mathbf{x}})||\Delta x_i|$$

Combining the above inequality with the expression for Δf, we get

$$\Delta f \leq \sum_{i=1}^{n}|f'(\bar{\mathbf{x}})||\Delta x_i|$$

where the right-hand side provides the approximate, maximum function error, i.e.,

$$\Delta \hat{f} = \sum_{i=1}^{n}|f'(\bar{\mathbf{x}})||\Delta x_i| \tag{6.21}$$

Example 6.1.2

The half-life period of thermal decomposition of a species is estimated from

$$t_{1/2} = \frac{a}{P^b}\exp\left(\frac{c}{T}\right) \quad [\text{s}]$$

where P is pressure in Pa, T is temperature in K, and a, b and c are constants. Their numerical values are $a = 1.5 \times 10^{-8}$, $b = 0.7$, and $c = 3 \times 10^4$.

Estimate the maximum error in $t_{1/2}$ at $\bar{P} = 4 \times 10^4$ Pa and $\bar{T} = 800$ K if there is 10% uncertainty in both P and T measurements.

Solution
From Equation (6.21) above, the maximum error in $t_{1/2}$ is given by

$$\Delta \hat{t}_{1/2} = \left(\left|\frac{\partial t_{1/2}}{\partial P}\right||\Delta P| + \left|\frac{\partial t_{1/2}}{\partial T}\right||\Delta T|\right)_{\bar{P},\bar{T}}$$

$$= \left|-\frac{b}{P}t_{1/2}\right|_{\bar{P},\bar{T}} \times 0.01\bar{P} + \left|-\frac{c}{T^2}t_{1/2}\right|_{\bar{P},\bar{T}} \times 0.01\bar{T}$$

$$= 6.65 \times 10^4 \text{ s}$$

The half-life at the given $\bar{P}$ and $\bar{T}$ is 1.74×10^5 s. Hence, the maximum relative error in the estimate of half-life period, i.e., $\Delta \hat{t}_{1/2}/t_{1/2}$, is about 0.38. This implies that the estimate is sensitive to the uncertainties in P and T measurements.

❑

6.2   Model Approximation

Without using fundamental laws, we can still develop an approximate model of a process based on experimental data. We can either figure out a relation using dimensional analysis, or simply postulate a relation for the involved variables. In either case, the relation when fitted on the data becomes the model of the process. This model is also known as empirical correlation.

We will first present dimensional analysis, and then present techniques for model fitting.

6.2.1 *Dimensional Analysis*

Dimensional analysis is the derivation of dimensionless combinations of the variables involved in a process. These combinations are called **dimensionless numbers**, which can be related with each other to form the model of the process.

Consider the vector $\mathbf{y} = \begin{bmatrix} y_0 & y_1 & \cdots & y_{n-1} \end{bmatrix}^{\mathsf{T}}$ of n variables in a system, which are to be related as

$$y_0 = f(y_1, y_2, \ldots, y_{n-1})$$

If m is the number of the involved dimensions (e.g., mass, length and time) then according to the Buckingham pi theorem, the above relation is equivalent to

$$\Pi_0 = F(\Pi_1, \Pi_2, \ldots, \Pi_{n-m-1}) \tag{6.22}$$

where Π_is are the dimensionless numbers given by

$$\Pi_i = \frac{p_i}{q_1^{a_1} q_2^{a_2} \cdots q_m^{a_m}}; \qquad i = 0, 1, \ldots, n - m - 1 \tag{6.23}$$

with the following provisions:

1. A q_i is an element of the variable vector $\mathbf{y}$, and carries the ith dimension D_i.

2. No two q_is are the same.

3. The p_is are also the elements of $\mathbf{y}$, but are other than the q_is.

The proof of the theorem is provided in Appendix 6.B on p. 221.

The benefit of the Buckingham pi theorem is that its application reduces the number of variables in a functional relation by the number of involved dimensions. This is advantageous when experiments need to be done to determine the relation between the variables. Consider for example, five variables involving three dimensions of mass, length and time. To determine the relation

$$y_0 = f(y_1, y_2, y_3, y_4)$$

we need to perform experiments with different values of dependent variables, and obtain the corresponding values of y_0. Suppose the requirement is to do experiments for five different values of each dependent variable. Then the total number of experiments to be done is $5^4 = 625$. However, the application of the Buckingham pi theorem yields an equivalent relation

$$\Pi_0 = F(\Pi_1)$$

whose determination for the same experimental requirement needs only five experiments as opposed to 625.

Example 6.2.1

Find the dimensionless numbers involved in the transfer of a species between a bubble of gas and liquid. The variables for this process are bubble diameter, density, viscosity, and approach velocity of liquid; and the mass transfer coefficient, and gas-phase diffusivity of the species.

Solution

The six process variables along with their units, and three dimensions are as follows:

$$\underset{\text{bubble diameter}}{d} \equiv \mathrm{m} \equiv \mathcal{L} \qquad\qquad \underset{\substack{\text{liquid velocity}\\\text{approaching the bubble}}}{v} \equiv \mathrm{m}\cdot\mathrm{s}^{-1} \equiv \mathcal{L}\mathcal{T}^{-1}$$

$$\underset{\substack{\text{liquid}\\\text{density}}}{\rho} \equiv \mathrm{kg}\cdot\mathrm{m}^{-3} \equiv \mathcal{M}\mathcal{L}^{-3} \qquad\qquad \underset{\substack{\text{species mass}\\\text{transfer coefficient}}}{k} \equiv \mathrm{m}\cdot\mathrm{s}^{-1} \equiv \mathcal{L}\mathcal{T}^{-1}$$

$$\underset{\substack{\text{liquid}\\\text{viscosity}}}{\mu} \equiv \mathrm{kg}\cdot\mathrm{m}^{-1}\cdot\mathrm{s}^{-1} \equiv \mathcal{M}\mathcal{L}^{-1}\mathcal{T}^{-1} \qquad\qquad \underset{\substack{\text{species diffusivity}\\\text{in gas}}}{D} \equiv \mathrm{m}^{2}\cdot\mathrm{s}^{-1} \equiv \mathcal{L}^{2}\mathcal{T}^{-1}$$

In the above equations, $\mathcal{M}$, $\mathcal{L}$ and $\mathcal{T}$ denote, respectively, the dimensions of mass, length and time. With $n = 6$ and $m = 3$, we seek the relation [see Equations (6.22) and (6.23), previous page]

$$\Pi_0 = F(\Pi_1, \Pi_2) \tag{6.24}$$

where

$$\Pi_i = \frac{p_i}{q_1^{a_1} q_2^{a_2} q_3^{a_3}}; \qquad i = 0, 1, 2$$

are the dimensionless numbers to be determined in terms of the six variables. We begin by selecting

$$q_1 = d, \quad q_2 = v, \quad \text{and} \quad q_3 = \rho$$

each of which involves one of the three dimensions, namely, $\mathcal{M}$, $\mathcal{L}$ and $\mathcal{T}$. The remaining variables are

$$p_0 = k, \quad p_1 = \mu, \quad \text{and} \quad p_2 = D$$

Determination of Π_0

The zeroth dimensionless number is given by

$$\Pi_0 = \frac{p_0}{q_1^{a_1} q_2^{a_2} q_3^{a_3}} = \frac{k}{d^{a_1} v^{a_2} \rho^{a_3}}$$

Using square brackets to denote dimensions, we have

$$[\Pi_0] = \left[\frac{k}{d^{a_1} v^{a_2} \rho^{a_3}} \right]$$

which means that

$$\mathcal{M}^0 \mathcal{L}^0 \mathcal{T}^0 = \frac{\mathcal{L}\mathcal{T}^{-1}}{(\mathcal{L})^{a_1} (\mathcal{L}\mathcal{T}^{-1})^{a_2} (\mathcal{M}\mathcal{L}^{-3})^{a_3}}$$

Equating the powers of $\mathcal{M}$, $\mathcal{L}$, and $\mathcal{T}$ in the last equation, we obtain, respectively,

$$0 = -a_3, \qquad 0 = 1 - a_1 - a_2 + 3a_3, \qquad \text{and} \qquad 0 = -1 + a_2.$$

From these relations, $a_1 = a_3 = 0$, and $a_2 = 1$. As a result,

$$\Pi_0 = \frac{k}{v}$$

Determination of Π_1

The next dimensionless number is given by

$$\Pi_1 = \frac{p_1}{q_1^{a_1} q_2^{a_2} q_3^{a_3}} = \frac{\mu}{d^{a_1} v^{a_2} \rho^{a_3}}$$

As before,

$$[\Pi_1] = \left[\frac{\mu}{d^{a_1} v^{a_2} \rho^{a_3}} \right]$$

which means that

$$\mathcal{M}^0 \mathcal{L}^0 \mathcal{T}^0 = \frac{\mathcal{M} \mathcal{L}^{-1} \mathcal{T}^{-1}}{(\mathcal{L})^{a_1} (\mathcal{L} \mathcal{T}^{-1})^{a_2} (\mathcal{M} \mathcal{L}^{-3})^{a_3}}$$

Equating the powers of $\mathcal{M}$, $\mathcal{L}$, and $\mathcal{T}$ in the above equation, we obtain, respectively,

$$0 = 1 - a_3, \qquad 0 = -1 - a_1 - a_2 + 3a_3, \qquad \text{and} \qquad 0 = -1 + a_2.$$

From these relations, $a_1 = a_2 = a_3 = 1$. As a result,

$$\Pi_1 = \frac{\mu}{dv\rho}$$

Determination of Π_2

The last dimensionless number is given by

$$\Pi_2 = \frac{p_2}{q_1^{a_1} q_2^{a_2} q_3^{a_3}} = \frac{D}{d^{a_1} v^{a_2} \rho^{a_3}}$$

As before,

$$[\Pi_2] = \left[\frac{D}{d^{a_1} v^{a_2} \rho^{a_3}} \right]$$

which means that

$$\mathcal{M}^0 \mathcal{L}^0 \mathcal{T}^0 = \frac{\mathcal{L}^2 \mathcal{T}^{-1}}{(\mathcal{L})^{a_1} (\mathcal{L} \mathcal{T}^{-1})^{a_2} (\mathcal{M} \mathcal{L}^{-3})^{a_3}}$$

Equating the powers of $\mathcal{M}$, $\mathcal{L}$, and $\mathcal{T}$ in the above equation, we obtain, respectively,

$$0 = -a_3, \qquad 0 = 2 - a_1 - a_2 + 3a_3, \qquad \text{and} \qquad 0 = -1 + a_2.$$

From these relations, $a_1 = a_2 = 1$, and $a_3 = 0$. As a result,

$$\Pi_2 = \frac{D}{dv}$$

Recombination

The dimensionless numbers are often recombined to obtain alternate dimensionless numbers that allow for easy physical interpretation and quantification. For example,

1. The ratio

$$\frac{\Pi_0}{\Pi_2} = \frac{kd}{D} \equiv \mathrm{Sh}$$

   where Sh is called Sherwood number. It is the ratio of the convective rate of mass transfer to the rate of diffusive mass transfer.

2. The inverse of Π_1, i.e.,

$$\frac{1}{\Pi_1} = \frac{dv\rho}{\mu} \equiv \mathrm{Re}$$

   is called Reynolds number. It is the ratio of force arising from momentum to that from fluid viscosity. This ratio helps quantify the relative contribution of the two forces.

3. The ratio

$$\Pi_2\Pi_0 = \frac{\mu}{\rho D} \equiv \mathrm{Sc}$$

   where Sc is called Schmidt number. It is the ratio of kinematic viscosity (μ/ρ) to diffusivity.

Thus, we can replace Π_0, Π_1 and Π_2 with, respectively, Sh, Re and Sc. Instead of determining the relation given by Equation (6.24) on p. 202, we can determine

$$\mathrm{Sh} = g(\mathrm{Re}, \mathrm{Sc})$$

from experiments. With this approach, we can obtain realistic interpretation of the process through the numerical values of the Sh, Re and Sc. For example, a low value of Sh would mean the dominance of diffusion over convection in the process. This information is not revealed by Π_0, Π_1 or Π_2 as such.

From experimental studies, several correlations of the form

$$\mathrm{Sh} = a\,\mathrm{Re}^b\,\mathrm{Sc}^c$$

with fitted values of parameters a, b and c are available.

❑

6.2.2   *Model Fitting*

Given a set of experimental y versus x data points – $\{(x_1, y_1) \quad (x_2, y_2) \quad \cdots \quad (x_n, y_n)\}$ – where x and y are process variables, we can find the unknown coefficients of a specified relation, e.g., a, b and c in the relation

$$y = ax^2 + bx + c$$

such that it predicts y values from x values satisfying certain criterion of accuracy. This procedure is known as model fitting. The relation and fitted coefficients constitute the fitted process model.

There are two main criteria for model fitting:

1. minimum least-squares error

2. perfect agreement of predictions at data points

Using the first criterion, a model is fitted such that the square of the errors between experimental data points, and the corresponding model predictions is minimum. Model fitting based on this criterion is known as **least-squares regression**.

The second criterion makes it necessary for the model prediction to match exactly with each data point. Model fitting based on this criterion is called **interpolation**.

Least-Squares Regression

We will first describe least-squares regression to fit a straight line, and a polynomial to a set of experimental y versus x data points: $\{(x_1, y_1) \quad (x_2, y_2) \quad \cdots \quad (x_n, y_n)\}$. Then we will present a generalized method for least-squares regression.

Statistical Terminology

For the aforementioned set of y versus x data points, the mean of y_is is defined as

$$\bar{y} = \frac{1}{n} \sum_{i=1}^{n} y_i$$

The sum of the squared errors around the mean is given by

$$S = \sum_{i=1}^{n} (\bar{y} - y_i)^2$$

The variance of y_i-values is defined as

$$s_y^2 = \frac{S}{n-1} = \frac{1}{n-1} \sum_{i=1}^{n} (\bar{y} - y_i)^2$$

where $(n-1)$ is called the degree of freedom. The standard deviation in y_i-values is s_y, i.e., the square root of variance. Finally, the coefficient of variation in y_i-values is the ratio, $s_y/\bar{y}$.

For the model $y = y(x)$ fitted on the data set, the sum of the squared errors is given by

$$E = \sum_{i=1}^{n} (\mathring{y}_i - y_i)^2$$

where $\mathring{y}_i$ is the model prediction for x_i. Then the coefficient of determination is defined as

$$r^2 = \frac{S - E}{S}$$

The square root of the coefficient of determination (i.e., r) is called correlation coefficient. It relates to the improvement in prediction by the model $y = y(x)$ over that by $\bar{y}$. For perfect fit, $E = 0$ and $r^2 = r = 1$.

Least-Squares Regression to Fit a Straight Line

The objective here is to fit to the data set, a straight line

$$y = a_1 + a_2 x \quad \Rightarrow \quad \mathring{y}_i = a_1 + a_2 x_i$$

such that the sum of the squared errors,

$$E = \sum_{i=1}^{n} (\mathring{y}_i - y_i)^2 = \sum_{i=1}^{n} (a_1 + a_2 x_i - y)^2$$

is minimum. The necessary conditions for the minimum are

$$\frac{\partial E}{\partial a_1} = 2 \sum_{i=1}^{n} (a_1 + a_2 x_i - y_i) = 0 \quad \text{and}$$

$$\frac{\partial E}{\partial a_2} = 2 \sum_{i=1}^{n} [(a_1 + a_2 x_i - y_i) x_i] = 0$$

It is easy to check that the sufficient conditions ($\partial^2 E / \partial a_1{}^2 > 0$ and $\partial^2 E / \partial a_2{}^2 > 0$) are satisfied for the minimum of E. Thus, the minimum corresponds to the solution of the necessary conditions, which yield

$$a_1 = \bar{y} - a_2 \bar{x} \quad \text{and} \quad a_2 = \frac{n \sum_{i=1}^{n} x_i y_i - \sum_{i=1}^{n} x_i \sum_{i=1}^{n} y_i}{n \sum_{i=1}^{n} x_i^2 - \left(\sum_{i=1}^{n} x_i \right)^2}$$

as the fitted coefficients of the straight line.

Least-Squares Regression to Fit a Quadratic

In this case, the objective is to fit to the data set, a quadratic

$$y = a_1 + a_2 x + a_3 x^2 \quad \Rightarrow \quad \mathring{y}_i = a_1 + a_2 x_i + a_3 x_i^2$$

such that the sum of the squared errors,

$$E = \sum_{i=1}^{n} (\mathring{y}_i - y_i)^2 = \sum_{i=1}^{n} (a_1 + a_2 x_i + a_3 x_i^2 - y_i)^2$$

is minimum. The necessary conditions for the minimum are

$$\frac{\partial E}{\partial a_1} = 2\sum_{i=1}^{n}(a_1 + a_2 x_i + a_3 x_i^2 - y_i) = 0$$

$$\frac{\partial E}{\partial a_2} = 2\sum_{i=1}^{n}[(a_1 + a_2 x_i + a_3 x_i^2 - y_i)x_i] = 0$$

$$\frac{\partial E}{\partial a_3} = 2\sum_{i=1}^{n}[(a_1 + a_2 x_i + a_3 x_i^2 - y_i)x_i^2] = 0$$

It is easy to check that the sufficient conditions (i.e., $\partial^2 E/\partial a_1{}^2 > 0$, $\partial^2 E/\partial a_2{}^2 > 0$, and $\partial^2 E/\partial a_3{}^2 > 0$) are satisfied for the minimum of E. Thus, the minimum corresponds to the solution of the necessary conditions recast as

$$
\begin{bmatrix}
n & \sum_{i=1}^{n} x_i & \sum_{i=1}^{n} x_i^2 \\
\sum_{i=1}^{n} x_i & \sum_{i=1}^{n} x_i^2 & \sum_{i=1}^{n} x_i^3 \\
\sum_{i=1}^{n} x_i^2 & \sum_{i=1}^{n} x_i^3 & \sum_{i=1}^{n} x_i^4
\end{bmatrix}
\begin{bmatrix}
a_1 \\
a_2 \\
a_3
\end{bmatrix}
=
\begin{bmatrix}
\sum_{i=1}^{n} y_i \\
\sum_{i=1}^{n} x_i y_i \\
\sum_{i=1}^{n} x_i^2 y_i
\end{bmatrix}
\tag{6.25}
$$

whose solution provides the fitted coefficients (namely, a_1, a_2 and a_3) of the quadratic.

Generalized Least-Squares Regression

Consider in general, the least square fit of a function

$$y = y(\underbrace{a_1, a_2, \ldots, a_m}_{\equiv \, \mathbf{a}^\mathsf{T}}, \underbrace{x_1, x_2, \ldots, x_l}_{\equiv \, \mathbf{x}^\mathsf{T}}) \equiv y(\mathbf{a}, \mathbf{x})$$

to the experimental set of n data points: $\{(\mathbf{x}_1, y_1) \quad (\mathbf{x}_2, y_2) \quad \cdots \quad (\mathbf{x}_n, y_n)\}$.

In the above equation, $\mathbf{a}$ is the coefficient vector, and $\mathbf{x}$ is the variable vector. Its ith experimental value is $\mathbf{x}_i = \begin{bmatrix} x_{1i} & x_{2i} & \cdots & x_{li} \end{bmatrix}^\mathsf{T}$ corresponding to y_i. The objective here is to minimize the sum of the squared errors given by

$$E(\mathbf{a}) = \sum_{i=1}^{n} (\underbrace{\mathring{y}_i - y_i}_{\equiv \, e_i(\mathbf{a})})^2 = \sum_{i=1}^{n}[e_i(\mathbf{a})]^2 \tag{6.26}$$

where $\mathring{y}_i = y(\mathbf{a}, \mathbf{x}_i)$ is the predicted value of y corresponding to $\mathbf{x}_i$.

From Newton's optimization method [see Appendix 6.C, p. 223], the algorithm to find the $\mathbf{a}$ that minimizes E is

$$\mathbf{a}^{(k+1)} = \mathbf{a}^{(k)} - \underbrace{\left[\frac{\partial}{\partial \mathbf{a}}\left(\frac{\partial E}{\partial \mathbf{a}}\right)\right]^{-1}_{\mathbf{a}^{(k)}}}_{\{\mathbf{H}[\mathbf{a}^{(k)}]\}^{-1}} \underbrace{\left[\frac{\partial E}{\partial \mathbf{a}}\right]_{\mathbf{a}^{(k)}}}_{\nabla E[\mathbf{a}^{(k)}]} = \mathbf{a}^{(k)} - \left\{\mathbf{H}[\mathbf{a}^{(k)}]\right\}^{-1}\nabla E[\mathbf{a}^{(k)}]$$

$$k = 0, 1, 2, \ldots$$

where ∇E and $\mathbf{H}$ are, respectively, the gradient and Hessian of E with respect to $\mathbf{a}$. Using Equation (6.26) on the previous page,

$$
\nabla E = 2 \sum_{i=1}^{n} e_i(\mathbf{a}) \frac{\partial e_i}{\partial \mathbf{a}} = 2 \underbrace{\begin{bmatrix} e_1 & e_2 & \cdots & e_n \end{bmatrix}}_{\mathbf{e}^{\mathsf{T}}} \underbrace{\begin{bmatrix} \dfrac{\partial e_1}{\partial a_1} & \dfrac{\partial e_1}{\partial a_2} & \cdots & \dfrac{\partial e_1}{\partial a_m} \\[2ex] \dfrac{\partial e_2}{\partial a_1} & \dfrac{\partial e_2}{\partial a_2} & \cdots & \dfrac{\partial e_2}{\partial a_m} \\[2ex] \vdots & \vdots & \ddots & \vdots \\[2ex] \dfrac{\partial e_n}{\partial a_1} & \dfrac{\partial e_n}{\partial a_2} & \cdots & \dfrac{\partial e_n}{\partial a_m} \end{bmatrix}}_{\mathbf{J}}
$$

$$
= 2\mathbf{J}^{\mathsf{T}}\mathbf{e}
$$

where $\mathbf{J}$ is the Jacobian of the error vector $\mathbf{e}$. Moreover,

$$
\mathbf{H} = \frac{\partial}{\partial \mathbf{a}}\left[2\sum_{i=1}^{n} e_i(\mathbf{a}) \frac{\partial e_i}{\partial \mathbf{a}} \right] = 2\left[\underbrace{\sum_{i=1}^{n}\left(\frac{\partial e_i}{\partial \mathbf{a}}\right)^2}_{\mathbf{J}^{\mathsf{T}}\mathbf{J}} + \sum_{i=1}^{n} e_i \frac{\partial^2 e_i}{\partial \mathbf{a}^2} \right]
$$

Gauss–Newton Algorithm

In this algorithm, the second order derivatives are discarded from the last equation, and the Hessian is approximated as $\mathbf{H} \approx 2\mathbf{J}^{\mathsf{T}}\mathbf{J}$. An initial guess vector $\mathbf{a}^{(0)}$ is then iteratively improved using,

$$
\mathbf{a}^{(k+1)} = \mathbf{a}^{(k)} - \left[\mathbf{J}^{\mathsf{T}}\mathbf{J}\right]^{-1}\mathbf{J}^{\mathsf{T}}\mathbf{e}; \quad k = 0, 1, 2, \ldots \tag{6.27}
$$

Example 6.2.2

Fit a quadratic to the following data of carbon dioxide diffusivity (D) versus its concentration (c) in polypropylene at 190°C and 2.67 MPa:[1]

c (kg m^{-3})	0	1.0847	2.1694	3.2541	4.3389	5.4236	6.5083	7.5536
$D \times 10^9$ (m^2 s^{-1})	3.9004	6.8729	7.0911	6.8354	6.3765	5.7881	5.0680	4.2123

Next, use the Gauss–Newton algorithm to fit the following logarithmic function:

$$
D = a_1 + a_2 \log_{10}(1 + c) + a_3 \left[\log_{10}(1 + c) \right]^2
$$

Solution

To fit a quadratic, we use Equation (6.25) on p. 207, which results in

$$
\begin{bmatrix}
8 & 30.3325 & 164.1269 \\
30.3325 & 164.1269 & 993.8141 \\
164.1269 & 993.8141 & 6404.8981
\end{bmatrix}
\begin{bmatrix}
a_1 \\
a_2 \\
a_3
\end{bmatrix}
=
\begin{bmatrix}
46.1447 \\
168.9424 \\
859.1468
\end{bmatrix}
$$

The solution of the above equation provides $a_1 = 4.7260$, $a_2 = 1.2737$ and $a_3 = -0.1846$. Hence, the fitted quadratic is

$$ D = 4.7260 + 1.2737c - 0.1846c^2 \tag{6.28} $$

Next, we apply Gauss–Newton algorithm to fit the given logarithmic function. The ith error, and its partial derivatives with respect to the fitting coefficients (a_1, a_2 and a_3) are given by

$$ e_i = a_1 + a_2 \log_{10}(1 + c_i) + a_3 \left[\log_{10}(1 + c_i) \right]^2 - D_i $$

$$ \frac{\partial e_i}{\partial a_1} = 1; \quad \frac{\partial e_i}{\partial a_2} = \log_{10}(1 + c_i); \quad \frac{\partial e_i}{\partial a_3} = \left[\log_{10}(1 + c_i) \right]^2; \quad i = 1, 2, \ldots, 8 $$

The above expressions form the error and Jacobian, respectively, as

$$
\mathbf{e} =
\begin{bmatrix}
e_1 \\
e_2 \\
\vdots \\
e_8
\end{bmatrix}
\quad \text{and} \quad
\mathbf{J} =
\begin{bmatrix}
\dfrac{\partial e_1}{\partial a_1} & \dfrac{\partial e_1}{\partial a_2} & \dfrac{\partial e_1}{\partial a_3} \\[2mm]
\dfrac{\partial e_2}{\partial a_1} & \dfrac{\partial e_2}{\partial a_2} & \dfrac{\partial e_2}{\partial a_3} \\[2mm]
\vdots & \vdots & \vdots \\[2mm]
\dfrac{\partial e_8}{\partial a_1} & \dfrac{\partial e_8}{\partial a_2} & \dfrac{\partial e_8}{\partial a_3}
\end{bmatrix}
$$

which are needed in the algorithm [see Equation (6.27), previous page].

With the initial guess of $\mathbf{a}^{(0)} = \begin{bmatrix} 1, 1, 1 \end{bmatrix}$, the algorithm converges in the second iteration to $a_1 = 3.9036$, $a_2 = 13.6387$ and $a_3 = -14.1466$. Hence, the fitted logarithmic function is

$$ D = 3.9036 + 13.6387 \log_{10}(1 + c) - 14.1466 \left[\log_{10}(1 + c) \right]^2 \tag{6.29} $$

The two fitted functions along with data points are plotted in Figure 6.2 on the next page. It is observed from the figure that the logarithmic function fits the data significantly better than the quadratic function.

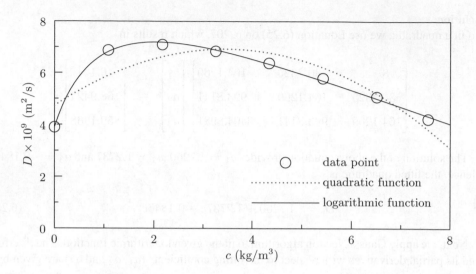

Figure 6.2   The fitted quadratic and logarithmic functions [Equations (6.28) and (6.29) on the previous page] along with data points

Table 6.2 below shows the sum of the squared errors (E), and the coefficient of determination (r^2) for the two fitted functions. With the logarithmic function $r^2 > 0.99$, and E is two orders of magnitude less than that with the quadratic function.

Table 6.2   Quality of the fitting functions

fitted function	E	r^2
quadratic, Equation (6.28)	2.3831	0.7810
logarithmic, Equation (6.29)	0.0432	0.9921

❑

Interpolation

Interpolation is typically carried out with polynomials. In this section, we present interpolation with polynomials and splines where the latter involve low-order polynomials.

We can express a polynomial of degree n, which is given by

$$y(x) \; = \; \alpha_0 + \alpha_1 x + \alpha_2 x^2 + \ldots + \alpha_n x^n$$

for a set of $(n+1)$ experimental data points: $\{(x_0, y_0) \quad (x_1, y_1) \quad \cdots \quad (x_n, y_n)\}$. The solution of the resulting equations yields the $(n+1)$ coefficients $(\alpha_i s)$ that are needed to fit the polynomial so that its prediction matches each data point exactly.

The coefficients can be determined systematically using Newton's interpolation method, or Lagrange's interpolation formula.

Newton's Interpolation Method

In this method, the nth degree polynomial is expressed involving the experimental x_is as

$$y(x) = a_0 + a_1(x - x_0) + a_2(x - x_0)(x - x_1) + a_3(x - x_0)(x - x_1)(x - x_2) + \cdots$$
$$+ a_n(x - x_0)(x - x_1) \cdots (x - x_{n-1}) \qquad (6.30)$$

The unknown coefficients a_is are found by substituting x_is for x in the above equation. Substituting x_0 for x in the above equation yields

$$a_0 = y(x_0)$$

Substituting x_1 for x in Equation (6.30) above yields

$$a_1 = \frac{y(x_1) - a_0}{x_1 - x_0} = \frac{y(x_1) - y(x_0)}{x_1 - x_0} \equiv \underbrace{y(x_1, x_0)}_{\text{divided difference}}$$

where multiple function arguments as in $y(x_1, x_0)$ imply a divided difference expression. Substituting x_2 for x in Equation (6.30) yields

$$a_2 = \frac{y(x_2) - \overbrace{a_0}^{y(x_0)} - a_1(x_2 - x_0)}{(x_2 - x_0)(x_2 - x_1)} = \frac{y(x_2) - y(x_1) + y(x_1) - y(x_0) - a_1(x_2 - x_0)}{(x_2 - x_0)(x_2 - x_1)}^{\overbrace{}^{a_1(x_1-x_0)}}$$

$$= \frac{y(x_2) - y(x_1) - a_1(x_2 - x_1)}{(x_2 - x_0)(x_2 - x_1)} = \frac{1}{x_2 - x_0}\left[\underbrace{\frac{y(x_2) - y(x_1)}{x_2 - x_1}}_{y(x_2,x_1)} - \underbrace{\frac{y(x_1) - y(x_0)}{x_1 - x_0}}_{a_1}\right]$$

$$= \frac{y(x_2, x_1) - y(x_1, x_0)}{x_2 - x_0} \equiv y(x_2, x_1, x_0)$$

Substituting x_3 for x in Equation (6.30) similarly yields

$$a_3 = \frac{y(x_3, x_2, x_1) - y(x_2, x_1, x_0)}{x_3 - x_0} \equiv y(x_3, x_2, x_1, x_0)$$

Thus, in general

$$a_i = \frac{y(x_i, x_{i-1}, \ldots, x_1) - y(x_{i-1}, x_{i-2}, \ldots, x_0)}{x_i - x_0} \equiv y(x_i, x_{i-1}, \ldots, x_0)$$

$$i = 1, 2, \ldots, n \qquad (6.31)$$

which involves the divided difference formula,

$$y(x_j, x_{j-1}, \ldots, x_{k-1}, x_k) = \frac{y(x_j, x_{j-1}, \ldots, x_{k-1}) - y(x_{j-1}, x_{j-2}, \ldots, x_k)}{x_j - x_k} \qquad (6.32)$$

Example 6.2.3

Find the interpolating polynomial for the following data using Newton's interpolation:

i	0	1	2	3
x_i	1.5	3	7	9
y_i	1	5	2	3.5

Solution

Since the number of data points is four, we need the third degree polynomial, i.e.,

$$y(x) \;=\; a_0 + a_1(x - x_0) + a_2(x - x_0)(x - x_1) + a_3(x - x_0)(x - x_1)(x - x_2)$$

where the coefficients a_is are obtained as follows.

$$a_0 \;=\; y(x_0) \;=\; y(1) \;=\; 1$$

For the remaining a_is, we use Equations (6.31) and (6.32) sequentially.

$$a_1 \;=\; y(x_1, x_0) \;=\; \frac{y(x_1) - y(x_0)}{x_1 - x_0} \;=\; 2.6667$$

$$a_2 \;=\; y(x_2, x_1, x_0) \;=\; \frac{y(x_2, x_1) - y(x_1, x_0)}{x_2 - x_0}$$

$$y(x_2, x_1) \;=\; \frac{y(x_2) - y(x_1)}{x_2 - x_1} \;=\; -0.7500$$

$$\Rightarrow \quad a_2 \;=\; -0.6212$$

$$a_3 \;=\; y(x_3, x_2, x_1, x_0) \;=\; \frac{y(x_3, x_2, x_1) - y(x_2, x_1, x_0)}{x_3 - x_0}$$

$$y(x_3, x_2, x_1) \;=\; \frac{y(x_3, x_2) - y(x_2, x_1)}{x_3 - x_1}$$

$$y(x_3, x_2) \;=\; \frac{y(x_3) - y(x_2)}{x_3 - x_2} \;=\; 0.7500$$

$$\Rightarrow \quad y(x_3, x_2, x_1) \;=\; 0.2500$$

$$y(x_2, x_1, x_0) \;=\; \frac{y(x_2, x_1) - y(x_1, x_0)}{x_2 - x_0} \;=\; -0.6212$$

$$\Rightarrow \quad a_3 \;=\; 0.1162$$

Thus, the interpolating polynomial is

$$y(x) \;=\; 1 + 2.6667(x - 1.5) - 0.6212(x - 1.5)(x - 3)$$
$$+\, 0.1162(x - 1.5)(x - 3)(x - 7)$$

The interpolating polynomial along with the data points is plotted in Figure 6.3 below.

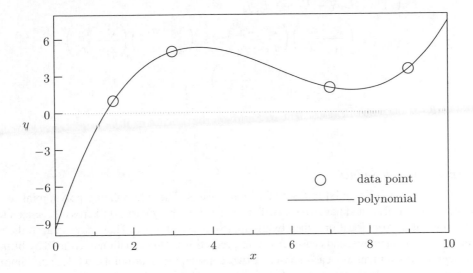

Figure 6.3   The fitted polynomial along with data points

❑

Lagrange's Interpolation Formula

This is a straightforward formula for the interpolating polynomial, and is given by

$$y(x) \; = \; \sum_{i=0}^{n} L_i(x) y_i \quad \text{where} \quad L_i(x) \; = \; \prod_{\substack{j=0 \\ j \neq i}}^{n} \frac{x - x_j}{x_i - x_j} \tag{6.33}$$

Example 6.2.4

Find the first and second order interpolating polynomials using Lagrange's interpolation method.

Solution

For the first order polynomial,

$$y(x) \; = \; L_0 y_0 + L_1 y_1$$

$$= \; \left(\frac{x - x_1}{x_0 - x_1} \right) y_0 + \left(\frac{x - x_0}{x_1 - x_0} \right) y_1$$

For the second order polynomial,

$$y(x) = L_0 y_0 + L_1 y_1 + L_2 y_2$$

$$= \left(\frac{x - x_1}{x_0 - x_1} \right) \left(\frac{x - x_2}{x_0 - x_2} \right) y_0 + \left(\frac{x - x_0}{x_1 - x_0} \right) \left(\frac{x - x_2}{x_1 - x_2} \right) y_1$$

$$+ \left(\frac{x - x_0}{x_2 - x_0} \right) \left(\frac{x - x_1}{x_2 - x_1} \right) y_2$$

❑

Interpolation with Splines

Consider fitting a polynomial to a set of 11 data points. The interpolating polynomial $y(x)$ will be of 10th degree, and have 10 roots. This means that the y-curve will cross the x-axis up to 10 times at the root locations, thereby giving rise to oscillations. They may occur in the x-interval of the data points, and cause erroneous predictions. This situation is avoided by fitting low-degree polynomials for each subset of two consecutive data points, while maintaining continuity at the internal data points. These polynomials are called splines.

We will derive linear, quadratic and cubic splines for the n intervals of the set of $(n + 1)$ experimental data points: $\{(x_0, y_0) \quad (x_1, y_1) \quad \cdots \quad (x_n, y_n)\}$.

Linear Splines

These are equations of a straight line for each subset of two consecutive data points, or the corresponding x-interval. Over the ith interval, $[x_{i-1}, x_i]$, linear splines are simply

$$y_i(x) = m_i(x - x_{i-1}) + y_{i-1}, \qquad x_{i-1} \le x \le x_i; \qquad i = 1, 2, \ldots, n$$

where m_i is the slope in the interval, and is determined from

$$m_i = \frac{y_i - y_{i-1}}{x_i - x_{i-1}}$$

To find y at a given x, the surrounding j^{th} interval $[x_{j-1}, x_j]$ is identified at first. The y-value is then obtained from the corresponding j^{th} spline, i.e., $y_j(x)$.

Quadratic Splines

These are quadratic polynomials for each subset of two consecutive data points, or the corresponding x-interval. Over the ith interval, $[x_{i-1}, x_i]$, quadratic splines are given by

$$y_i(x) = a_i x^2 + b_i x + c_i, \qquad x_{i-1} \le x \le x_i; \qquad i = 1, 2, \ldots, n$$

where a_i, b_i and c_i are $3n$ unknown coefficients. They are determined based on the following considerations:

Function Equality at Intermediate Points At an intermediate x_i, both the ith and $(i + 1)^{\text{th}}$ splines should yield the same y_i. This condition provides the following $2(n - 1)$ equations:

$$a_i x_i^2 + b_i x_i + c_i = a_{i+1} x_i^2 + b_{i+1} x_i + c_{i+1} = y_i; \qquad i = 1, 2, \ldots, n - 1$$

Derivative Equality at Intermediate Points Again, at an intermediate x_i, both the ith and $(i+1)^{th}$ splines should yield the same derivative with respect to x. This condition provides the following $(n-1)$ equations:

$$2a_i x_i + b_i = 2a_{i+1} x_i + b_{i+1}; \qquad i = 1, 2, \ldots, n-1$$

Terminal Function Values At the terminal points, we have the following two equations from the corresponding splines:

$$a_1 x_0^2 + b_1 x_0 + c_1 = y_0$$
$$a_n x_n^2 + b_n x_n + c_n = y_n$$

Zero Second Derivative Lastly, we specify that $d^2 y/dx^2 = 0$ at x_0. This specification yields

$$\left. \frac{d^2 y_1}{dx^2} \right|_{x_0} = 2a_1 = 0 \quad \Rightarrow \quad a_1 = 0$$

The $3n$ equations above appear as the following matrix equation:

$$
\begin{bmatrix}
x_0^2 & x_0 & 1 & 0 & 0 & 0 & 0 & 0 & 0 & 0 & \cdots & 0 \\
x_1^2 & x_1 & 1 & 0 & 0 & 0 & 0 & 0 & 0 & 0 & \cdots & 0 \\
0 & 0 & 0 & x_1^2 & x_1 & 1 & 0 & 0 & 0 & 0 & \cdots & 0 \\
\vdots & \vdots & \ddots & \ddots & \ddots & \ddots & \ddots & \ddots & \ddots & \ddots & \vdots & \vdots \\
0 & 0 & 0 & 0 & 0 & 0 & x_{n-1}^2 & x_{n-1} & 1 & 0 & \cdots & 0 \\
0 & 0 & 0 & 0 & 0 & 0 & 0 & \cdots & 0 & x_{n-1}^2 & x_{n-1} & 1 \\
0 & 0 & 0 & 0 & 0 & 0 & 0 & \cdots & 0 & x_n^2 & x_n & 1 \\
2x_1 & 1 & 0 & -2x_1 & -1 & 0 & 0 & 0 & 0 & 0 & \cdots & 0 \\
0 & 0 & 0 & 2x_2 & 1 & 0 & -2x_2 & -1 & 0 & 0 & \cdots & 0 \\
\vdots & \vdots & \ddots & \ddots & \ddots & \ddots & \ddots & \ddots & \ddots & \ddots & \vdots & \vdots \\
0 & 0 & \cdots & 0 & 2x_{n-2} & 1 & 0 & -2x_{n-2} & -1 & 0 & 0 & 0 \\
0 & 0 & 0 & 0 & \cdots & 0 & 2x_n & 1 & 0 & -2x_n & -1 & 0 \\
1 & 0 & 0 & 0 & 0 & 0 & 0 & 0 & 0 & 0 & \cdots & 0
\end{bmatrix}
\begin{bmatrix}
a_1 \\ b_1 \\ c_1 \\ a_2 \\ b_2 \\ c_2 \\ \vdots \\ \vdots \\ a_{n-1} \\ b_{n-1} \\ c_{n-1} \\ a_n \\ b_n \\ c_n
\end{bmatrix}
=
\begin{bmatrix}
y_0 \\ y_1 \\ y_1 \\ y_2 \\ y_2 \\ \vdots \\ y_{n-1} \\ y_{n-1} \\ y_n \\ 0 \\ 0 \\ \vdots \\ 0
\end{bmatrix}
$$

$$(6.34)$$

The solution of Equation (6.34) on the previous page provides the $3n$ unknowns, namely, a_i, b_i, c_i for $i = 1, 2, \ldots, n$. The value of y at a given x is obtained in the surrounding j^{th} interval $[x_{j-1}, x_j]$ from the corresponding quadratic spline, i.e., $y_j(x)$.

Example 6.2.5

Fit a set of quadratic splines to the following data:

x	0	1	2	3	4
y	2	3	1	5	4

Use the splines to calculate $y(2.5)$.

Solution

Applying Equation (6.34) on the previous page to the above data for $n = 4$, we obtain

$$
\begin{bmatrix}
0 & 0 & 1 & 0 & 0 & 0 & 0 & 0 & 0 & 0 & 0 & 0 \\
1 & 1 & 1 & 0 & 0 & 0 & 0 & 0 & 0 & 0 & 0 & 0 \\
0 & 0 & 0 & 1 & 1 & 1 & 0 & 0 & 0 & 0 & 0 & 0 \\
0 & 0 & 0 & 4 & 2 & 1 & 0 & 0 & 0 & 0 & 0 & 0 \\
0 & 0 & 0 & 0 & 0 & 0 & 4 & 2 & 1 & 0 & 0 & 0 \\
0 & 0 & 0 & 0 & 0 & 0 & 9 & 3 & 1 & 0 & 0 & 0 \\
0 & 0 & 0 & 0 & 0 & 0 & 0 & 0 & 0 & 9 & 3 & 1 \\
0 & 0 & 0 & 0 & 0 & 0 & 0 & 0 & 0 & 16 & 4 & 1 \\
2 & 1 & 0 & -2 & -1 & 0 & 0 & 0 & 0 & 0 & 0 & 0 \\
0 & 0 & 0 & 4 & 1 & 0 & -4 & -1 & 0 & 0 & 0 & 0 \\
0 & 0 & 0 & 0 & 0 & 0 & 6 & 1 & 0 & -6 & -1 & 0 \\
1 & 0 & 0 & 0 & 0 & 0 & 0 & 0 & 0 & 0 & 0 & 0
\end{bmatrix}
\begin{bmatrix}
a_1 \\ b_1 \\ c_1 \\ a_2 \\ b_2 \\ c_2 \\ a_3 \\ b_3 \\ c_3 \\ a_4 \\ b_4 \\ c_4
\end{bmatrix}
=
\begin{bmatrix}
2 \\ 3 \\ 3 \\ 1 \\ 1 \\ 5 \\ 5 \\ 4 \\ 0 \\ 0 \\ 0 \\ 0
\end{bmatrix}
$$

The solution of the above equation is

$$
\begin{bmatrix} a_1 \\ b_1 \\ c_1 \end{bmatrix} = \begin{bmatrix} 0 \\ 1 \\ 2 \end{bmatrix}, \quad
\begin{bmatrix} a_2 \\ b_2 \\ c_2 \end{bmatrix} = \begin{bmatrix} -3 \\ 7 \\ -1 \end{bmatrix}, \quad
\begin{bmatrix} a_3 \\ b_3 \\ c_3 \end{bmatrix} = \begin{bmatrix} 9 \\ -41 \\ 47 \end{bmatrix}, \quad \text{and} \quad
\begin{bmatrix} a_4 \\ b_4 \\ c_4 \end{bmatrix} = \begin{bmatrix} -14 \\ 97 \\ -160 \end{bmatrix}
$$

$$\underbrace{}_{\text{for } x \in [0, 1]} \qquad \underbrace{}_{\text{for } x \in [1, 2]} \qquad \underbrace{}_{\text{for } x \in [2, 3]} \qquad \underbrace{}_{\text{for } x \in [3, 4]}$$

Since, $x = 2.5$ lies in the third interval $[2, 3]$,

$$y(2.5) = \underbrace{9}_{a_3} \times 2.5^2 + \underbrace{(-41)}_{b_3} \times 2.5 + \underbrace{7}_{c_3} = 0.75.$$

□

Note that in quadratic splines, the specification of zero derivative results in a straight line interpolation between the first two data points. This is obviated in cubic splines.

Cubic Splines

These are cubic polynomials for each subset of two consecutive data points, or the corresponding x-interval. Over the ith interval, $[x_{i-1}, x_i]$, cubic splines are given by

$$y_i(x) = a_i x^3 + b_i x^2 + c_i x + d_i, \qquad x_{i-1} \le x \le x_i; \qquad i = 1, 2, \ldots, n$$

where a_i, b_i, c_i and d_i are $4n$ unknown coefficients. Their determination requires $4n$ equations, which can be obtained from the same procedure as followed above for quadratic splines. There is an efficient procedure, however, which reduces the number of equations. That procedure is as follows.

Let p and q denote, respectively, the first and second derivatives of y with respect to x. Then the equations for cubic spline interpolation are obtained in the following three steps:

Step 1

We express q over the ith interval, $[x_{i-1}, x_i]$ using Lagrange's interpolation formula [see Equation (6.33), p. 213] for $n = 1$. Thus,

$$q_i(x) = \frac{x - x_i}{x_{i-1} - x_i} q(x_{i-1}) + \frac{x - x_{i-1}}{x_i - x_{i-1}} q(x_i), \qquad x_{i-1} \le x \le x_i; \qquad i = 1, 2, \ldots, n$$

Step 2

Next, we integrate the above equations twice. The integration constants are resolved using the condition $y_i(x) = y(x_i)$ at each intermediate x_i. This results in

$$y_i(x) = \frac{q(x_{i-1})}{6(x_i - x_{i-1})} (x_i - x)^3 + \frac{q(x_i)}{6(x_i - x_{i-1})} (x - x_{i-1})^3$$

$$+ \left[\frac{y(x_{i-1})}{x_i - x_{i-1}} - \frac{q(x_{i-1})(x_i - x_{i-1})}{6} \right] (x_i - x)$$

$$+ \left[\frac{y(x_i)}{x_i - x_{i-1}} - \frac{q(x_i)(x_i - x_{i-1})}{6} \right] (x - x_{i-1}); \qquad i = 1, 2, \ldots, n \qquad (6.35)$$

The above equation has unknown qs at x_{i-1} and x_i. They are found in the next step.

Step 3

We differentiate the last set of equations to obtain $p_i(x)$ for $i = 1, 2, \ldots, n$. Considering the equality of dy/dx at x_i obtained from $p_i(x)$ and $p_{i+1}(x)$, we get

$$(x_i - x_{i-1})q(x_{i-1}) + 2(x_{i+1} - x_{i-1})q(x_i) + (x_{i+1} - x_i)q(x_{i+1})$$

$$= \frac{6}{x_{i+1} - x_i}[y(x_{i+1}) - y(x_i)] + \frac{6}{x_i - x_{i-1}}[y(x_{i-1}) - y(x_i)]$$

$$i = 1, 2, \ldots, n-1$$

With the specification of $q(x_0) = q(x_n) = 0$, the above equations can be expressed as

$$
\begin{bmatrix}
2(x_2-x_0) & (x_2-x_1) & 0 & 0 & \cdots & & \cdots & & 0 \\
(x_2-x_1) & 2(x_3-x_1) & (x_3-x_2) & 0 & \cdots & & \cdots & & 0 \\
0 & (x_3-x_2) & 2(x_4-x_2) & (x_4-x_3) & \cdots & & \cdots & & \vdots \\
0 & 0 & (x_4-x_3) & 2(x_5-x_3) & (x_5-x_4) & & \cdots & & \vdots \\
\vdots & \vdots & \vdots & \vdots & \ddots & & \vdots & & \vdots \\
\vdots & \vdots & \vdots & 0 & (x_{n-2}-x_{n-3}) & 2(x_{n-1}-x_{n-3}) & (x_{n-1}-x_{n-2}) \\
0 & 0 & 0 & \cdots & 0 & (x_{n-1}-x_{n-2}) & 2(x_n-x_{n-2})
\end{bmatrix}
\begin{bmatrix}
q(x_1) \\
q(x_2) \\
\vdots \\
\vdots \\
\vdots \\
q(x_{n-2}) \\
q(x_{n-1})
\end{bmatrix}
$$

$$
= 6
\begin{bmatrix}
\dfrac{y(x_2)-y(x_1)}{x_2-x_1} + \dfrac{y(x_0)-y(x_1)}{x_1-x_0} \\[2ex]
\dfrac{y(x_3)-y(x_2)}{x_3-x_2} + \dfrac{y(x_1)-y(x_2)}{x_2-x_1} \\[2ex]
\vdots \\[2ex]
\vdots \\[2ex]
\vdots \\[2ex]
\dfrac{y(x_{n-1})-y(x_{n-2})}{x_{n-1}-x_{n-2}} + \dfrac{y(x_{n-3})-y(x_{n-2})}{x_{n-2}-x_{n-3}} \\[2ex]
\dfrac{y(x_n)-y(x_{n-1})}{x_n-x_{n-1}} + \dfrac{y(x_{n-2})-y(x_{n-1})}{x_{n-1}-x_{n-2}}
\end{bmatrix}
\tag{6.36}
$$

Equation (6.36) above is solved once to obtain the $q(x_i)$s. The value of y at a given x is then obtained in the surrounding j^{th} interval $[x_{j-1}, x_j]$ from Equation (6.35) on the previous page for $i = j$.

Example 6.2.6

Fit a set of cubic splines to the data of Example 6.2.5 on p. 216. Compare the fit with that of the quadratic splines.

Solution

Applying Equation (6.36) on the previous page to those data, we obtain

$$\begin{bmatrix} 4 & 1 & 0 \\ 1 & 4 & 1 \\ 0 & 1 & 4 \end{bmatrix} \begin{bmatrix} q(1) \\ q(2) \\ q(3) \end{bmatrix} = \begin{bmatrix} -18 \\ 36 \\ -30 \end{bmatrix}$$

The solution of the above equation is

$$\begin{bmatrix} q(1) & q(2) & q(3) \end{bmatrix} = \begin{bmatrix} -7.93 & 13.71 & -10.93 \end{bmatrix}$$

Note that $q(0) = q(4) = 0$. With these $q(x_i)$-values, Equation (6.35) on p. 217 enables the calculation of y at a given value of x. Thus, $y(2.5) = 2.83$ from that equation for $i = 3$, i.e., the third x-interval.

Comparison with Quadratic Splines

Figure 6.4 below shows the y-values obtained from cubic splines of this example, and quadratic splines of Example 6.2.5 on p. 216, along with data points. The y-values outside the x-interval $[0, 4]$ are linear extrapolations from the terminal splines.

From the figure, it is observed that cubic splines provide a tighter fit to the data than quadratic splines. The latter have oscillations that grow beyond $x = 2$. No such oscillations are present in cubic splines. The value of $y(2.5)$ is 0.75 from quadratic splines as opposed to 2.83 from cubic splines.

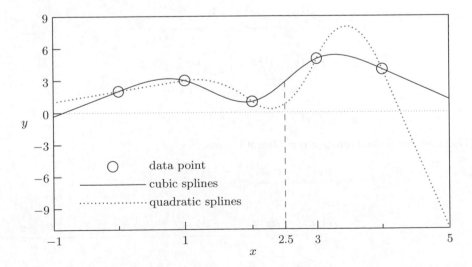

Figure 6.4 The y-values obtained from cubic splines of this example, and quadratic splines of Example 6.2.5, along with data points

6.A   Linear Function

Let f be a function of the variable vector* $\mathbf{x} = \begin{bmatrix} x_1 & x_2 & \cdots & x_n \end{bmatrix}^\mathsf{T}$. Then f is a linear function if it satisfies the superposition principle, i.e.,

$$f[\alpha_1 \mathbf{x}^{(1)} + \alpha_2 \mathbf{x}^{(2)}] = \alpha_1 f[\mathbf{x}^{(1)}] + \alpha_2 f[\mathbf{x}^{(2)}]$$

for all $\mathbf{x}^{(1)}$ and $\mathbf{x}^{(2)}$ in the x-domain, and all scalars α_1 and α_2.

Example 6.A.1

Determine whether the following functions:

   i. $f(\mathbf{x}) = x_1 + x_2$

   ii. $g(\mathbf{x}) = x_1 + x_2^2$

are linear with respect to $\mathbf{x} = \begin{bmatrix} x_1 & x_2 \end{bmatrix}^\mathsf{T}$.

Solution

In order to be linear, $f(\mathbf{x})$ and $g(\mathbf{x})$ must satisfy the superposition principle.
   Given two arbitrary variable vectors in the x-domain,

$$\mathbf{x}^{(1)} = \begin{bmatrix} x_1^{(1)} \\ x_2^{(1)} \end{bmatrix} \quad \text{and} \quad \mathbf{x}^{(2)} = \begin{bmatrix} x_1^{(2)} \\ x_2^{(2)} \end{bmatrix}$$

the first function is linear because

$$f[\alpha_1 \mathbf{x}^{(1)} + \alpha_2 \mathbf{x}^{(2)}] = [\underbrace{\alpha_1 x_1^{(1)} + \alpha_2 x_1^{(2)}}_{x_1}] + [\underbrace{\alpha_1 x_2^{(1)} + \alpha_2 x_2^{(2)}}_{x_2}]$$

$$= \alpha_1 [\underbrace{x_1^{(1)} + x_2^{(1)}}_{f[\mathbf{x}^{(1)}]}] + \alpha_2 [\underbrace{x_1^{(2)} + x_2^{(2)}}_{f[\mathbf{x}^{(2)}]}]$$

$$= \alpha_1 f[\mathbf{x}^{(1)}] + \alpha_2 f[\mathbf{x}^{(2)}]$$

However, the second function is not linear because

$$g[\alpha_1 \mathbf{x}^{(1)} + \alpha_2 \mathbf{x}^{(2)}] = [\underbrace{\alpha_1 x_1^{(1)} + \alpha_2 x_1^{(2)}}_{x_1}] + [\underbrace{\alpha_1 x_2^{(1)} + \alpha_2 x_2^{(2)}}_{x_2}]^2$$

$$= \alpha_1 \{ \underbrace{x_1^{(1)} + \alpha_1 [x_2^{(1)}]^2 + 2\alpha_2 x_2^{(1)} x_2^{(2)}}_{\neq g[\mathbf{x}^{(1)}]} \} + \alpha_2 \{ \underbrace{x_1^{(2)} + \alpha_2 [x_2^{(2)}]^2}_{\neq g[\mathbf{x}^{(2)}]} \}$$

$$\neq \alpha_1 g[\mathbf{x}^{(1)}] + \alpha_2 g[\mathbf{x}^{(2)}]$$

❑

---

*It could be a scalar with just one element.

General Form

The general form of a linear function of $\mathbf{x} = \begin{bmatrix} x_1 & x_2 & \cdots & x_n \end{bmatrix}^\top$ is

$$f(\mathbf{x}) = \mathbf{a} \cdot \mathbf{x} + b$$

where $\mathbf{a} = \begin{bmatrix} a_1 & a_2 & \cdots & a_n \end{bmatrix}^\top$ is a vector of constants, and b is another constant.

Linearity Check

A quick way to determine whether a function is linear is to check the interaction of the variables. A function is linear if its variables satisfy *all* of the following conditions:

1. There is no multiplication or division between variables.

2. No variable is raised to a power other than unity.

3. No variable is an argument to a transcendental function (such as exponential, logarithmic and trigonometric functions).

Put differently, a function is non-linear if any of the above conditions is violated.

6.B   Proof of Buckingham Pi Theorem

Consider n variables, y_0 to y_{n-1}, which can be related as

$$y_0 = f(y_1, y_2, \ldots, y_{n-1}) \tag{6.37}$$

Let the set of y_is carry m dimensions, D_1 to D_m. Then according to the Buckingham pi theorem, the above relation can be expressed as

$$\Pi_0 = F(\Pi_1, \Pi_2, \ldots, \Pi_{n-m-1})$$

where Π_is are dimensionless numbers.

Scaling Transformation

Note that the dimensions of y_i are given by

$$[y_i] = D_1^{\alpha_{i1}} \times D_2^{\alpha_{i2}} \times \cdots \times D_m^{\alpha_{im}} ; \qquad i = 0, 1, \ldots, n-1$$

Thus, when the first dimension is scaled by a factor e^β to give the transformed dimension,

$$\tilde{D}_1 = e^\beta D_1$$

the variables get transformed accordingly to

$$\tilde{y}_i = (e^\beta)^{\alpha_{i1}} y_i; \qquad i = 0, 1, \ldots, n-1$$

Because of dimensional consistency on both sides of Equation (6.37) above, we have

$$\tilde{y}_0 = f(\tilde{y}_1, \tilde{y}_2, \ldots, \tilde{y}_{n-1})$$

Elimination of D_1

Let y_j for $j > 0$ be the variable depending on D_1, i.e., with α_{j1} not zero. Then we define

$$z_i = y_i y_j^{-\alpha_{i1}/\alpha_{j1}}; \qquad i = 0, 1, \ldots, j-1$$

$$z_{n-1} = y_j$$

$$z_{i-1} = y_i y_j^{-\alpha_{i1}/\alpha_{j1}}; \qquad i = j+1, j+2, \ldots, n-1$$

Let the z_is be related to each other as

$$z_0 = g(z_1, z_2, \ldots, z_{n-1}) \tag{6.38}$$

Then if z_is undergo the aforementioned scaling transformation then

$$\tilde{z}_0 = g(\tilde{z}_1, \tilde{z}_2, \ldots, \tilde{z}_{n-1}) \tag{6.39}$$

Invariance

It is interesting to note that z_is other than z_{n-1} are invariant with respect to the dimension D_1 since they are not affected by the scaling transformation. For example,

$$\tilde{z}_i = e^{\beta \alpha_{i1}} y_i (e^{\beta \alpha_{j1}} y_j)^{-\alpha_{i1}/\alpha_{j1}} = y_i y_j^{-\alpha_{i1}/\alpha_{j1}}$$

$$= z_i; \qquad i = 0, 1, \ldots, n-2$$

Thus, we can write Equation (6.39) above as

$$z_0 = g(z_1, z_2, \ldots, z_{n-2}, e^{\beta \alpha_{j1}} z_{n-1})$$

Comparison of the last equation with Equation (6.38) above shows that g is independent of z_{n-1}. Scaling it does not change z_0. Consequently, we can write

$$z_0 = g(z_1, z_2, \ldots, z_{n-2})$$

where z_is are invariant with respect to D_1 as shown above. Moreover, the number of arguments of g is reduced by one.

Repeated Elimination

Repeating the above steps, we can eliminate the remaining dimensions one by one using a new y_j each time, and obtain the final relation

$$\Pi_0 = F(\Pi_1, \Pi_2, \ldots, \Pi_{n-m-1})$$

where

$$\Pi_i \equiv \frac{y_i}{y_j^{\alpha_{i1}/\alpha_{j1}} \times y_{j+1}^{\alpha_{i2}/\alpha_{(j+1),2}} \times \cdots \times y_{j+m-1}^{\alpha_{im}/\alpha_{(j+m-1),m}}}; \quad i = 0, 1, \ldots, n-m-1$$

is invariant with respect to *all* dimensions, or is a dimensionless number.

Simplification of Π_i

In terms of

$$q_k \equiv y_{j+k-1}, \qquad a_k \equiv \frac{\alpha_{ik}}{\alpha_{(j+k-1),k}}; \qquad k = 1, 2, \ldots, m$$

$$\text{and} \quad p_i \equiv y_i; \qquad i = 0, 1, \ldots, n - m - 1$$

we can write

$$\Pi_i = \frac{p_i}{q_1^{a_1} q_2^{a_2} \cdots q_m^{a_m}}; \qquad i = 0, 1, \ldots, n - m - 1$$

where

1. a q_i is a variable y_i, which carries the ith dimension D_i,
2. no two q_is are the same, and
3. the p_is are also the y_is, but are other than the q_is.

6.C Newton's Optimization Method

This method is used to find the root of the derivative of a twice-differentiable function, $f(\mathbf{x})$. At that root, the function satisfies the necessary condition to be optimum.

The second order Taylor expansion of $f(\mathbf{x})$ at $\mathbf{x}^{(k+1)}$ near a reference point $\mathbf{x}^{(k)}$ is given by

$$f[\mathbf{x}^{(k+1)}] = f[\mathbf{x}^{(k)}] + \mathbf{f}'[\mathbf{x}^{(k)}]^{\top} \left[\mathbf{x}^{(k+1)} - \mathbf{x}^{(k)} \right]$$

$$+ \frac{1}{2} \left[\mathbf{x}^{(k+1)} - \mathbf{x}^{(k)} \right]^{\top} \mathbf{H}[\mathbf{x}^{(k)}] \left[\mathbf{x}^{(k+1)} - \mathbf{x}^{(k)} \right] \qquad (6.40)$$

where $\mathbf{x} = \begin{bmatrix} x_1 & x_2 & \cdots & x_n \end{bmatrix}^{\top}$, $\mathbf{f}'[\mathbf{x}^{(k)}]$ is the vector of partial derivatives of f (i.e., the gradient of f) evaluated at $\mathbf{x}^{(k)}$, and is given by

$$\mathbf{f}'[\mathbf{x}^{(k)}] = \begin{bmatrix} \left.\frac{\partial f}{\partial x_1}\right|_{\mathbf{x}^{(k)}} & \left.\frac{\partial f}{\partial x_2}\right|_{\mathbf{x}^{(k)}} & \cdots & \left.\frac{\partial f}{\partial x_n}\right|_{\mathbf{x}^{(k)}} \end{bmatrix}^{\top}$$

and $\mathbf{H}[\mathbf{x}^{(k)}]$ is the matrix of second order partial derivatives of $f(\mathbf{x})$, or Hessian, evaluated at $\mathbf{x}^{(k)}$, i.e.,

$$
\mathbf{H}[\mathbf{x}^{(k)}] = \begin{bmatrix}
\dfrac{\partial^2 f}{\partial x_1{}^2}\bigg|_{\mathbf{x}^{(k)}} & \dfrac{\partial^2 f}{\partial x_1 \partial x_2}\bigg|_{\mathbf{x}^{(k)}} & \cdots & \dfrac{\partial^2 f}{\partial x_1 \partial x_n}\bigg|_{\mathbf{x}^{(k)}} \\[2ex]
\dfrac{\partial^2 f}{\partial x_2 \partial x_1}\bigg|_{\mathbf{x}^{(k)}} & \dfrac{\partial^2 f}{\partial x_2{}^2}\bigg|_{\mathbf{x}^{(k)}} & \cdots & \dfrac{\partial^2 f}{\partial x_2 \partial x_n}\bigg|_{\mathbf{x}^{(k)}} \\[2ex]
\vdots & \vdots & \ddots & \vdots \\[2ex]
\dfrac{\partial^2 f}{\partial x_n \partial x_1}\bigg|_{\mathbf{x}^{(k)}} & \dfrac{\partial^2 f}{\partial x_n \partial x_2}\bigg|_{\mathbf{x}^{(k)}} & \cdots & \dfrac{\partial^2 f}{\partial x_n{}^2}\bigg|_{\mathbf{x}^{(k)}}
\end{bmatrix}
$$

Differentiating Equation (6.40) on the previous page throughout with respect to $\mathbf{x}^{(k+1)}$, and assuming that it is the desired root of the derivative of $f(\mathbf{x})$, we obtain

$$
\underbrace{\mathbf{f}'[\mathbf{x}^{(k+1)}]}_{=0} = \mathbf{f}'[\mathbf{x}^{(k)}] + \mathbf{H}[\mathbf{x}^{(k)}]\left[\mathbf{x}^{(k+1)} - \mathbf{x}^{(k)}\right]
$$

The above equation upon rearrangement yields the algorithm of Newton's optimization method, i.e.,

$$
\mathbf{x}^{(k+1)} = \mathbf{x}^{(k)} - \left\{\mathbf{H}[\mathbf{x}^{(k)}]\right\}^{-1}\mathbf{f}'[\mathbf{x}^{(k)}]; \qquad k = 0, 1, \ldots
$$

References

[1] P. Kundra et al. "Experimental Determination of Concentration-Dependent Diffusivity of Carbon Dioxide in Polypropylene". In: *Journal of Chemical & Engineering Data* 56 (2011), pp. 21–26.

Bibliography

[1] W.B. Krantz. *Scaling Analysis in Modeling Transport and Reaction Processes: A Systematic Approach to Model Building and the Art of Approximation.* Hoboken, NJ: John Wiley & Sons, Inc., 2007.

[2] G.W. Bluman and S. Kumei. *Symmetries and Differential Equations.* 2nd edition. Chapter 1. Berlin: Springer-Verlag, 1989.

[3] C.C. Lin and L.A. Segel. *Mathematics Applied to Deterministic Problems in the Natural Sciences.* Chapter 6. Philadelphia: Society for Industrial and Applied Mathematics, 1988.

[4] M.M. Denn. *Process Modeling.* Chapter 6. Massachusetts: Pitman Publishing Inc., 1986.

[5] S.C. Chapra and R.P. Canale. *Numerical Methods for Engineers*. 6th edition. Chapters 17,18. New York: McGraw-Hill, 2010.

Exercises

6.1 The steady state reactant concentration c in a fluid flowing in a circular pipe is given by

$$v \frac{\partial c}{\partial z} = \frac{D}{r} \frac{\partial}{\partial r} \left(r \frac{\partial c}{\partial r} \right) \qquad v = \frac{3\bar{v}}{2} \left[1 - \left(\frac{r}{R} \right)^2 \right]$$

$$z = 0 : \quad c = c_0 \quad \forall \ 0 < r \leq R$$

$$r = \left\{ \begin{array}{ll} 0 : & \dfrac{\partial c}{\partial r} = 0 \\[3mm] R : & kc = -D \dfrac{\partial c}{\partial r} \end{array} \right\} \quad \forall \ 0 \leq z \leq L$$

where v is fluid velocity under laminar flow condition with an average of $\bar{v}$, z is axial variable, r is radial variable, D is reactant diffusivity in liquid, R is inside pipe radius, c_0 is inlet reactant concentration, k is the rate coefficient of the reaction at pipe wall, and L is pipe length.

Simplify the above model for **(a)** $kR/D \ll 1$ and **(b)** $kR/D \gg 1$.

6.2 Among the T and P measurements of Example 6.1.2 on p. 200, which one needs to be taken more accurately. Justify your answer with calculations.

6.3 Heat transfer in a turbulent fluid in a pipe involves its diameter (d) as well as the velocity (v), density (ρ), viscosity (μ), thermal conductivity (k), heat capacity ($\hat{C}_{\text{P}}$), and heat transfer coefficient (h) of the fluid. Apply Buckingham pi theorem to relate these variables through dimensionless numbers.

Recombine these numbers as necessary into Nusselt, Prandtl, and Reynold numbers, which are defined, respectively, as

$$\text{Nu} = \frac{hd}{k}, \qquad \text{Pr} = \frac{\mu \hat{C}_{\text{P}}}{k} \qquad \text{and} \qquad \text{Re} \equiv \frac{dv\rho}{\mu}$$

6.4 For the least-squares fit of a plane

$$z = ax + by + c$$

on the data points:

$$(x_1, y_1, z_1) \quad (x_2, y_2, z_2) \quad \ldots \quad (x_n, y_n, z_n)$$

derive the equations to determine the coefficients a, b and c.

6.5 Derive the divided difference formula for the polynomial interpolation of equispaced data points.

6.6 Write a computer program to generate quadratic splines. Enable linear extrapolation outside the end points of a given data set. Use the program to fit the following data:

x	3	4.5	7	9	11	13
y	2	30	4	50	6	7

6.7 Solve the last problem using cubic splines. Compare the results with those obtained with quadratic splines in that problem.

7

Process Simulation

Process simulation is the solution of process models. While analytical methods provide solutions as explicit analytical expressions for the unknowns, the equations involved in most process models are too difficult to be solved in this manner. Numerical methods circumvent this problem by approximating model equations with Taylor expansions, and often involve iterative calculations. They require initial guesses, which improve and converge to solutions. Their accuracies therefore depend upon the convergence criteria, and the approximations they utilize.

In this chapter, we focus on simple and effective numerical methods that are widely used in process simulation to solve algebraic equations, ordinary differential equations, and partial differential equations.

7.1 Algebraic Equations

In process modeling, algebraic equations typically result from the steady state description of processes with lumped parameters. These equations also result from the finite difference approximations of differential equations.

We first describe the numerical solution of linear algebraic equations, which forms the basis for the solution of non-linear algebraic equations.

7.1.1 Linear Algebraic Equations

In general, a set of linear algebraic equations can be expressed as

$$a_{00}x_0 + a_{01}x_1 + \cdots + a_{0,n-1}x_{n-1} = b_0$$

$$a_{10}x_0 + a_{11}x_1 + \cdots + a_{1,n-1}x_{n-1} = b_1$$

$$\vdots \qquad\qquad \vdots \;\; \vdots$$

$$a_{n-1,0}x_0 + a_{n-1,1}x_1 + \cdots + a_{n-1,n-1}x_{n-1} = b_{n-1}$$

Process Modeling and Simulation for Chemical Engineers: Theory and Practice, First Edition. Simant Ranjan Upreti.
© 2017 John Wiley & Sons Ltd. Published 2017 by John Wiley & Sons Ltd.
Companion website: www.wiley.com/go/upreti/pms_for_chemical_engineers

where a_{ij}s and b_is are constants, and x_is are variables. In matrix notation,

$$\overbrace{\begin{bmatrix} a_{00} & a_{01} & \cdots & a_{0,n-1} \\ a_{10} & a_{01} & \cdots & a_{0,n-1} \\ \vdots & \vdots & \ddots & \vdots \\ a_{n-1,0} & a_{n-1,1} & \cdots & a_{n-1,n-1} \end{bmatrix}}^{\mathbf{A}} \overbrace{\begin{bmatrix} x_0 \\ x_1 \\ \vdots \\ x_{n-1} \end{bmatrix}}^{\mathbf{x}} = \overbrace{\begin{bmatrix} b_0 \\ b_1 \\ \vdots \\ b_{n-1} \end{bmatrix}}^{\mathbf{b}}$$

where the function $\mathbf{f(x)} \equiv \mathbf{Ax} - \mathbf{b}$ is linear with respect to $\mathbf{x}$. The vector $\mathbf{x}$ for which $\mathbf{f(x)} = \mathbf{0}$ is the solution of the equations. It is also called the root of $\mathbf{f}$.

To solve linear algebraic equations, we present the method of lower-upper decomposition, followed by forward and back substitution. This method is illustrated for a set of three equations as follows.

Lower-Upper Decomposition

This is the first step in which a matrix $\mathbf{A}$ is decomposed into lower and upper triangular matrices, which are, respectively,

$$\mathbf{L} = \begin{bmatrix} 1 & 0 & 0 \\ l_{10} & 1 & 0 \\ l_{20} & l_{21} & 1 \end{bmatrix} \quad \text{and} \quad \mathbf{U} = \begin{bmatrix} u_{00} & u_{01} & u_{02} \\ 0 & u_{11} & u_{12} \\ 0 & 0 & u_{22} \end{bmatrix}$$

such that

$$\mathbf{A} = \mathbf{LU} \tag{7.1}$$

Forward Substitution

This is the second step in which an intermediate vector $\mathbf{d} = \begin{bmatrix} d_0 & d_1 & d_2 \end{bmatrix}^{\top}$ is obtained by solving the set of equations

$$\mathbf{Ld} = \mathbf{b} \tag{7.2}$$

starting from the first equation. Note that the first equation immediately yields d_0, which when substituted forward into the second equation yields d_1. Finally, d_0 and d_1 when substituted forward into the third equation yield d_3.

Backward Substitution

This is the last step in which the solution $\mathbf{x}$ is obtained by solving the set of equations

$$\mathbf{Ux} = \mathbf{d} \tag{7.3}$$

starting from the last equation. Note that the last equation straightaway yields x_2, which when substituted backward into the previous (i.e., second) equation yields x_1. Finally, x_2 and x_1 when substituted backward into the first equation yield x_0.

Proof

The solution vector $\mathbf{x}$ satisfies Equation (7.3) on the previous page. Multiplying both sides of the equation by $\mathbf{L}$, we get,

$$\underbrace{\mathbf{LU}}_{\mathbf{A}}\,\mathbf{x} \;=\; \underbrace{\mathbf{Ld}}_{\mathbf{b}}$$

In the above equation, the coefficient of $\mathbf{x}$ on the left-hand side is $\mathbf{A}$ from Equation (7.1) on the previous page. Also, the right-hand side is $\mathbf{b}$ from Equation (7.2) on the previous page. In other words, $\mathbf{x}$ is the solution of

$$\mathbf{A}\mathbf{x} \;=\; \mathbf{b}$$

Method for Lower-Upper Decomposition

Based on Gaussian elimination, this method employs elementary row operations systematically to yield triangular matrices $\mathbf{L}$ and $\mathbf{U}$ such that their product is $\mathbf{A}$. The operations are presented in three steps. The goal is to eliminate the elements below the diagonal in

$$\mathbf{A} \;=\; \begin{bmatrix} a_{00} & a_{01} & a_{02} \\ a_{10} & a_{11} & a_{12} \\ a_{20} & a_{21} & a_{22} \end{bmatrix}$$

Step 1 The first row of $\mathbf{A}$ is selected, and called the *pivotal row*. Its element that is on the diagonal, a_{00}, is the *pivotal element*. Its column is the *pivotal column*. To minimize round-off errors in operations that will follow, the pivotal row is switched with another suitable row to obtain a pivotal element with maximum absolute value in the pivotal column. At the same time, the identical rows of $\mathbf{b}$ are switched so that the equations in the set, $\mathbf{A}\mathbf{x} = \mathbf{b}$, do not change. This process is known as **partial pivoting**.

    a. The first row is multiplied throughout by the factor $l_{10} = a_{10}/a_{00}$, and subtracted from the second row. The result is the modified second row in the matrix shown below with a_{10} eliminated.

$$\begin{bmatrix} a_{00} & a_{03} & a_{02} \\ 0 & a_{11}^{(1)} & a_{12}^{(1)} \\ a_{20} & a_{21} & a_{22} \end{bmatrix}$$

    The superscript '$^{(1)}$' denotes the modified elements.

    b. Similarly, the third row is modified. The 1st row is multiplied by the factor $l_{20} = a_{20}/a_{00}$, and subtracted from the third row to obtain the modified third row as shown below with a_{20} eliminated.

$$\begin{bmatrix} a_{00} & a_{03} & a_{02} \\ 0 & a_{11}^{(1)} & a_{12}^{(1)} \\ 0 & a_{21}^{(1)} & a_{22}^{(1)} \end{bmatrix}$$

At this point, all elements below the pivotal element in the same column are eliminated.

Step 2 The second row is selected as the pivotal row with the diagonal element, $a_{11}^{(1)}$, as the pivotal element. If partial pivoting is done, the l_{ij}s corresponding to the two switched rows are also interchanged.

    a. The second row is multiplied by the factor $l_{21} = a_{21}^{(1)}/a_{11}^{(1)}$, and subtracted from the third row to obtain the modified third row with $a_{21}^{(1)}$ eliminated.

    At this point, all elements below the diagonal are eliminated, and the resulting matrix is the upper triangular matrix,

$$\mathbf{U} = \begin{bmatrix} a_{00} & a_{03} & a_{02} \\ 0 & a_{11}^{(1)} & a_{12}^{(1)} \\ 0 & 0 & a_{22}^{(2)} \end{bmatrix}$$

    The superscript '(2)' denotes the modified element.

Step 3 The factors l_{ij}s used in the previous steps are collected according to their subscripts to form the lower triangular matrix,

$$\mathbf{L} = \begin{bmatrix} 1 & 0 & 0 \\ l_{10} & 1 & 0 \\ l_{20} & l_{21} & 1 \end{bmatrix}$$

Remarks

Using matrix multiplication, and expanding l_{ij}s, readers can verify that the multiplication of **L** and **U** yields a matrix, say, $\mathbf{A}'$, which is **A**, but after the rows switches that happen due to any partial pivoting. Note that the above steps apply the same row switches in **b** during partial pivoting, and modify **b** to, say, $\mathbf{b}'$. Thus, eventually when any i^{th} row of **A** becomes the kth row in $\mathbf{A}'$, b_i becomes b_k'. As a result, the solution of $\mathbf{Ax} = \mathbf{b}$ is equivalent to that of $\mathbf{A}'\mathbf{x} = \mathbf{b}'$. It is this last set of equations that is solved using the decomposed **L** and **U** in forward and backward substitution. In the absence of partial pivoting, $\mathbf{A}' = \mathbf{A}$ and $\mathbf{b}' = \mathbf{b}$.

Example 7.1.1

Solve the following equation using the method of lower-upper decomposition, and forward and backward substitution:

$$
\underbrace{\begin{bmatrix} 10^{-9} & 3 & -2 \\ -5 & 4 & 8 \\ 3 & -20 & 4 \end{bmatrix}}_{A} \underbrace{\begin{bmatrix} x_0 \\ x_1 \\ x_2 \end{bmatrix}}_{x} = \underbrace{\begin{bmatrix} 1 \\ 2 \\ 3 \end{bmatrix}}_{b} \tag{7.4}
$$

Solution

We begin with lower-upper decomposition of A.

Lower-Upper Decomposition

The steps of the method, which was described earlier in Section 7.1.1 on p. 229, are as follows.

Step 1

The first row is the pivotal row, and the pivotal element is $a_{00} = 10^{-9}$, which is not the absolute maximum in its column. Therefore, we need to do partial pivoting, i.e., switch appropriate rows of A in order to have the maximum absolute value for a_{00} in the first column. Note that the same rows of b need to be switched simultaneously.

The following operations are carried out in this step:

$$
\begin{bmatrix} 10^{-9} & 3 & -2 \\ -5 & 4 & 8 \\ 3 & -20 & 4 \end{bmatrix} \xrightarrow[\text{(and corresponding } b_0 \text{ and } b_1)]{\textbf{pivot: switch top two rows}} \begin{bmatrix} -5 & 4 & 8 \\ 10^{-9} & 3 & -2 \\ 3 & -20 & 4 \end{bmatrix}
$$

$$
\xrightarrow[\text{second row} - \underbrace{\left(\dfrac{10^{-9}}{-5}\right)}_{l_{10}} \times \text{ first row}]{\text{replace second row with:}} \begin{bmatrix} -5 & 4 & 8 \\ 0 & 3 & -2 \\ 3 & -20 & 4 \end{bmatrix} \xrightarrow[\text{third row} - \underbrace{\left(\dfrac{3}{-5}\right)}_{l_{20}} \times \text{ first row}]{\text{replace third row with:}}
$$

$$
\longrightarrow \begin{bmatrix} -5 & 4 & 8 \\ 0 & 3 & -2 \\ 0 & -17.6 & 8.8 \end{bmatrix}
$$

Step 2

Here, the second row is the pivotal row, and a_{11} is the pivotal element. The operations for this step are as follows:

$$\xrightarrow[\text{(and corresponding } b_1 \text{ and } b_2, \text{ and } l_{10} \text{ and } l_{20})]{\textbf{pivot}: \text{ switch bottom two rows}} \begin{bmatrix} -5 & 4 & 8 \\ 0 & -17.6 & 8.8 \\ 0 & 3 & -2 \end{bmatrix} \longrightarrow$$

$$\xrightarrow[\text{third row} - \underbrace{\left(\dfrac{3}{17.6}\right)}_{l_{21}} \times \text{ second row}]{\text{replace third row with:}} \begin{bmatrix} -5 & 4 & 8 \\ 0 & -17.6 & 8.8 \\ 0 & 0 & -0.5 \end{bmatrix} = \mathbf{U}$$

Step 3

In the final step, we gather the l_{ij}s to form

$$\mathbf{L} = \begin{bmatrix} 1 & 0 & 0 \\ l_{10} & 1 & 0 \\ l_{20} & l_{21} & 0 \end{bmatrix} = \begin{bmatrix} 1 & 0 & 0 \\ -0.6 & 1 & 0 \\ -2 \times 10^{-10} & -0.1705 & 1 \end{bmatrix}$$

Effect of Partial Pivoting

Note that during partial pivoting in Step 2 above, l_{10} and l_{20} were swapped. Now the product of $\mathbf{L}$ and $\mathbf{U}$ gives

$$\begin{bmatrix} -5 & 4 & 8 \\ 3 & -20 & 4 \\ 10^{-9} & 3 & -2 \end{bmatrix} \equiv \mathbf{A}'$$

which is $\mathbf{A}$ *after* applying the row switches of Steps 1 and 2. This $\mathbf{A}'$ corresponds to $\mathbf{b}'$, which is $\mathbf{b}$ with its elements swapped in Steps 1 and 2. Thus, the equation to be solved is

$$\underbrace{\begin{bmatrix} -5 & 4 & 8 \\ 3 & -20 & 4 \\ 10^{-9} & 3 & -2 \end{bmatrix}}_{\mathbf{A}'} \underbrace{\begin{bmatrix} x_0 \\ x_1 \\ x_2 \end{bmatrix}}_{\mathbf{x}} = \underbrace{\begin{bmatrix} 2 \\ 3 \\ 1 \end{bmatrix}}_{\mathbf{b}'}$$

Note that the above equation is equivalent to Equation (7.4) on the previous page.

Forward Substitution

We find **d** in this step as follows:

$$\mathbf{Ld} = \mathbf{b'} \Rightarrow \begin{bmatrix} 1 & 0 & 0 \\ -0.6 & 1 & 0 \\ -2 \times 10^{-10} & -0.1705 & 1 \end{bmatrix} \begin{bmatrix} d_0 \\ d_1 \\ d_2 \end{bmatrix} = \begin{bmatrix} 2 \\ 3 \\ 1 \end{bmatrix}$$

$$\begin{aligned} d_0 &= 2 \\ \Rightarrow \quad d_1 &= 3 - l_{10}d_0 = 4.2 \\ d_2 &= 1 - l_{20}d_0 - l_{21}d_1 = 1.7159 \end{aligned} \qquad \Rightarrow \quad \mathbf{d} = \begin{bmatrix} 2 \\ 4.2 \\ 1.7159 \end{bmatrix}$$

Backward Substitution

Finally, we find **x** as follows:

$$\mathbf{Ux} = \mathbf{d} \Rightarrow \begin{bmatrix} -5 & 4 & 8 \\ 0 & -17.6 & 8.8 \\ 0 & 0 & -0.5 \end{bmatrix} \begin{bmatrix} x_0 \\ x_1 \\ x_2 \end{bmatrix} = \begin{bmatrix} 2 \\ 4.2 \\ 1.7159 \end{bmatrix}$$

$$\begin{aligned} x_2 &= \frac{1.7159}{-0.5} = -3.4318 \\ \Rightarrow \quad x_1 &= \frac{4.2 - 8.8x_2}{-17.6} = -1.9546 \quad \Rightarrow \quad \mathbf{x} = \begin{bmatrix} -7.4546 \\ -1.9546 \\ -3.4318 \end{bmatrix} \\ x_0 &= \frac{2 - 4x_1 - 8x_2}{2} = -7.4546 \end{aligned}$$

❑

Inverse of Square Matrix

The above method to solve linear algebraic equations also enables the determination of the inverse of a square matrix. Given a square matrix $\mathbf{A}$, its inverse $\mathbf{A}^{-1}$ is the solution of the following equation:

$$\underbrace{\begin{bmatrix} a_{00} & a_{01} & \cdots & a_{0,n-1} \\ a_{10} & a_{11} & \cdots & a_{0,n-1} \\ \vdots & \vdots & \cdots & \vdots \\ a_{n-1,0} & a_{n-1,1} & \cdots & a_{n-1,n-1} \end{bmatrix}}_{\mathbf{A}} \underbrace{\begin{bmatrix} \tilde{a}_{00} & \tilde{a}_{01} & \cdots & \tilde{a}_{0,n-1} \\ \tilde{a}_{10} & \tilde{a}_{01} & \cdots & \tilde{a}_{0,n-1} \\ \vdots & \vdots & \ddots & \vdots \\ \tilde{a}_{n-1,0} & \tilde{a}_{n-1,1} & \cdots & \tilde{a}_{n-1,n-1} \end{bmatrix}}_{\mathbf{A}^{-1}} = \underbrace{\begin{bmatrix} 1 & 0 & \cdots & 0 \\ 0 & 1 & \cdots & 0 \\ \vdots & \vdots & \ddots & \vdots \\ 0 & 0 & \cdots & 1 \end{bmatrix}}_{\mathbf{I}}$$

where the ith column of $\mathbf{A}^{-1}$ satisfies

$$\mathbf{A} \underbrace{\begin{bmatrix} \tilde{a}_{0i} \\ \tilde{a}_{1i} \\ \vdots \\ \tilde{a}_{n-1,i} \end{bmatrix}}_{\mathbf{x}_i} = \underbrace{\begin{bmatrix} b_0 \\ b_1 \\ \vdots \\ b_{n-1} \end{bmatrix}}_{\mathbf{b}_i} ; \qquad i = 0, 1, \ldots, n-1 \qquad (7.5)$$

with all elements of $\mathbf{b}_i$ zero except the ith one, which is unity. Observe that the ith column of $\mathbf{A}^{-1}$ is the solution of the above equation. Hence, the inverse of a square matrix is determinable as follows:

1. $\mathbf{A}$ is decomposed into $\mathbf{L}$ and $\mathbf{U}$.

2. To obtain the ith column of $\mathbf{A}^{-1}$, the vector $\mathbf{b}_i$ is obtained by setting all its elements to zero except b_i, which is set to unity.

3. Using forward substitution, $\mathbf{d}$ is obtained from $\mathbf{Ld} = \mathbf{b}_i$.

4. Using backward substitution, $\mathbf{x}_i$ is obtained from $\mathbf{Ux}_i = \mathbf{d}$. This $\mathbf{x}_i$ is the ith column of $\mathbf{A}^{-1}$.

5. Steps 2–4 are repeated to obtain the rest of the columns of $\mathbf{A}^{-1}$.

If during the decomposition of $\mathbf{A}$, its ith row eventually became the kth row of $\mathbf{A}'$ (due to one or more instances of partial pivoting) then i is set to k in Step 2 above. See Appendix 7.A on p. 281 for the explanation.

Example 7.1.2

Find the inverse of $\mathbf{A}$ in Example 7.1.1 on p. 231.

Solution

We will make use of $\mathbf{L}$ and $\mathbf{U}$ that were determined in that example. Due to partial pivoting, $\mathbf{LU} = \mathbf{A}'$, which is different from the original $\mathbf{A}$. Similarly, $\mathbf{b}'$ is different from the original $\mathbf{b}$. Compared to the originals, the rows in $\mathbf{A}'$ and $\mathbf{b}'$ are switched. Specifically,

$$b_2' = b_0, \qquad b_0' = b_1 \qquad \text{and} \qquad b_1' = b_2 \qquad (7.6)$$

First column of $\mathbf{A}^{-1}$

To find this column, we set the first element of $\mathbf{b}$ to unity, and rest of the elements to zero, i.e.,

$$\mathbf{b} = \begin{bmatrix} 1 & 0 & 0 \end{bmatrix}^{\top}$$

The above $\mathbf{b}$ from Equation (7.6) above corresponds to

$$\mathbf{b}' = \begin{bmatrix} 0 & 0 & 1 \end{bmatrix}^{\top}$$

We use this $\mathbf{b}'$, and solve equation $\mathbf{A}'\mathbf{x}_1 = \mathbf{b}'$ using the lower-upper decomposition, and forward and backward substitution:

$$\mathbf{Ld} = \mathbf{b}' \quad \Rightarrow \quad \mathbf{d} = \begin{bmatrix} 0 & 0 & 1 \end{bmatrix}^\mathsf{T}$$

$$\mathbf{Ux}_1 = \mathbf{d} \quad \Rightarrow \quad \mathbf{x}_1 = \begin{bmatrix} -4 & -1 & -2 \end{bmatrix}^\mathsf{T}$$

The solution $\mathbf{x}$ is the first column of $\mathbf{A}^{-1}$.

Second column of $\mathbf{A}^{-1}$

To find the second column of $\mathbf{A}^{-1}$, we set the second element of $\mathbf{b}$ to unity, and the rest of the elements to zero. This $\mathbf{b}$ from Equation (7.6) on the previous page corresponds to

$$\mathbf{b}' = \begin{bmatrix} 1 & 0 & 0 \end{bmatrix}$$

With this $\mathbf{b}'$, we solve equation $\mathbf{A}'\mathbf{x}_2 = \mathbf{b}'$ as before to obtain

$$\mathbf{x}_2 = \begin{bmatrix} -0.6364 & -0.1364 & -0.2045 \end{bmatrix}^\mathsf{T}$$

which is the second column of $\mathbf{A}^{-1}$.

Third column of $\mathbf{A}^{-1}$

Finally, to find the third column of $\mathbf{A}^{-1}$, we set the third element of $\mathbf{b}$ to unity, and the rest of the elements to zero. This $\mathbf{b}$ from Equation (7.6) on the previous page corresponds to

$$\mathbf{b}' = \begin{bmatrix} 0 & 1 & 0 \end{bmatrix}$$

With this $\mathbf{b}'$, we solve equation $\mathbf{A}'\mathbf{x}_3 = \mathbf{b}'$ as before to obtain

$$\mathbf{x}_3 = \begin{bmatrix} -0.7273 & -0.2273 & -0.3409 \end{bmatrix}^\mathsf{T}$$

which is the third column of $\mathbf{A}^{-1}$.

Putting the three columns together, we obtain,

$$\mathbf{A}^{-1} = \begin{bmatrix} -4 & -0.6364 & -0.7273 \\ -1 & -0.1364 & -0.2273 \\ -2 & -0.2045 & -0.3409 \end{bmatrix}$$

❑

7.1.2 Non-Linear Algebraic Equations

A set of non-linear algebraic equations can be written in general as

$$f_0(x_0, x_1, \ldots, x_{n-1}) = 0$$
$$f_1(x_0, x_1, \ldots, x_{n-1}) = 0$$
$$\vdots \qquad \vdots \ \vdots$$
$$f_{n-1}(x_0, x_1, \ldots, x_{n-1}) = 0$$

or, in vector notation, $$\mathbf{f}(\mathbf{x}) = \mathbf{0}$$

where $\mathbf{f}$ is a non-linear function of $\mathbf{x}$. These equations are solvable using the Newton–Raphson method.

Newton–Raphson Method

This method involves a computational algorithm to solve non-linear algebraic equations using derivative information.

For a single equation, $f(x) = 0$, Figure 7.1 below illustrates the steps of the algorithm. The tangent $f'[x^{(0)}]$ to the function at the initial guess $x^{(0)}$ provides an improved estimate $x^{(1)}$ for the root, i.e., the solution of $f(x) = 0$. Another tangent at $x^{(1)}$ does the same by providing a further improved estimate $x^{(2)}$. This procedure is repeated until the function value at the final estimate is sufficiently close to zero at the root x_r.

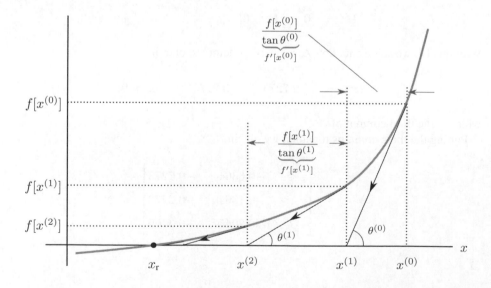

Figure 7.1   Steps of the Newton–Raphson method for a single equation, $f(x) = 0$

Thus, the Newton–Raphson method has the following algorithm:

$$x^{(k+1)} = x^{(k)} - \frac{f[x^{(k)}]}{f'[x^{(k)}]}; \qquad k = 0, 1, 2, \ldots \tag{7.7}$$

For a set of simultaneous algebraic equations, $\mathbf{f}(\mathbf{x}) = \mathbf{0}$, the algorithm becomes

$$\mathbf{x}^{(k+1)} = \mathbf{x}^{(k)} - \mathbf{J}^{-1}[\mathbf{x}^{(k)}]\,\mathbf{f}[\mathbf{x}^{(k)}]; \qquad k = 0, 1, 2, \ldots \tag{7.8}$$

where $\mathbf{J}[\mathbf{x}^{(k)}]$ is the matrix of partial derivatives of $\mathbf{f}$, or Jacobian, evaluated at $\mathbf{x}^{(k)}$, i.e.,

$$\mathbf{J}[\mathbf{x}^{(k)}] = \begin{bmatrix} \left.\frac{\partial f_0}{\partial x_0}\right|_{\mathbf{x}^{(k)}} & \left.\frac{\partial f_0}{\partial x_1}\right|_{\mathbf{x}^{(k)}} & \cdots & \left.\frac{\partial f_0}{\partial x_{n-1}}\right|_{\mathbf{x}^{(k)}} \\[2ex] \left.\frac{\partial f_1}{\partial x_0}\right|_{\mathbf{x}^{(k)}} & \left.\frac{\partial f_1}{\partial x_1}\right|_{\mathbf{x}^{(k)}} & \cdots & \left.\frac{\partial f_1}{\partial x_{n-1}}\right|_{\mathbf{x}^{(k)}} \\[2ex] \vdots & \vdots & \ddots & \vdots \\[2ex] \left.\frac{\partial f_{n-1}}{\partial x_0}\right|_{\mathbf{x}^{(k)}} & \left.\frac{\partial f_{n-1}}{\partial x_1}\right|_{\mathbf{x}^{(k)}} & \cdots & \left.\frac{\partial f_{n-1}}{\partial x_{n-1}}\right|_{\mathbf{x}^{(k)}} \end{bmatrix}$$

Appendix 7.B on p. 281 provides the derivation of the Newton–Raphson method. Essentially, the method uses $\mathbf{x}^{(0)}$ to calculate the left-hand side of Equation (7.8) above, which is $\mathbf{x}^{(1)}$ – an improvement of $\mathbf{x}^{(0)}$.

In the next iteration, the method improves $\mathbf{x}^{(1)}$ to $\mathbf{x}^{(2)}$. As the iterations continue, the left-hand side of the equation improves progressively, and converges to the root. The iterations are stopped when any of the following happens:

1. The improvement in $\mathbf{x}^{(k)}$ becomes negligible.

   For example, given the vector of absolute relative errors,

$$\mathbf{e} = \begin{bmatrix} \left|1 - \dfrac{x_0^{(k)}}{x_0^{(k+1)}}\right| & \left|1 - \dfrac{x_1^{(k)}}{x_1^{(k+1)}}\right| & \cdots & \left|1 - \dfrac{x_{n-1}^{(k)}}{x_{n-1}^{(k+1)}}\right| \end{bmatrix}^{\top}$$

   either the magnitude, or the largest element of $\mathbf{e}$ becomes less than a specified, small positive number (i.e., accuracy).

2. The function vector at $\mathbf{x}^{(k)}$ gets sufficiently close to $\mathbf{0}$.

   For example, either the magnitude of $\mathbf{f}[\mathbf{x}^{(k)}]$, or the largest absolute value among the elements of $\mathbf{f}[\mathbf{x}^{(k)}]$ becomes less than a specified, small positive number (or accuracy).

3. The Jacobian becomes singular.

4. The number of iterations exceeds a specified maximum number.

Example 7.1.3

Obtain the molar volume V of n-butane at temperature $T = 510.24$ K, and different values of pressure (P) from 1 atm to the critical pressure P_c using Redlich–Kwong equation of state, i.e.,

$$z^3 - z^2 + (A - B - B^2)z - AB = 0$$

where

$$z = \frac{PV}{RT}, \quad A = 0.42748\frac{P_r}{T_r}F, \quad P_r = \frac{P}{P_c},$$

$$T_r = \frac{T}{T_c}, \quad F = \frac{1}{\sqrt{T_r^3}}, \quad \text{and} \quad B = 0.08664\frac{P_r}{T_r}.$$

The parameters in the above equations are $R = 82.06 \times 10^{-3}$ atm m^3 kmol^{-1} K^{-1}, $P_c = 37.45$ atm, $T_c = 425.15$ K, and $\omega = 0.193$.

Solution

For a given P in the specified range, we need to find the root z_r of the equation of state, and use the root to determine V from

$$V = \frac{z_r RT}{P}$$

For $f(z) \equiv z^3 - z^2 + (A - B - B^2)z - AB = 0$, the Newton–Raphson algorithm to find the root is [see Equation (7.7), previous page]

$$z^{(k+1)} = z^{(k)} - \frac{f[z^{(k)}]}{f'[z^{(k)}]}; \quad k = 0, 1, 2, \ldots$$

where $f'[z^{(k)}]$ is the derivative of f with respect to z, and evaluated at $z = z^{(k)}$. For $P = 1$ atm, Table 7.1 below shows the results of the Newton–Raphson iterations starting with an initial guess of $z_0 = 1.2$, and the accuracy of 10^{-8} in the absolute relative change in z_i, or $f(z_i)$. At convergence, $z = 0.9947$, which corresponds to $V = 41.6477$ m^3 kmol^{-1}.

Table 7.1 Iterations of the Newton–Raphson method for $P = 1$ atm

i	z_i	z_{i+1}	$\lvert 1 - z_i/z_{i+1}\rvert$	$f(z_i)$	$f'(z_i)$
0	1.2000	1.0471	1.5288×10^{-1}	2.9435×10^{-1}	1.9253
1	1.0471	0.9995	4.7648×10^{-2}	5.7198×10^{-2}	1.2004
2	0.9995	0.9947	4.7384×10^{-3}	4.7535×10^{-3}	1.0032
3	0.9947	0.9947	4.5477×10^{-5}	4.4763×10^{-5}	0.9843
4	0.9947	0.9947	4.1698×10^{-9}	4.1035×10^{-9}	0.9841

To obtain V for the next P, say, 1.1 atm we use the converged $z = 0.9947$ at 1 atm as the guess for the next round of the Newton–Raphson iterations. They converge to the new root of z, which yields the corresponding $\underline{V}$. These steps are thus repeated for increasing P. The results are plotted in Figure 7.2 below.

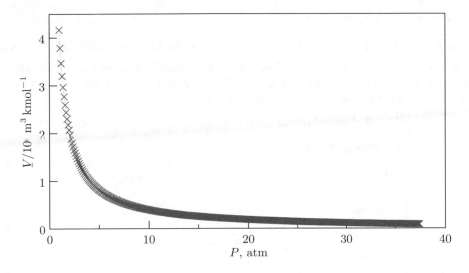

Figure 7.2   Molar volume of n-butane versus pressure at 510.24 K

❏

Example 7.1.4

Find the steady state concentrations of species in the CSTR described by Equations (4.1)–(4.4) on p. 82. Use the parameters given in Table 7.2 below. The volume of reactants in the reactor is 9.5 m³.

Table 7.2   Values of parameters for the CSTR in Example 7.1.4 above

parameter	value	parameter	value
c_{Af}	$2\ \text{kmol}\,\text{m}^{-3}$	c_{Bf}	$3\ \text{kmol}\,\text{m}^{-3}$
k_1	$3\ \text{m}^3\,\text{kmol}^{-1}\,\text{s}^{-1}$	k_2	$6\ \text{m}^3\,\text{kmol}^{-1}\,\text{s}^{-1}$
F_{Af}	$2\times10^{-3}\ \text{m}^3\,\text{s}^{-1}$	F_{Bf}	$2\times10^{-3}\ \text{m}^3\,\text{s}^{-1}$

Solution

Setting the time derivatives to zero in Equations (4.1)–(4.4), we obtain

$$f_0 \equiv \frac{F_{0f}c_{0f} - Fc_0}{V} - k_1 c_0 c_1 \equiv 0$$

$$f_1 \equiv \frac{F_{1f}c_{1f} - Fc_1}{V} - k_1 c_0 c_1 - k_2 c_1 c_2 \equiv 0$$

$$f_2 \equiv -\frac{F c_2}{V} + k_1 c_0 c_1 - k_2 c_1 c_2 \equiv 0$$

$$f_3 \equiv -\frac{F c_3}{V} + k_1 c_1 c_2 \equiv 0$$

or $\mathbf{f}(\mathbf{c}) = \mathbf{0}$ where $\mathbf{f} = \begin{bmatrix} f_0 & f_1 & f_2 & f_3 \end{bmatrix}^\top$ is the function vector, which depends upon $\mathbf{c} = \begin{bmatrix} c_0 & c_1 & c_2 & c_3 \end{bmatrix}^\top$. The latter is the variable vector of the concentration of species A, B, C and D denoted by subscripts '0', '1', '2' and '3', respectively. The Newton–Raphson algorithm for the above set of equations is

$$\mathbf{c}^{(k+1)} = \mathbf{c}^{(k)} - \mathbf{J}^{-1}[\mathbf{c}^{(k)}]\, \mathbf{f}[\mathbf{c}^{(k)}]; \quad k = 0, 1, 2, \dots$$

where $\mathbf{J}$ is the Jacobian given by

$$\mathbf{J} = \begin{bmatrix} \underbrace{-\dfrac{F}{V} - k_1 c_1}_{\partial f_0/\partial c_0} & \underbrace{- k_1 c_0}_{\partial f_0/\partial c_1} & \underbrace{0}_{\partial f_0/\partial c_2} & \underbrace{0}_{\partial f_0/\partial c_3} \\[2em] \underbrace{- k_1 c_1}_{\partial f_1/\partial c_0} & \underbrace{-\dfrac{F}{V} - k_1 c_0 - k_2 c_2}_{\partial f_1/\partial c_1} & \underbrace{- k_2 c_1}_{\partial f_1/\partial c_2} & \underbrace{0}_{\partial f_1/\partial c_3} \\[2em] \underbrace{k_1 c_1}_{\partial f_2/\partial c_0} & \underbrace{k_1 c_0 - k_2 c_2}_{\partial f_2/\partial c_1} & \underbrace{-\dfrac{F}{V} - k_2 c_1}_{\partial f_2/\partial c_2} & \underbrace{0}_{\partial f_2/\partial c_3} \\[2em] \underbrace{0}_{\partial f_3/\partial c_0} & \underbrace{k_1 c_2}_{\partial f_3/\partial c_1} & \underbrace{k_1 c_1}_{\partial f_3/\partial c_2} & \underbrace{-\dfrac{F}{V}}_{\partial f_3/\partial c_3} \end{bmatrix}$$

Given an initial guess $\mathbf{c}^{(0)} = \begin{bmatrix} 0.1 & 0.1 & 0.1 & 0.1 \end{bmatrix}^\top$, the algorithm yields

$$\mathbf{c}^{(1)} = \begin{bmatrix} 0.1 \\ 0.1 \\ 0.1 \\ 0.1 \end{bmatrix} - \underbrace{\begin{bmatrix} -0.3004 & -0.3000 & 0.0000 & 0.0000 \\ -0.3000 & -0.9004 & -0.6000 & 0.0000 \\ 0.3000 & -0.3000 & -0.6004 & 0.0000 \\ 0.0000 & 0.3000 & 0.3000 & -0.0004 \end{bmatrix}^{-1}}_{\mathbf{J}[\mathbf{c}^{(0)}]} \underbrace{\begin{bmatrix} -0.0296 \\ -0.0894 \\ -0.0300 \\ 0.0300 \end{bmatrix}}_{\mathbf{f}[\mathbf{c}^{(0)}]}$$

$$= \begin{bmatrix} 0.1505 & -0.0493 & 0.1498 & 0.3499 \end{bmatrix}^\top$$

Applying the algorithm repeatedly we get $\mathbf{c}^{(2)}$, $\mathbf{c}^{(3)}$, etc. until convergence. It is attained in six iterations with the specified accuracy of (i) machine epsilon, of $O(10^{-16})$, for the maximum, absolute relative change in an element of $\mathbf{c}^{(k)}$, and (ii) 10^{-8} for the maximum absolute value among the elements of $\mathbf{f}[\mathbf{c}^{(k)}]$. At convergence

$$\mathbf{c} = \begin{bmatrix} 0.2049 & 0.0006 & 0.0908 & 0.3522 \end{bmatrix}^\top$$

which is the vector of species concentrations in the CSTR at steady state.

❑

Secant Method

This method implements the Newton–Raphson method with finite difference approximation of the involved derivatives. The derivatives required for the solution of $\mathbf{f}(\mathbf{x}) = \mathbf{0}$ are given by

$$\frac{\partial \mathbf{f}}{\partial x_i} = \frac{\mathbf{f}(\tilde{\mathbf{x}}, x_i + \Delta x_i) - \mathbf{f}(\mathbf{x})}{\Delta x_i}$$

where $\tilde{\mathbf{x}}$ is $\mathbf{x}$ excluding x_i, and $\Delta x_i = \epsilon x_i$ with ϵ as a small, positive fraction such as 10^{-4}.

The secant method is helpful when dealing with complicated functions whose derivatives are difficult to obtain analytically. For a single equation, $f(x) = 0$, Figure 7.3 below illustrates the secant method. It provides improved estimates of the root using a chord connecting two points on the function curve at a given $\mathbf{x}$, and its neighborhood.

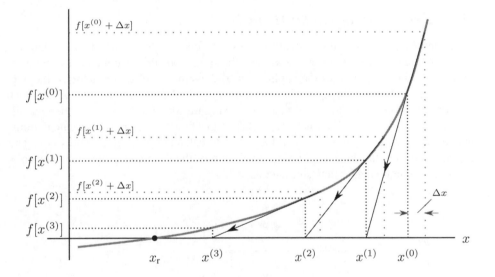

Figure 7.3   Steps of the secant method for a single equation, $f(x) = 0$

7.2   Differential Equations

Frequently encountered in process models, differential equations describe the change in system properties with space and time. A differential equation that involves one independent variable is called an **ordinary differential equation**. A differential equation with two or more independent variables is called a **partial differential equation**.

The analytical solution of a differential equation is the explicit, analytical expression for the dependent variable in terms of the involved independent variables. In numerical methods we obtain the set of values of the dependent variable for specific discrete values of the involved independent variables.

7.2.1 Ordinary Differential Equations

Ordinary differential equations typically describe temporal property changes in uniform, or lumped-parameter parameters. Under steady state conditions, these equations describe spatial property changes in distributed-parameter systems. These equations also result from finite difference approximation of partial differential equations.

Recall that the derivatives of order higher than one in differential equations can be expressed as additional dependent variables [see Section 5.4.1, p. 178]. Thus, a general form for ordinary differential equations is $dy/dt = f(t, y)$, where y is the vector of dependent variables, and t is the independent variable. We focus on the numerical solution of this form of equation with y specified at the initial t. Ordinary differential equations that are linear can be solved analytically using the well-developed theory [Section 8.13, p. 327], or Laplace transformation [Section 5.3, p. 161].

7.2.2 Explicit Runge–Kutta Methods

These methods have been used widely to solve ordinary differential equations. Utilizing a slope estimate at a given t where y is known, these methods find y at the next value of t.

Figure 7.4 below illustrates the application of the simplest of these methods, the explicit Euler's method, to solve $dy/dt = f(t, y)$. Given the initial condition, $y(0) = y_0$, this method estimates the slope of $y(t)$ at t_0 as $f(t_0, y_0)$. Using the slope, $y_1 \equiv y(t_1)$ is then calculated from $y_0 + f(t_0, y_0)(t_1 - t_0)$ where $(t_1 - t_0)$ is the first time step. These calculations are repeated at t_1 to calculate $y_2 \equiv y(t_2)$, then at t_2 to calculate $y_3 \equiv y(t_3)$, and so on. The set of y values thus obtained is the numerical solution of the differential equation. The solution is more accurate if smaller time steps are used.

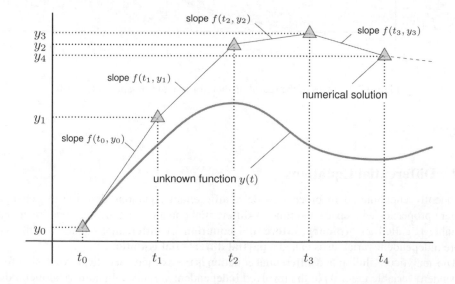

Figure 7.4   Application of the explicit Euler's method to solve $dy/dt = f(t, y)$ with $y(0) = y_0$

General Formulation

For the ordinary differential equation

$$\frac{d\mathbf{y}}{dt} = \mathbf{f}(t, \mathbf{y}), \quad \mathbf{y} = \mathbf{y}_0 \text{ at } t = t_0$$

The nth stage explicit Runge–Kutta method is given by

$$\mathbf{y}_{i+1} = \mathbf{y}_i + h \underbrace{\sum_{i=1}^{n} b_i \mathbf{k}_i}_{\text{slope estimate}}$$

where $\mathbf{y}_i \equiv \mathbf{y}(t_i)$, h is a constant interval between any t_i and t_{i+1}, and $\mathbf{k}_i$s are the different slopes, which are given by

$$\mathbf{k}_1 = \mathbf{f}(t_i, \mathbf{y}_i)$$
$$\mathbf{k}_2 = \mathbf{f}(t_i + c_2 h, \mathbf{y}_i + a_{21}\mathbf{k}_1 h)$$
$$\mathbf{k}_3 = \mathbf{f}[t_i + c_3 h, (\mathbf{y}_i + a_{31}\mathbf{k}_1 + a_{32}\mathbf{k}_2)h]$$
$$\vdots = \vdots$$
$$\mathbf{k}_n = \mathbf{f}[t_i + c_n h, (\mathbf{y}_i + a_{n1}\mathbf{k}_1 + a_{n2}\mathbf{k}_2 + \cdots + a_{nn}\mathbf{k}_n)h]$$

The coefficients c_is, b_is and a_{ij}s in the above equations are conveniently tabulated in a form called a *Butcher tableau*, which is shown in Table 7.3 below. Depending on the formulation, some of the coefficients may need to be specified.

Table 7.3 Butcher tableau for the nth order explicit Runge–Kutta method

0						
c_2	a_{21}					
c_3	a_{31}	a_{32}				
c_4	a_{41}	a_{42}	a_{43}			
$\vdots$	$\vdots$	$\vdots$	$\vdots$			
c_n	a_{n1}	a_{n2}	a_{n3}	$\cdots$	$a_{n,n-1}$	
	b_1	b_2	b_3	$\cdots$	b_{n-1}	b_n

For example, the formulation for the second stage explicit Runge–Kutta methods is

$$\mathbf{y}_{i+1} = \mathbf{y}_i + (b_1\mathbf{k}_1 + b_2\mathbf{k}_2)h$$
$$\mathbf{k}_1 = \mathbf{f}(t_i, \mathbf{y}_i)$$
$$\mathbf{k}_2 = \mathbf{f}(t_i + c_2 h, \mathbf{y}_i + a_{21}\mathbf{k}_1 h)$$

Using the first order Taylor expansion [Equation (8.12) on p. 322 for $n = 1$] for the right-hand side of the last equation, we get

$$\mathbf{k}_2 = \mathbf{f}(t_i, \mathbf{y}_i) + c_2 h \frac{\partial \mathbf{f}}{\partial t}\Big|_{t_i, \mathbf{y}_i} + a_{21} \mathbf{k}_1 h \frac{\partial \mathbf{f}}{\partial \mathbf{y}}\Big|_{t_i, \mathbf{y}_i} + O(h^2)$$

Combining the above equation with the previous ones for $\mathbf{k}_1$ and $\mathbf{y}_{i+1}$, we obtain

$$\mathbf{y}_{i+1} = \mathbf{y}_i + (b_1 + b_2)\mathbf{f}(t_i, \mathbf{y}_i)h + \left[b_2 c_2 \frac{\partial \mathbf{f}}{\partial t}\Big|_{t_i, \mathbf{y}_i} + b_2 a_{21} \mathbf{f}(t_i, \mathbf{y}_i) \frac{\partial \mathbf{f}}{\partial \mathbf{y}}\Big|_{t_i, \mathbf{y}_i} \right] h^2$$

$$+ O(h^3) \tag{7.9}$$

But, from the second order Taylor expansion [Equation (8.12) for $n = 2$]

$$\mathbf{y}_{i+1} = \mathbf{y}_i + \frac{d\mathbf{y}}{dt}\Big|_{t_i, \mathbf{y}_i} h + \frac{d^2\mathbf{y}}{dt^2}\Big|_{t_i, \mathbf{y}_i} \frac{h^2}{2!} + O(h^3)$$

$$= \mathbf{y}_i + \mathbf{f}(t_i, \mathbf{y}_i)h + \frac{d\mathbf{f}}{dt}\Big|_{t_i, \mathbf{y}_i} \frac{h^2}{2!} + O(h^3)$$

$$= \mathbf{y}_i + \mathbf{f}(t_i, \mathbf{y}_i)h + \left[\frac{\partial \mathbf{f}}{\partial t} + \frac{\partial \mathbf{f}}{\partial \mathbf{y}} \underbrace{\frac{d\mathbf{y}}{dt}}_{\mathbf{f}} \right]_{t_i, \mathbf{y}_i} \frac{h^2}{2!} + O(h^3) \tag{7.10}$$

Comparison of Equations (7.9) and (7.10) above results in three equations

$$b_1 + b_2 = 1 \qquad b_2 c_2 = 0.5 \qquad b_2 a_{21} = 0.5$$

whereas there are four unknowns, namely, c_2, a_{21}, b_1 and b_2. Thus, one unknown needs to be specified. Specification of $b_2 = 1$ results in the *midpoint method* with $c_2 = 0.5$, $a_{21} = 0.5$ and $b_1 = 0$. For *Heun's* and *Ralston's* methods, $b_2 = 0.5$ and $b_2 = 2/3$, respectively.

Note that the order of a Runge-Kutta method is the order of the derivative of $\mathbf{f}$ (not $\mathbf{y}$) that would appear in the first term of the truncation error. For the midpoint, Heun's and Ralston's methods, the truncation error is of $O(h^3)$, which means that these are second order methods.

Fourth Order Runge–Kutta Method

Higher order methods are derived in a similar manner as above. They have better accuracy but at the cost of increased function evaluations. A balanced choice for simple ordinary differential equations is the classical, fourth-order Runge–Kutta method, which is given by

$$\mathbf{y}_{i+1} = \mathbf{y}_i + \frac{h}{6}(\mathbf{k}_1 + 2\mathbf{k}_2 + 2\mathbf{k}_3 + \mathbf{k}_4) \tag{7.11}$$

$$\mathbf{k}_1 = \mathbf{f}(t_i, \mathbf{y}_i) \tag{7.12}$$

$$\mathbf{k}_2 = \mathbf{f}(t_i + 0.5h, \mathbf{y}_i + 0.5\mathbf{k}_1 h) \tag{7.13}$$

$$\mathbf{k}_3 = \mathbf{f}(t_i + 0.5h, \mathbf{y}_i + 0.5\mathbf{k}_2 h) \tag{7.14}$$

$$\mathbf{k}_4 = \mathbf{f}(t_i + h, \mathbf{y}_i + \mathbf{k}_3 h) \tag{7.15}$$

Example 7.2.1

Using the fourth order Runge–Kutta method with a step size of 0.1, integrate the following equations:

$$\frac{dy_0}{dt} = e^{-at} + \frac{y_0^2}{y_1} - y_0, \quad y_0(0) = 1 \tag{7.16}$$

$$\frac{dy_1}{dt} = y_0^2 - by_1, \qquad y_1(0) = 1 \tag{7.17}$$

up to $t = 1$ for $a = 0.1$ and $b = 1$.

Solution

Given a time instant, and the known values of $\mathbf{y}$ at that instant, we apply Equations (7.12)–(7.15) on the previous page to obtain the values of $\mathbf{y}$ from Equation (7.11) at the next time instant. At the ith time instant, the right-hand side of the differential equations is given by

$$\mathbf{f}(t_i, \mathbf{y}_i) = \mathbf{f}(t_i, y_{0i}, y_{1i}) = \begin{bmatrix} e^{-at_i} + \dfrac{y_{0i}^2}{y_{1i}} - y_{0i} \\[2mm] y_{0i}^2 - by_{1i} \end{bmatrix}$$

We begin with the subscript counter $i = 0$ when the time is $t_0 = 0$. From the initial condition,

$$\mathbf{y}_0 = \begin{bmatrix} 1 \\ 1 \end{bmatrix}$$

$$\mathbf{k}_1 = \mathbf{f}(t_0, \mathbf{y}_0) = \mathbf{f}(0, 1, 1) = \begin{bmatrix} 1 \\ 0 \end{bmatrix}$$

$$\mathbf{k}_2 = \mathbf{f}(t_0 + 0.5h, \mathbf{y}_0 + 0.5\mathbf{k}_1 h) = \mathbf{f}(0.05, 1.05, 1) = \begin{bmatrix} 1.0475 \\ 0.1025 \end{bmatrix}$$

$$\mathbf{k}_3 = \mathbf{f}(t_0 + 0.5h, \mathbf{y}_0 + 0.5\mathbf{k}_2 h) = \mathbf{f}(0.05, 1.0524, 1.0051) = \begin{bmatrix} 1.0445 \\ 0.1024 \end{bmatrix}$$

$$\mathbf{k}_4 = \mathbf{f}(t_0 + h, \mathbf{y}_0 + \mathbf{k}_3 h) = \mathbf{f}(0.1, 1.1044, 1.0102) = \begin{bmatrix} 1.0930 \\ 0.2096 \end{bmatrix}$$

$$\mathbf{y}_1 = \mathbf{y}_0 + \frac{h}{6}(\mathbf{k}_1 + 2\mathbf{k}_2 + 2\mathbf{k}_3 + \mathbf{k}_4) = \begin{bmatrix} 2.0462 \\ 1.1032 \end{bmatrix}$$

For the next step, we increment the counter i by one. Then the time is $t_1 = t_0 + h = 0.1$. We recalculate $\mathbf{k}_1$, $\mathbf{k}_2$, $\mathbf{k}_3$ and $\mathbf{k}_4$ to obtain $\mathbf{y}_2$. This procedure is repeated until $\mathbf{y}$ is obtained at $t = 1$. Figure 7.5 below plots the solution, i.e., values of $\mathbf{y}$ at the discrete time instants.

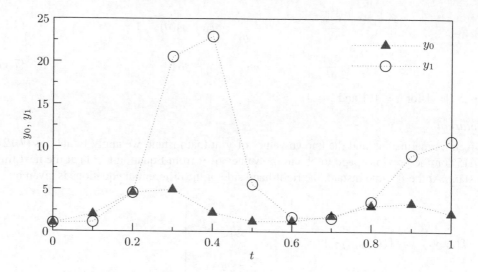

Figure 7.5 Solution of Equations (7.16) and (7.17) on the previous page using the fourth order Runge–Kutta method with a step size of 0.1

❑

7.2.3 *Step-Size Control*

When dependent variables change rapidly, the magnitudes of their derivatives are large, and lead to large truncation errors. To reduce them to maintain a desired accuracy, we must reduce the size of integration steps. This increases computations, which is necessary though. On the other hand, when the dependent variables change slowly, the derivatives are small. Then we can afford to use large step sizes, yet maintaining the accuracy with reduced computations. This is the concept of step-size control in numerical integration.

Since the analytical solution of a differential equation would not necessarily exist, we have to rely on an approximate but more accurate solution to determine the errors. This additional solution can be obtained by using multiple, smaller integration steps, or a higher order method in parallel. The last approach is efficiently implemented by embedded Runge–Kutta methods, which provide a nominal and a higher order (more accurate) solution simultaneously with common function evaluations.

Implementation

The step size of numerical integration is controlled using the procedure that is explained next. The error in an integration step is calculated from

$$\mathbf{e} = |\tilde{\mathbf{y}} - \mathbf{y}|$$

where $\tilde{\mathbf{y}}$ and $\mathbf{y}$ are, respectively, the more and less accurate solutions of the differential equation $d\mathbf{y}/dt = \mathbf{f}(\mathbf{y})$ at the next step. Each element of $\mathbf{e}$, i.e., e_i is normalized by a scaled absolute value $\hat{y}_i$ for the corresponding dependent variable, y_i. Then for a given accuracy ϵ, one requirement is*

$$\frac{e_i}{\hat{y}_i} \leq \epsilon \quad \text{or} \quad \frac{e_i}{\epsilon \hat{y}_i} \leq 1; \quad i = 0, 1, \ldots, n - 1$$

Let E be the maximum among all values of $e_i/(\epsilon \hat{y}_i)$. Then for the mth order method,

$$E \propto h^{m+1}$$

If integration carried out with the step size h_{new} results in E_{new} then

$$E_{\text{new}} \propto h_{\text{new}}^{m+1} \quad \Rightarrow \quad \frac{E_{\text{new}}}{E} = \left(\frac{h_{\text{new}}}{h}\right)^{m+1} \quad \Rightarrow \quad h_{\text{new}} = h\left(\frac{E_{\text{new}}}{E}\right)^{\frac{1}{m+1}}$$

Note that if h_{new} enables integration in the limit of the given accuracy then $E_{\text{new}} = 1$. Substitution of this result in the last equation yields

$$h_{\text{new}} = h\left(\frac{1}{E}\right)^{\frac{1}{m+1}} \tag{7.18}$$

which is the step size at the limit of the given accuracy.

Thus, the step-size control procedure at each integration step is as follows:

1. Obtain E based on a standard and more accurate solution.

2. If $E > 1$ then calculate h_{new} from Equation (7.18).

3. Repeat calculations until $E \leq 1$.

When the procedure is successful, the more accurate solution is used to advance the solution. This approach is known as *local extrapolation*.

The aforementioned procedure requires the specification for ϵ as well as $\hat{y}_i$s. A general specification for the latter is

$$\hat{y}_i = |y_i| + \left|h\frac{dy_i}{dt}\right|$$

7.2.4   Stiff Equations

Accurate solutions of certain differential equations require very small step sizes with explicit methods in general. Such equations are called stiff equations. The reason for stiffness is the presence of terms that cause rapid accumulation of truncation errors as numerical integration progresses.

---

*Using the norm of the normalized error is another choice, which is less stringent.

Consider, for example, the ordinary differential equation,

$$\frac{dy}{dt} = \underbrace{-\alpha y}_{f}, \qquad y(0) = 1 \tag{7.19}$$

where α is a positive constant. The above equation has the analytical solution, $y = e^{-\alpha t}$. We examine the numerical solutions obtained from the explicit and implicit Euler's methods given, respectively, by

$$y_{i+1} = y_i + f(t_i, y_i)h \quad \text{and} \quad y_{i+1} = y_i + f(t_{i+1}, y_{i+1})h; \qquad i = 0, 1, \ldots$$

The last method is implicit with respect to y_{i+1} because it uses f at the next time instant.

From the explicit Euler's method,

$$y_i = (1 - \alpha h)^i; \qquad i = 0, 1, \ldots$$

where h is a fixed time step. In order for the numerical solution to attain a stable value of y at infinite t,

$$\lim_{i \to \infty} y_i = \lim_{i \to \infty} (1 - \alpha h)^i = \lim_{t \to \infty} y(t) = \lim_{t \to \infty} e^{-\alpha t} = 0$$

This result in conjunction with Equation (7.19) above implies that

$$|(1 - \alpha h)| \leq 1 \qquad \text{or} \qquad h \leq \frac{2}{\alpha}$$

for a stable numerical solution.

The explicit Euler's method, on the other hand, yields the solution

$$y_i = \frac{1}{(1 + \alpha h)^i}; \qquad i = 0, 1, \ldots$$

which is stable regardless of the step size h because

$$\lim_{i \to \infty} y_i = \lim_{i \to \infty} \frac{1}{(1 + \alpha h)^i} = 0$$

Thus, for a stable solution when $\alpha = 2 \times 10^5$, the explicit method requires a step size less than 10^{-5}. It comes down to excessive computational overhead of more than 10^5 function evaluations over the $[0, 1]$ t-interval. In contrast, the implicit Euler's method does not have such restriction. It generates good quality solutions with significantly larger step sizes, and less computations than those with the explicit method.

Truncation Errors

Using Taylor series, the accumulated truncation errors in y_n calculated from the explicit and implicit Euler's methods are, respectively,

$$\epsilon_n = \left[(1 - \alpha h)^{n-1} e^{-\alpha t_0} + (1 - \alpha h)^{n-2} e^{-\alpha t_1} + \cdots + e^{-\alpha t_{n-1}} \right] C$$

$$\text{where} \quad C = \frac{(\alpha h)^2}{2!} - \frac{(\alpha h)^3}{3!} + \frac{(\alpha h)^4}{4!} - \cdots$$

and

$$\tilde{\epsilon}_n = \left[\frac{\tilde{e}^{-\alpha t_1}}{(1+\alpha h)^{n-1}} + \frac{\tilde{e}^{-\alpha t_2}}{(1+\alpha h)^{n-2}} + \cdots + \tilde{e}^{-\alpha t_n} \right] \tilde{C}$$

$$\text{where} \quad \tilde{C} = -\frac{(\alpha h)^2}{2!} - \frac{(\alpha h)^3}{3!} - \frac{(\alpha h)^4}{4!} - \cdots$$

Comparing term-wise with ϵ_n, the ith exponential term of $\tilde{\epsilon}_n$ has a more negative index, and $(1+\alpha h) > 1$ in the denominator. Moreover, $\tilde{C} < 0$. The implication is that for the same h, the truncation errors with the implicit method

1. are smaller than those with the explicit method, and

2. being always negative, cannot cause oscillations in the solution that may happen with the explicit method due to alternating signs of the terms in C.

Figure 7.6 below compares the numerical solutions for $\alpha = 17$ and $h = 0.125$. It is observed that the solution from the implicit method follows the true solution closely. On the other hand, the solution from the explicit method oscillates and diverges. A good quality solution from this method would require h considerably smaller than $2/\alpha$.

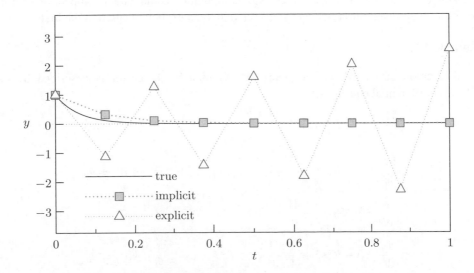

Figure 7.6　Comparison of the true solution of Equation (7.19) on the previous page with those obtained from implicit and explicit Euler's methods for $\alpha = 17$ and $h = 0.125$

In general, implicit integration methods are effective in solving stiff differential equations. The differential equations, if non-linear, must be linearized at each integration step when using these methods. This is explained next.

Linearization

Consider an implicit integration method of the form

$$\mathbf{y}_{i+1} = \mathbf{y}_i + h\mathbf{f}(\mathbf{y}_{i+1}); \qquad i = 0, 1, 2, \ldots$$

If $\mathbf{f}(\mathbf{y})$ is non-linear then it needs to be linearized before an explicit expression for $\mathbf{y}_{i+1}$ could be obtained.

Using the first order Taylor expansion for $\mathbf{f}$ at $\mathbf{y}_i$ [see Section 8.8.2, p. 323], we get

$$\mathbf{y}_{i+1} = \mathbf{y}_i + h\left[\mathbf{f}(\mathbf{y}_i) + \mathbf{J}_i(\mathbf{y}_{i+1} - \mathbf{y}_i)\right]; \qquad i = 0, 1, 2, \ldots$$

where $\mathbf{J}_i$ is the Jacobian of $\mathbf{f}$ with respect to $\mathbf{y}$, and is evaluated at $\mathbf{y}_i$. The above equation, upon rearrangement, yields

$$\mathbf{y}_{i+1} = \mathbf{y}_i + \left[\frac{\mathbf{I}}{h} - \mathbf{J}_i\right]^{-1}\mathbf{f}(\mathbf{y}_i); \qquad i = 0, 1, 2, \ldots$$

where $\mathbf{I}$ is the identity matrix.

Thus, to solve a stiff set of ordinary differential equations

$$\frac{d\mathbf{y}}{dt} = \mathbf{f}(t, \mathbf{y})$$

we need to provide the Jacobian of $\mathbf{f}$. The calculation of the new $\mathbf{y}$ at each step requires the inversion of the matrix $(\mathbf{I}/h - \mathbf{J})$. This extra computation is more than compensated for by the ability to take significantly larger steps relative to an explicit integration method.

Example 7.2.2

Polymerization of methyl methacrylate (MMA) in a batch reactor is described by the following differential equations:[1]

$$\frac{dV}{dt} = -K_p m \lambda V M_m\left[\frac{1}{\rho_p} - \frac{1}{\rho_m}\right], \qquad V(0) = V_0$$

$$\frac{dm}{dt} = -K_p m \lambda - \frac{m}{V}\frac{dV}{dt}, \qquad m(0) = m_0$$

$$\frac{di}{dt} = K_d i - \frac{i}{V}\frac{dV}{dt}, \qquad i(0) = i_0$$

$$\frac{ds}{dt} = -K_s s \lambda - \frac{s}{V}\frac{dV}{dt}, \qquad s(0) = s_0$$

$$\frac{dz}{dt} = -K_z z \lambda - \frac{z}{V}\frac{dV}{dt}, \qquad z(0) = z_0$$

$$\frac{dT}{dt} = \frac{-\Delta H K_p m \lambda}{\rho_m \hat{C}_P} - \frac{UA(T - T_j)}{V \rho_m \hat{C}_P}, \qquad T(0) = T_0$$

$$\frac{d\lambda}{dt} = 2f K_d i - K_t \lambda^2 - K_z z \lambda - \frac{\lambda}{V}\frac{dV}{dt}, \qquad \lambda(0) = \lambda_0$$

where (i) V is the volume of reactants, (ii) m, i, s and z, are, respectively, the concentrations of monomer (MMA), initiator, solvent and inhibitor, (iii) T is the temperature of reactants, (iv) λ is the zeroth moment of the polymer radical, (v) K_p, K_d, K_s, K_z and K_t are, respectively, the rate coefficients for propagation, initiation, solvent chain transfer, inhibitor chain transfer, and termination, (vi) M_m is the monomer molecular weight, (vii) ρ_m and ρ_p are, respectively, the densities of monomer and polymer, (viii) $-\Delta H$ is the heat of reaction, (ix) U and A are, respectively, the heat transfer coefficient, and area, (x) T_j is the coolant temperature in the reactor jacket, (xi) $\hat{C}_P$ is the specific heat capacity of the reaction mixture, and (xii) f is initiator efficiency. Expressions for the relevant variables are as follows.

$$K_p = K_{p1} K_{p2} \quad [\text{L mol}^{-1} \text{min}^{-1}]$$

$$\text{where} \quad K_{p1} = 2.95 \times 10^7 \exp\left[\frac{-4350}{R(T + 273.15)}\right] \quad [\text{L mol}^{-1} \text{min}^{-1}]$$

$$K_{p2} = \begin{cases} 7.1 \times 10^{-5} \exp(171.53 V_f), & \text{if } V_f < 0.05 \\ 1, & \text{if } V_f \geq 0.05 \end{cases}$$

$$V_f = 0.025 + \frac{\alpha_M (T - T_{gM}) m M_{wM}}{\rho_M} + \frac{\alpha_P (T - T_{gP})(m_0 - m) M_{wM}}{\rho_P}$$
$$+ \frac{\alpha_S (T - T_{gS}) s M_{wS}}{\rho_S} \quad [\text{L}]$$

$$K_t = K_{t1} K_{t2} \quad [\text{L mol}^{-1} \text{min}^{-1}]$$

$$\text{where} \quad K_{t1} = 3.12 \times 10^{10} \exp\left[\frac{-1394}{R(T + 273.15)}\right] \quad [\text{L mol}^{-1} \text{min}^{-1}]$$

$$K_{t2} = \begin{cases} 2.3 \times 10^{-6} \exp(75 V_f), & \text{if } V_f < V_{ft} \\ 0.10575 \exp(17.15 V_f - 0.01715T), & \text{if } V_f \geq V_{ft} \end{cases}$$

$$V_{ft} = 0.1856 - 2.965 \times 10^{-4} T \quad [\text{L}]$$

$$K_d = 1.014 \times 10^{16} \exp\left[\frac{-3 \times 10^4}{R(T + 273.15)}\right] \quad [\text{min}^{-1}]$$

$$K_s = 1.58 \times 10^{-5} K_p \quad \text{L mol}^{-1} \text{min}^{-1}$$

$$K_z = 10^6 K_p \quad \text{L mol}^{-1} \text{min}^{-1}$$

$$\rho_m = 309.85 \times 0.25357^{-[1-(T+273.15)/564]^{0.28571}} \quad [\text{g L}^{-1}]$$

$$\rho_p = 1.18 \times 10^3 - (T + 273.15) \quad [\text{g L}^{-1}]$$

$$\rho_s = 300.9 \times 0.2677^{-[1-(T+273.15)/562.16]^{0.2818}} \quad [\text{g L}^{-1}]$$

$$\hat{C}_P = 0.10113342 + 2.575065 \times 10^{-3}(T + 273.15) - 7.53159 \times 10^{-6}(T + 273.15)^2$$
$$+ 9.0138 \times 10^{-9}(T + 273.15)^3 \quad [\text{cal g MMA}^{-1} \, {}^\circ\text{C}^{-1}]$$

For the parameters given in Table 7.4 below, solve the above differential equations to obtain the dependent variables versus time for one hour of reactor operation. Analyze the effect of increase in the initial temperature of reactants.

Table 7.4   Parameters for the polymerization of MMA in a batch reactor

parameter	value	parameter	value
V_0	1 L	λ_0	0
m_0	$1.7648 \text{ mol L}^{-1}$	f	0.6
i_0	$4.13 \times 10^{-2} \text{ mol L}^{-1}$	M_m	$100.12 \text{ g mol}^{-1}$
s_0	$8.4692 \text{ mol L}^{-1}$	M_s	78.11 g mol^{-1}
z_0	$10^{-5} \text{ mol L}^{-1}$	UA	$156.49 \text{ cal min}^{-1}\,{}^\circ\text{C}^{-1}$
T_0, T_j	$90\,{}^\circ\text{C}$	$-\Delta H$	$13.2553 \times 10^3 \text{ cal mol}^{-1}$
T_{gM}	$-106\,{}^\circ\text{C}$	α_S, α_M	10^{-3}
T_{gP}	$114\,{}^\circ\text{C}$	α_P	4.8×10^{-4}
T_{gS}	$-102\,{}^\circ\text{C}$	R	$1.987 \text{ cal mol}^{-1}\,\text{K}^{-1}$

Solution

The above differential equations are very stiff. We present the results obtained using (i) the explicit, fifth order Runge–Kutta-Fehlberg (RKF) method with adaptive step-size control, and (ii) a stiff solver[2] with analytical expressions for the Jacobian elements. Figure 7.7 on the next page shows the dependent variables thus obtained with the integration accuracy, $\epsilon = 10^{-10}$ [see Section 7.2.3, p. 246]. The stiffness of the differential equations can be ascertained from the small steps [of $O(10^{-4})$ min] taken by the explicit method to meet the accuracy requirement [see Figure 7.8, next page]. In contrast, the stiff solver took considerably larger steps – as large as 27 min – to integrate the differential equations.

Computational Overhead

Figures 7.9 and 7.10 on p. 254 show computational overheads of the two integration methods for different initial temperatures, T_0s. Compared to the explicit method, the stiff solver

1. needed considerably less number of steps and derivative evaluations – at least by three orders of magnitude, and

2. notwithstanding the Jacobian evaluation at each step, used computation time, which was 10^3 to 10^4 times less, and mostly decreased with the rise in stiffness with T_0.

Effect of Temperature

The high initial temperature of reactants (T_0) contributes to the stiffness of the above differential equations. The reason is that temperature appears as $e^{-\alpha/T}$ in the kinetic rate expressions used in the differential equations. The terms of the form $e^{-\alpha/T}$ persist in the derivatives that make up the truncation errors of the numerical method. As a result, an increase in T, which means an increase in $e^{-\alpha/T}$, raises the truncation errors. To curb them, the method has to use smaller steps, and consequently do more computations. This is evident for the explicit method, especially when T_0 is increased from 120°C to 130°C. The effect of increase in T_0 on the stiff solver is relatively very mild.

❑

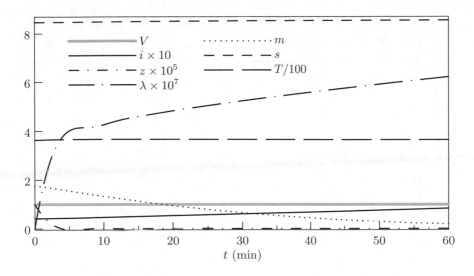

Figure 7.7   Dependent variables versus time in Example 7.2.2 on p. 250

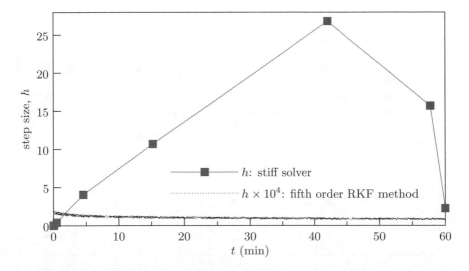

Figure 7.8   Sizes of steps taken by the stiff solver, and the explicit, fifth order RKF method in Example 7.2.2

7.3   Partial Differential Equations

A partial differential equation is a differential equation with more than one independent variable. These equations are frequently encountered in engineering practice. Those with time as independent variable describe unsteady state phenomena, which under steady state are described by ordinary differential equations.

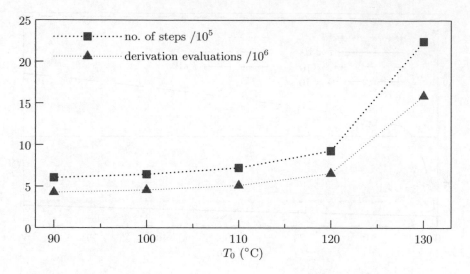

Figure 7.9 Computational overhead versus T_0 with the explicit, fifth order RKF method in Example 7.2.2 on p. 250

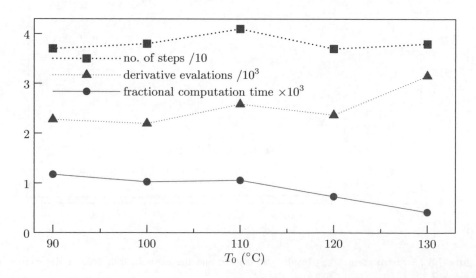

Figure 7.10 Computational overhead versus T_0 with the stiff solver in Example 7.2.2. Fractional computation time is the ratio of time taken by this solver to that by the explicit, fifth order RKF method

The majority of process models involving partial differential equations are expressed in Cartesian, cylindrical, or spherical coordinate system. In the simple geometries of these coordinates systems, the **method of finite differences** is an effective means for the numerical solution of partial differential equations. This method is illustrated next.

7.3.1   Finite Difference Method

Basically, this method replaces the derivatives in a partial differential equation with their finite difference approximations, or formulas. With systematic replacement of derivatives, this method converts

1. a partial differential equation into a set of simultaneous, ordinary differential equations, and

2. an ordinary differential equation into a set of simultaneous algebraic equations.

Either the intermediate ordinary differential equations or the final algebraic equations can be solved to obtain the solution of the original PDEs. The former approach is called the **method of lines**. It is often preferable because ordinary differential equations can be efficiently solved with adaptive step-size control, thereby reducing the amount of computation.

Finite Difference Formulas

These formulas are approximations of derivatives using discrete data points. Consider for example, the first order Taylor expansion [Equation (8.12) on p. 322 for $n = 1$] of a function $f(x)$ at a reference point $x = x_i$ for $n = 1$. With slight rearrangement, the series yields the finite difference formula for the first derivative of the function, i.e.,

$$\underbrace{\left[\frac{\mathrm{d}f}{\mathrm{d}x}\right]_{x_i}}_{f^{(1)}(x_i)} = \frac{f(\overbrace{x_i + h}^{\equiv\, x_{i+1}}) - f(x_i)}{h} - \underbrace{\frac{h}{2}\left[\frac{\mathrm{d}^2 f}{\mathrm{d}x^2}\right]_{\zeta}}_{O(h)}, \qquad x_i < \zeta < x_{i+1}$$

where $h = (x_{i+1} - x_i)$ is the step size. In the limit of $h \to 0$,

$$f^{(1)}(x_i) = \frac{f(x_{i+1}) - f(x_i)}{h}$$

The above equation is the *forward, first order finite difference formula* for the first derivative of $f(x)$. For a finite step size, the formula has the truncation error of $O(h)$.

In the same manner as above for a step size $-h$, we obtain the $O(h)$-accurate *backward, first order finite difference formula*

$$f^{(1)}(x_i) = \frac{f(x_i) - f(x_{i-1})}{h}$$

where $x_{i-1} = x_i - h$.

To obtain the *centered, first order finite difference formula*, we subtract Equation (8.12) for $n = 2$ and step size $-h$ from the same equation, but for step size h. This results in the $O(h^2)$-accurate formula,

$$f^{(1)}(x_i) = \frac{f(x_i + h) - f(x_i - h)}{2h} + O(h^2)$$

A general derivation of finite difference formulas is provided in Appendix 7.C on p. 284. Higher order formulas have higher accuracy but require more function values, and arithmetic operations. In most problems, the second order formulas provide a good balance between accuracy, and computational overhead. Appendix 7.C.4 on p. 289 provides the first and second order-accurate finite difference formulas for the derivatives of a function up to the fourth order.

In a given application of the finite difference method, the finite difference formulas of the same order of accuracy are used consistently. When a finite difference formula is needed at the 'left' boundary of the interval of an independent variable, a forward formula is used since it does not depend on function values at the backward points in the interval. Similarly, at the 'right' boundary of the interval, a backward finite difference formula is applicable.

Solution Procedure

The procedure to solve partial differential equations using the finite difference method is as follows:

1. An independent variable is selected, and its interval is discretized using grid points.

2. The differential equations are expressed at the internal grid points using finite difference formulas for the derivatives with respect to the selected variable.

3. The associated initial, or boundary conditions are expressed similarly at the first and last grid points. At either of them, if the dependent variable is not specified then the differential equation is first simplified using the associated condition, and then expressed as in Step 2 above.

   The outcome is a set of finite-differenced equations at each grid point, without the partial derivatives with respect to the selected independent variable. If the equations are ordinary differential equations then they can be integrated to obtain the solution, the approach being that of the **method of lines**.

4. The next independent variable is then selected. Starting at each existing grid point, the interval of the newly selected variable is discretized using grid points.

5. Steps 2–4 above are repeated for all independent variables that are involved.

   The outcome is a set of algebraic equations. They can be solved using a relevant method such as the Newton–Raphson method. The solution yields the values of dependent variables at each grid point.

Strategy

The quality of solution obtained with the above procedure depends on the number of grid points employed to obtain finite-differenced equations. More grid points mean smaller step sizes in finite difference formulas, thereby leading to smaller truncation errors, and a more accurate solution. However, there is a caveat. Increasing the number of grid points also results in more equations, which need more computation time for their solution. In addition, the arithmetic operations increase, which increase round-off errors. The latter begin to offset the decrease in truncation errors, and may reduce the solution accuracy beyond a certain number of grid points.

Therefore, a recommended strategy is to solve the equations with an increasing number of grid points until the change in the solution is either negligible, or small enough to be acceptable within a reasonable computational time limit.

Example 7.3.1

Oxygen concentration in a blood capillary is described by

$$\frac{\partial c}{\partial t} = D\left(\frac{\partial^2 c}{\partial r^2} + \frac{1}{r}\frac{\partial c}{\partial r}\right) - r_{O_2} \tag{7.20}$$

with the following initial and boundary conditions:

$$c(r,0) = \begin{cases} \bar{c} & \forall \; R_1 < r \le R_2 \\ c_f & \text{at} \; r = R_1 \end{cases} \tag{7.21}$$

$$\left.\begin{aligned} c(r,t) &= c_f \; \text{at} \; r = R_1 \\ \frac{\partial c}{\partial r} &= 0 \; \text{at} \; r = R_2 \end{aligned}\right\} \; \forall \; 0 < t \le t_f \tag{7.22} \atop (7.23)$$

Solve the above model using the method of finite differences.

Solution

We apply the solution procedure outlined on p. 256. The steps are as follows:

Step 1

The model has c as dependent variable, and t and r as independent variables. We first select r, and discretize its interval using N_i equispaced grid points, as shown in Figure 7.11a on the next page. The N_i grid points divide the r-interval into $(N_i - 1)$ sub-intervals. At the ith grid point, the radial distance, and the dependent variable are given by, respectively, $r_i = (r_0 + i\Delta r)$ and $c(r_i, t) \equiv c_i$ where $\Delta r = (R_2 - R_1)/(N_i - 1)$ is the size of sub-intervals, or the distance between any two consecutive grid points. Note that c_i depends on the remaining independent variable, i.e., t. Thus, $c_i = c_i(t)$.

Step 2

Next, we express Equation (7.20) above at each grid point. This equation has first and second order derivatives with respect to r. We replace them with centered, first and second order finite difference formulas, respectively. Thus, for the internal grid points, we obtain

$$\frac{dc_i}{dt} = D\left(\frac{c_{i+1} - 2c_i + c_{i-1}}{\Delta r^2} + \frac{1}{r_i}\frac{c_{i+1} - c_{i-1}}{2\Delta r}\right) - r_{O_2}, \qquad i = 1, 2, \ldots, (N_i - 2) \tag{7.24}$$

Step 3

At the first grid point, the concentration is specified to be c_f by the initial and boundary conditions, i.e., Equations (7.21) and (7.22). Thus, we have at all times

$$c_0 = c_f \tag{7.25}$$

At the last grid point, r_{N-1}, the concentration is not specified. At that point, we have the second boundary condition given by Equation (7.23) above. Because of this condition, Equation (7.20) above at $r = R_2$ simplifies to

$$\frac{\partial c}{\partial t} = D\frac{\partial^2 c}{\partial r^2} - r_{O_2}$$

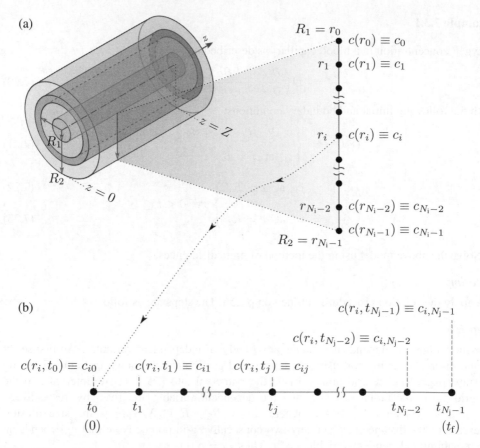

Figure 7.11 Grid points for oxygen concentration in the capillary along (a) the radial direction from R_1 to R_2, and (b) the time axis from 0 to t_f, at a radial position r_i

In the last equation, using the centered, second order finite difference formula for the second derivative, we get

$$\frac{dc_{N_i-1}}{dt} = D\frac{c_{N_i} - 2c_{N_i-1} + c_{N_i-2}}{\Delta r^2} - r_{O_2} \qquad (7.26)$$

Note that in the above equation, $c_{N_i} \equiv c(r_{N_i})$ is the concentration at the grid point r_{N_i} outside the interval $[r_0, r_{N_i-1}]$. This concentration gets resolved through the second boundary condition. Using the centered, second order finite difference formula for the first derivative, that condition is expressible as

$$\frac{c_{N_i} - c_{N_i-2}}{2\Delta r} = 0 \quad \Rightarrow \quad c_{N_i} = c_{N_i-2}$$

With the help of the above result, Equation (7.26) above becomes

$$\frac{dc_{N_i-1}}{dt} = 2D\frac{c_{N_i-2} - c_{N_i-1}}{\Delta r^2} - r_{O_2} \qquad (7.27)$$

Now the remaining independent variable is t, for which we have the initial condition, Equation (7.21) on p. 257. In terms of c_i, this condition becomes

$$c_i(0) = \bar{c}; \qquad i = 1, 2, \ldots, N_i - 1 \tag{7.28}$$

At this juncture, we have a set of ordinary differential equations, Equation (7.24) on p. 257 and Equation (7.27) on the previous page, with one less independent variable. These equations can be integrated using the initial conditions given by Equation (7.28) above to obtain the solution. This approach is the method of lines. Alternatively, the equations can be discretized further as follows.

Step 4 – Selection of the Next Independent Variable

We select the next and the last independent variable, i.e., t, and discretize its interval using N_j equispaced grid points t_j. As shown in Figure 7.11b on the previous page, for each existing radial location r_i, we have N_j grid points, which divide the t-interval into $(N_j - 1)$ sub-intervals of the same duration. At the jth grid point, the time duration, and the dependent variable are, respectively, given by

$$t_j = t_0 + j\Delta t \quad \text{and} \quad c(r_i, t_j) \equiv c_{ij}$$

$$\text{where} \quad \Delta t = \frac{t_f - 0}{N_j - 1}$$

is the duration between any two consecutive time instants, and t_f is the final time.

Next, we repeat Steps 2–4 of the solution procedure outlined on p. 256.

Step 2

We express at each time instant Equations (7.24) and (7.25) on p. 257, and Equation (7.27) on the previous page.

Equation (7.24) and Equation (7.27) have first order derivatives with respect to t. We replace them with their centered, second order finite difference formulas. Thus, for the internal grid points in the t-interval, we obtain

$$\frac{c_{i,j+1} - c_{i,j-1}}{2\Delta t} = D\left(\frac{c_{i+1,j} - 2c_{i,j} + c_{i-1,j}}{\Delta r^2} + \frac{1}{r_i}\frac{c_{i+1,j} - c_{i-1,j}}{2\Delta r}\right) - r_{O_2}$$

$$i = 1, 2, \ldots, N_i - 2; \quad j = 1, 2, \ldots, N_j - 2 \tag{7.29}$$

$$\frac{c_{N_i-1,j+1} - c_{N_i-1,j-1}}{2\Delta t} = 2D\frac{c_{N_i-2,j} - c_{N_i-1,j}}{\Delta r^2} - r_{O_2}; \qquad j = 1, 2, \ldots, N_j - 2 \tag{7.30}$$

where from Equation (7.25) on p. 257, $c_{0j} = c_f$ for $j = 0, 1, \ldots, N_j - 1$.

Step 3

At the first grid point of the t-interval, the initial concentration at $r > R$ is specified to be $\bar{c}$ by Equation (7.21) on p. 257. Thus, we have $c_{i0} = \bar{c}$ for $i = 1, 2, \ldots, N_i - 1$.

We now need to discretize Equation (7.24) on p. 257, and Equation (7.27) on the previous page at the last grid point in the t-interval, i.e., t_{N_j-1}. Note that the centered, second order finite difference formula requires the out-of-interval concentrations, c_{iN_j}s, which cannot be resolved since there is no final condition. This situation is addressed by utilizing the backward, second order finite difference formula. With this approach, Equation (7.24) and Equation (7.27) get discretized at the last grid point (t_{N_j-1}) into the following equations:

$$\frac{3c_{i,N_j-1} - 4c_{i,N_j-2} + c_{i,N_j-3}}{2\Delta t} = D\left(\frac{c_{i+1,j} - 2c_{i,j} + c_{i-1,j}}{\Delta r^2} \right.$$

$$\left. + \frac{1}{r_i}\frac{c_{i+1,j} - c_{i-1,j}}{2\Delta r} \right) - r_{O_2}$$

$$i = 1, 2, \ldots, N_i - 2 \quad (7.31)$$

$$\frac{3c_{N_i-1,N_j-1} - 4c_{N_i-1,N_j-2} + c_{N_i-1,N_j-3}}{2\Delta t} = 2D\frac{c_{N_i-2,N_j-1} - c_{N_i-1,N_j-1}}{\Delta r^2} - r_{O_2} \quad (7.32)$$

Thus, the partial differential equation of the process model given by Equations (7.20)–(7.23) on p. 257 gets discretized to the set of Equations (7.29)–(7.32). Readers should verify that this set comprises $(N_i - 1)(N_j - 1)$ equations involving the same number of unknown c_{ij}s. Solution of the discretized equations will provide these unknown c_{ij}s, i.e., the values of oxygen concentration for $t > 0$ and $r > R_1$ at the grid points shown in Figure 7.12 below.

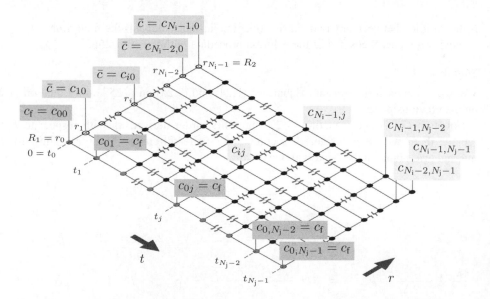

Figure 7.12 Oxygen concentrations at grid points in the r- and t- intervals in the capillary

Programming Discretized Equations

Because of the repeating patterns in discretized equations at intermediate grid points, it is easy to program the equations. For the method of lines, Figure 7.13 on the next page shows the

```
// Note: No equation is needed at the first grid point where concentration is already specified
// Thus, in the program, the total no. of differential equations (n_) is one less than the no. of grid points

const double rate = k_; // first order reaction rate

// Equation (7.24) on p. 257 for the second grid point at radial distance r_[0] ≡ R₁ + Δr
// '0' is the index of the first equation for the second grid point where concentration is c_[0]
dcdt[0] = D_*( (c[1]−2*c[0]+cf_)/drSqr_ + (c[1]−cf_)/dr2_/r_[0] ) − rate;
// drSqr_ ≡ Δr², dr2 ≡ 2Δr

// Equation (7.24) for the remaining grid points except the last one
for (unsigned i=1; i<n_−1; ++i) {
    dcdt[i] = D_*( (c[i+1]−2*c[i]+c[i−1])/drSqr_ + (c[i+1]−c[i−1])/dr2_/r_[i] ) − rate;
    // r_[i] ≡ r_[i-1] + Δr
}

// Equation (7.27) on p. 258 for the last grid point at radial distance R₂
// 'n_-1' is the index of the last (n_ᵗʰ) equation for the last or (n_+1)ᵗʰ grid point where concentration is c_[n_-1]
dcdt[n_−1] = 2*D_*(c[n_−2]−c[n_−1])/drSqr_ − rate;
```

Figure 7.13 Program code in C language for the ordinary differential equations [Equation (7.24), p. 257, and Equation (7.27), p. 258]

program code for the ordinary differential equations that can be integrated using a suitable solver. The program code for the algebraic equations [see Figure 7.14, next page] is more involved because they carry unknown concentrations in the two dimensions of r and t.

Solution of Ordinary Differential Equations

For the parameter values given in Table 7.5 below, Figure 7.15 on p. 263 shows the oxygen concentration at different grid points versus time obtained from the integration of the ordinary differential equations using the fifth order Runge–Kutta–Fehlberg method with adaptive steps. The concentration closer to the capillary rises faster to its oxygen concentration (c_f). A steady state is achieved in about 1.5 s.

Table 7.5 Values of parameters for the oxygen transfer process in a blood capillary

parameter	value	parameter	value
R_1	5×10^{-6} m	c_f	8×10^{-3} kg m^{-3}
R_2	2.25×10^{-5} m	$\bar{c}$	10^{-3} kg m^{-3}
k	9×10^{-6} kg m^{-3} s^{-1}	D	10^{-9} m^2 s^{-1}

Accuracy of Solution

Figure 7.16 on p. 263 shows the effect of doubling the number of radial grid points on oxygen concentration at the radial location furthest from the centre. The error relative to half as many

```
// Note: No equations are needed at R₁ where concentration is specified
// Radial index i = 0 corresponds to (R₁ + Δr), the second radial grid point
// No equations are needed at t = 0 when concentration is already specified
// Temporal index j = 0 corresponds to (0 + Δt), the second time instant

const double rate = k_; // first order reaction rate
// For intermediate time instants ———
unsigned i,j; double p,q;
for (j=0; j<nj_−1; ++j) { // loop over time
  // From Equation (7.29) on p. 259 for all radial grid points except the last one
  for (i=0; i<ni_−1; ++i) { // loop over radial distance
    if (i == 0) p = cf_; /* first boundary condition */ else p = c[i−1][j];
    if (j == 0) q = cs_; /* initial condition */ else q = c[i][j−1];
    f[i][j] = (c[i][j+1]−q)/dt2_ − D_*( (c[i+1][j]−2*c[i][j]+p)/drSqr_
                    + (c[i+1][j]−p)/dr2_/r_[i] ) + rate;
    // dt2_ ≡ 2Δt, drSqr_ ≡ Δr², dr2_ ≡ 2Δr, r_[i] ≡ r_[i-1] + Δr
  }
  // From Equation (7.30) on p. 259 for the last radial grid point
  i = ni_−1;
  if (j == 0) q = cs_; /* initial condition */ else q = c[i][j−1];
  f[i][j] = (c[i][j+1]−q)/dt2_ − 2*D_*(c[i−1][j]−c[i][j])/drSqr_ + rate;
}


// For the last time instant ———
j = nj_−1;
// From Equation (7.31) on p. 260 for all radial grid points except the last one
for (i=0; i<ni_−1; ++i) { // loop over radial distance
  if (i == 0) p = cf_; /* first boundary condition */ else p = c[i−1][j];
  f[i][j] = (3*c[i][j]−4*c[i][j−1]+c[i][j−2])/dt2_ − D_*( (c[i+1][j]−2*c[i][j]+p)/drSqr_
                  + (c[i+1][j]−p)/dr2_/r_[i] ) + rate;
}
// From Equation (7.32) on p. 260 for the last radial grid point
i = ni_−1;
f[i][j] = (3*c[i][j]−4*c[i][j−1]+c[i][j−2])/dt2_ − 2*D_*(c[i−1][j]−c[i][j])/drSqr_
                + rate;
}
```

Figure 7.14 Program code in C language for the algebraic equations given by Equations (7.29)–(7.32) on pp. 259 and 260

radial grid points decreases with the number of grid points. The maximum absolute error with eight grid points is 4.7%. The error goes down to 0.3% with 32 grid points.

Note that the errors peak close to the initial time. The reason is that during that time period, the oxygen concentration undergoes rapid changes. Its time derivatives are very large compared to those at a later time. Being part of the truncation errors of the numerical integration method, the larger derivatives result in larger errors.

❑

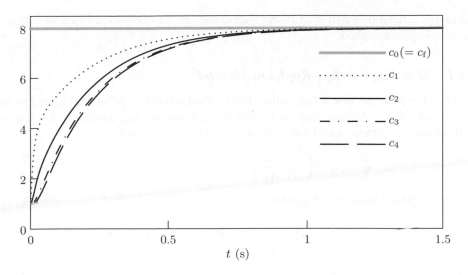

Figure 7.15 Oxygen concentration versus time with five grid points along the r-direction in Example 7.3.1 on p. 257

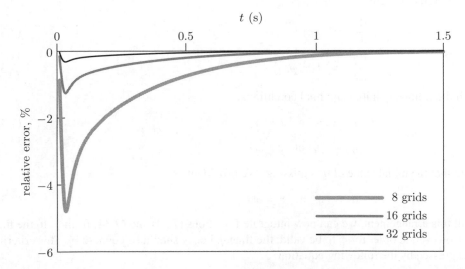

Figure 7.16 Error in $c(R_2, t)$ relative to that obtained with half as many radial grid points in Example 7.3.1

7.4 Differential Equations with Split Boundaries

Sometimes process models result in coupled ordinary differential equations with split boundary conditions. In other words, not all equations have conditions specified at the same boundary of the independent variable interval. For example, some equations have final conditions instead of initial conditions. Such equations can be solved using the shooting Newton–Raphson method.

7.4.1 Shooting Newton–Raphson Method

Based on the Newton–Raphson algorithm, this method solves the differential equations with guessed missing conditions. They are improved iteratively until all conditions are satisfied. To illustrate the method, consider the following differential equation:

$$\frac{d^2 p}{dx^2} = f\left(x, p, \frac{dp}{dx}\right)$$

with the following boundary conditions:

$$p = p_i \text{ at } x = x_i \quad \text{and} \quad \frac{dp}{dx} = p_f' \text{ at } x = x_f$$

With the help of

$$q \equiv \frac{dp}{dx} \quad \Rightarrow \quad \frac{dq}{dx} = \frac{d^2 p}{dx^2}$$

the differential equation can be expressed as the set of the following coupled ordinary differential equations:

$$\frac{dp}{dx} = q \tag{7.33}$$

$$\frac{dq}{dx} = f(x, p, q) \tag{7.34}$$

with the following initial and final conditions:

$$p = p_i \text{ at } x = x_i \tag{7.35}$$

$$q = p_f' \text{ at } x = x_f$$

Note that the initial value of q is missing. We provide it as

$$q = q_i \text{ at } x = x_i \tag{7.36}$$

With this arrangement, we can now integrate Equations (7.33) and (7.34) from x_i to the final x_f. However, in order for q_i to be valid, the final q, i.e., q_f should be equal to p_f'. This criterion is expressed by the following equation:

$$q_f(q_i) - p_f' = 0 \quad \text{or} \quad \underbrace{[q(q_i)]_{x_f} - p_f'}_{\equiv g(q_i)} = 0 \tag{7.37}$$

Observe that q_f is a function of q_i. The satisfaction of the above equation requires finding the root of the right-hand side denoted as the function $g(q_i)$. Thus, the Newton–Raphson algorithm to determine q_i is,

$$q_i^{(k+1)} = q_i^{(k)} - \left[\frac{g}{g'}\right]_{q_i^{(k)}} ; \qquad k = 0, 1, 2, \dots$$

where, on the basis of Equation (7.37) on the previous page,

$$g' = \frac{dg}{dq_i} = \left[\frac{\partial q}{\partial q_i}\right]_{x_f}$$

With the help of the last equation, the Newton–Raphson algorithm [see Equation (7.7), p. 237] is given by

$$q_i^{(k+1)} = q_i^{(k)} - \left\{\left[\frac{\partial q}{\partial q_i}\right]_{x_f}^{-1} g\right\}_{q_i^{(k)}} \quad ; \qquad k = 0, 1, 2, \ldots \qquad (7.38)$$

Differential Equations for Derivatives

To obtain the derivative $[\partial q/\partial q_i]_{x_f}$ for use in the above algorithm, we first need to differentiate Equations (7.33) and (7.34) on the previous page with respect to q_i as follows:

$$\frac{\partial}{\partial q_i}\left(\frac{dp}{dx}\right) = \frac{d}{dx}\left(\frac{\partial p}{\partial q_i}\right) = \frac{\partial q}{\partial q_i} \qquad (7.39)$$

$$\frac{\partial}{\partial q_i}\left(\frac{dq}{dx}\right) = \frac{d}{dx}\left(\frac{\partial q}{\partial q_i}\right) = \frac{\partial f}{\partial p}\frac{\partial p}{\partial q_i} + \frac{\partial f}{\partial q}\frac{\partial q}{\partial q_i} \qquad (7.40)$$

Equations (7.39) and (7.40) above need to be integrated simultaneously with Equations (7.33) and (7.34) to obtain $[\partial q/\partial q_i]_{x_f}$, which is required by the Newton–Raphson algorithm of Equation (7.38). The initial conditions for Equations (7.39) and (7.40) are

$$\frac{\partial p}{\partial q_i} = 0 \text{ at } x = x_i \qquad (7.41)$$

$$\frac{\partial q}{\partial q_i} = 1 \text{ at } x = x_i \qquad (7.42)$$

which are obtained, respectively, from Equations (7.35) and (7.36) on the previous page by differentiating them with respect to q_i.

Computational Algorithm

The shooting Newton–Raphson algorithm to solve the boundary value problem is as follows:

1. Set the iteration counter $k = 0$, and provide the initial guess $q_i^{(k)}$.

2. Set $q = q_i^{(k)}$ at $x = x_i$. Obtain $[\partial q/\partial q_i]_{x_f}$ by integrating Equations (7.33), (7.34), (7.39) and (7.40) using the initial conditions given by Equations (7.35), (7.36), (7.41) and (7.42).

3. Obtain the improved $q_i^{(k+1)}$ from the Newton–Raphson algorithm, i.e., Equation (7.38) above.

4. Assign $q_i^{(k+1)} \rightarrow q_i^{(k)}$, increment k by one, and go to Step 2 above until convergence when either the absolute relative change in $q_i^{(k)}$, or the absolute value of $g[q_i^{(k)}]$ becomes less than a specified, small positive number (i.e., accuracy).

Upon successful convergence, the above algorithm yields q_i and the desired solution, i.e., the values of p at different values of x.

Example 7.4.1

At steady state, the temperature in a tapered fin can be obtained from the unsteady state model [see Section 4.2.3, p. 96] as

$$\frac{d^2T}{dy^2} = \underbrace{\frac{1}{Z - ay}\left[a\frac{dT}{dy} + \frac{h}{k}(T - T_a)\right]}_{\equiv f}$$

with $z = (Z - ay)$ where a defines the taper, and the following boundary conditions:

$$T = T_w \text{ at } y = 0 \quad \text{and} \quad \frac{dT}{dy} = 0 \text{ at } y = Y$$

Convert the above steady state model into an initial value problem, and solve it using the shooting Newton–Raphson method for the parameter values given in Table 7.6 below.

Table 7.6   Values of parameters for the tapered fin

parameter	value	parameter	value
a	3×10^{-2}	h/k	0.125 m^{-1}
T_a	$23 \degree C$	T_w	$200 \degree C$
Z	$2 \times 10^{-3} \text{ m}$	Y	0.05 m

Solution

Comparing this problem to the general one in Section 7.4.1 on p. 264, we observe that $T \equiv p$, $y \equiv x$, and the right-hand side of the given differential equation is f.

Following the development of Section 7.4.1, and having $q \equiv dT/dy$, the set of equations for the initial value problem is

$$\frac{dT}{dy} = q \tag{7.43}$$

$$\frac{dq}{dy} = \frac{1}{Z - ay}\left[aq + \frac{h}{k}(T - T_a)\right] \tag{7.44}$$

$$\frac{d}{dy}\left(\frac{\partial T}{\partial q_i}\right) = \frac{\partial q}{\partial q_i} \tag{7.45}$$

$$\frac{d}{dy}\left(\frac{\partial q}{\partial q_i}\right) = \overbrace{\frac{h/k}{Z - ay}\frac{\partial T}{\partial q_i}}^{\partial f/\partial T} + \overbrace{\frac{a}{Z - ay}\frac{\partial q}{\partial q_i}}^{\partial f/\partial q} \tag{7.46}$$

with the following initial conditions:

$$T = T_\text{w}, \qquad q = q_\text{i}, \qquad \frac{\partial T}{\partial q_\text{i}} = 0, \quad \text{and} \quad \frac{\partial q}{\partial q_\text{i}} = 1$$

To solve this problem, we carry out the computational algorithm outlined on p. 265. Thus, we make a guess for q_i, i.e., provide $q_\text{i}^{(0)}$, and integrate Equations (7.43)–(7.46) on the previous page up to $y = Y$ in order to obtain $[\partial q/\partial q_\text{i}]_Y^{(0)}$. This derivative value is used in the following Newton–Raphson algorithm for $k = 0$ to improve $q_\text{i}^{(0)}$ to $q_\text{i}^{(1)}$:

$$q_\text{i}^{(k+1)} = q_\text{i}^{(k)} - \left\{ \left[\frac{\partial q}{\partial q_\text{i}} \right]_Y^{-1} g \right\}_{q_\text{i}^{(k)}} ; \qquad k = 0, 1, 2, \dots$$

where $g \equiv q_\text{f} - 0 = q(Y)$. Equations (7.43)–(7.46) on the previous page are integrated again with the improved $q_\text{i}^{(1)}$ to obtain $[\partial q/\partial q_\text{i}]_Y^{(1)}$, which is used in the above Newton–Raphson algorithm for $k = 1$ to obtain the further improved $q_\text{i}^{(2)}$. These calculations are repeated until convergence.

Figure 7.17 below shows the results for the parameters of Table 7.6 on the previous page. With $q_\text{i}^{(0)} = 1$, and the specified accuracy of (i) machine epsilon, of $O(10^{-16})$, for the maximum, absolute relative change in $q_\text{i}^{(k)}$, and (ii) 10^{-8} for the maximum absolute value of $g(q_\text{i}^{(k)})$, the algorithm converged in one iteration. Observe that the boundary condition of $\text{d}T/\text{d}y = 0$ is satisfied at the end of the fin.

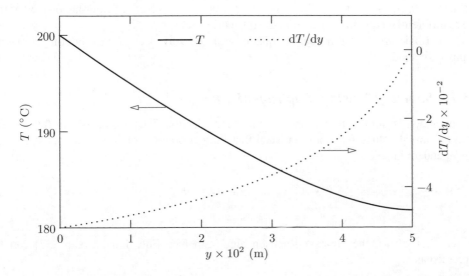

Figure 7.17   Steady state temperature distribution in the tapered fin at the convergence of the shooting Newton–Raphson algorithm

❑

Simplified Approach

The shooting Newton–Raphson method may be simplified by using the secant method. In this case, the derivative equations are not needed. Thus, in the above example, only Equations (7.33) and (7.34) need to be solved, albeit two times with two slightly different guesses for q_i in order to approximate the derivative in the Newton–Raphson algorithm.

7.5 Periodic Differential Equations

These differential equations describe periodic processes, which repeat after a time period. Nature abounds with periodic processes, such as photosynthesis, which follow the circadian rhythm. Man-made periodic processes are devised to achieve specialized objectives that are difficult to realize under the normal operations.

We can describe a periodic process in a uniform system as a set of ordinary differential equations

$$\frac{d\mathbf{y}}{dt} = \mathbf{f}(\mathbf{y}) \tag{7.47}$$

subject to the periodicity conditions

$$\mathbf{y}_0 = \mathbf{y}(\tau) \tag{7.48}$$

where (i) $\mathbf{y} = \begin{bmatrix} y_0 & y_1 & \cdots & y_{n-1} \end{bmatrix}^\top$ is the vector of dependent variables, (ii) $\mathbf{y}_0$ is $\mathbf{y}$ at time $t = 0$, (iii) $\mathbf{f} = \begin{bmatrix} f_0 & f_1 & \cdots & f_{n-1} \end{bmatrix}^\top$ is the vector of right-hand sides of the above differential equations, and (iv) τ is time period.

The periodic ordinary differential equations can be solved using the shooting Newton–Raphson method.

7.5.1 Shooting Newton–Raphson Method

The approach is similar to the one introduced in Section 7.4.1 on p. 264. In the present case, the initial values $\mathbf{y}_0$ should be such that the periodicity conditions are satisfied, i.e., the dependent function

$$\mathbf{g}(\mathbf{y}_0) \equiv \underbrace{\Big[\mathbf{y}(\mathbf{y}_0)\Big]_\tau}_{\text{y at } t\, =\, \tau} - \underbrace{\mathbf{y}_0}_{\text{initial y}} = \mathbf{0}$$

To zero out $\mathbf{g}(\mathbf{y}_0)$, the Newton–Raphson algorithm [see Equation (7.8), p. 237] can be expressed as

$$\mathbf{y}_0^{(k+1)} = \mathbf{y}_0^{(k)} - \big[\mathbf{J}^{-1}\mathbf{g}\big]_{\mathbf{y}_0^{(k)}}; \qquad k = 0, 1, 2, \ldots \tag{7.49}$$

where $\mathbf{J}$ is the Jacobian of $\mathbf{g}$ with respect to $\mathbf{y}_0$ at $t = \tau$, and is given by

$$
\mathbf{J} \;=\;
\begin{bmatrix}
\dfrac{\partial y_0}{\partial y_{00}} - 1 & \dfrac{\partial y_0}{\partial y_{10}} & \cdots & \dfrac{\partial y_0}{\partial y_{n-1,0}} \\[2.5ex]
\dfrac{\partial y_1}{\partial y_{00}} & \dfrac{\partial y_1}{\partial y_{10}} - 1 & \cdots & \dfrac{\partial y_1}{\partial y_{n-1,0}} \\[2.5ex]
\vdots & \vdots & \ddots & \vdots \\[2.5ex]
\dfrac{\partial y_{n-1}}{\partial y_{00}} & \dfrac{\partial y_{n-1}}{\partial y_{10}} & \cdots & \dfrac{\partial y_{n-1}}{\partial y_{n-1,0}} - 1
\end{bmatrix}_\tau
\tag{7.50}
$$

Determination of the Jacobian

Note that in Equation (7.50) above, $\mathbf{J}$ is made of partial derivatives evaluated for a given $y_0^{(k)}$ at $t = \tau$. Each of these derivatives is a dependent variable of a differential equation derivable from Equations (7.47) and (7.48) on the previous page. For example, differentiating the differential equation, and its initial condition for y_i, i.e.,

$$
\frac{dy_i}{dt} = f_i \quad \text{and} \quad y_i(0) = y_{i0}
$$

with respect to y_{j0}, we obtain the following differential equation for the derivative $\partial y_i / \partial y_{j0}$ along with the initial condition:

$$
\frac{d}{dt}\left(\frac{\partial y_i}{\partial y_{j0}}\right) = \sum_{k=1}^{n-1} \frac{\partial f_i}{\partial y_k}\frac{\partial y_k}{\partial y_{j0}}, \qquad \left[\frac{\partial y_i}{\partial y_{j0}}\right]_{t=0} = \begin{cases} 1 & \text{if } i = j \\ 0 & \text{if } i \neq j \end{cases}
\tag{7.51}
$$

Integration of the differential equations for all $\partial y_i / \partial y_{j0}$s along with Equation (7.47) up to $t = \tau$ provides the elements of $\mathbf{J}$.

Computational Algorithm

The shooting Newton–Raphson algorithm to solve the periodic, ordinary differential equations is as follows:

1. Set the iteration counter $k = 0$, and provide the initial guess $y_0^{(k)}$ for Equation (7.48) on the previous page.

2. Set $\mathbf{y}_0 = \mathbf{y}_0^{(k)}$. Obtain the elements of $\mathbf{J}$ by integrating up to $t = \tau$, Equation (7.47) on the previous page simultaneously with Equation (7.51) above for $i, j = 0, 1, \ldots, n-1$ (i.e., the differential equations for all $\partial y_i / \partial y_{j0}$s).

3. Obtain the improved $\mathbf{y}_0^{(k+1)}$ from the Newton–Raphson algorithm, i.e., Equation (7.49) on the previous page.

4. Assign $\mathbf{y}_0^{(k+1)} \to \mathbf{y}_0^{(k)}$, increment k by one, and go to Step 2 above until convergence when either the absolute relative change in $\mathbf{y}_0^{(k)}$, or the magnitude of $\mathbf{g}[\mathbf{y}_0^{(k)}]$ (or the maximum absolute value among its elements) becomes less than a specified, small positive number (i.e., accuracy).

Example 7.5.1

The process model of a CSTR in periodic operation is given by

$$\frac{dy_0}{dt} = \tilde{F}_0(y_f - y_0) - k_0 y_0^2 y_1 \tag{7.52}$$

$$\frac{dy_1}{dt} = \tilde{F}_1 - \tilde{F}_0 y_1 \tag{7.53}$$

subject to the periodicity conditions

$$y_0(0) = y_0(\tau) \quad \text{and} \quad y_1(0) = y_1(\tau)$$

In the model equations, (i) y_0 and y_1 are the concentrations of, respectively, the reactant and catalyst, (ii) y_f is the reactant concentration in the feed, (iii) τ is time period, and (iv) $\tilde{F}_0$ and $\tilde{F}_1$ are, respectively, the volumetric and catalyst mass flow rates per unit volume of the reaction mixture. The flow rates are periodic, i.e.,

$$\tilde{F}_0(0) = \tilde{F}_0(\tau) \quad \text{and} \quad \tilde{F}_1(0) = \tilde{F}_1(\tau)$$

For the parameter values provided in Table 7.7 below, use the shooting Newton–Raphson method to find the species concentrations versus time in the CSTR for $\tau = 10$ min.

Table 7.7　Values of parameters for the periodic CSTR

parameter	value	parameter	value
$\tilde{F}_0$	$2 + \sin(2\pi t/\tau)$ cm^3 min^{-1}	y_f	4 g cm^{-3}
k_0	5×10^7 cm^6 g^{-2} min^{-1}	$\tilde{F}_1$	$10 + 5\sin(2\pi t/\tau)$ g s^{-1}

Solution

In order to solve this problem using the shooting Newton–Raphson method, we need the derivative differential equations.

Differential Equations for Derivatives

These equations are obtained from Equation (7.51) on the previous page for $i, j = \{1, 2\}$, and f_0 and f_1 as the right-hand sides of Equations (7.52) and (7.53), respectively. The differential equations for derivatives are as follows:

$$\frac{d}{dt}\left(\frac{\partial y_0}{\partial y_{00}}\right) = -(\tilde{F}_0 + 2k_0 y_0 y_1)\frac{\partial y_0}{\partial y_{00}} - k_0 y_0^2 \frac{\partial y_1}{\partial y_{00}}, \qquad \left[\frac{\partial y_0}{\partial y_{00}}\right]_{t=0} = 1$$

$$\frac{d}{dt}\left(\frac{\partial y_0}{\partial y_{10}}\right) = -(\tilde{F}_0 + 2k_0 y_0 y_1)\frac{\partial y_0}{\partial y_{10}} - k_0 y_0^2 \frac{\partial y_1}{\partial y_{10}}, \qquad \left[\frac{\partial y_0}{\partial y_{10}}\right]_{t=0} = 0$$

$$\frac{d}{dt}\left(\frac{\partial y_1}{\partial y_{00}}\right) = -\tilde{F}_0 \frac{\partial y_1}{\partial y_{00}}, \qquad \left[\frac{\partial y_1}{\partial y_{00}}\right]_{t=0} = 0$$

$$\frac{d}{dt}\left(\frac{\partial y_1}{\partial y_{10}}\right) = -\tilde{F}_0 \frac{\partial y_1}{\partial y_{10}}, \qquad \left[\frac{\partial y_1}{\partial y_{10}}\right]_{t=0} = 1$$

Computational Algorithm
The following is the shooting Newton–Raphson algorithm to find the species concentrations:

1. Set the counter $k = 0$, and provide the initial guess $\mathbf{y}_0^k = \begin{bmatrix} y_{00}^k & y_{10}^k \end{bmatrix}^\top$.

2. Set $\mathbf{y}_0 = \mathbf{y}_0^{(k)}$. Obtain $\partial y_i/\partial y_{j0}$s by simultaneously integrating the above differential equations for derivatives, and Equations (7.52) and (7.53) on the previous page from $t = 0$ to τ.

3. Obtain the improved $\mathbf{y}_0^{(k+1)}$ by applying Equations (7.49) and (7.50) on p. 268–269, i.e.,

$$
\underbrace{\begin{bmatrix} y_{00}^{(k+1)} \\ y_{10}^{(k+1)} \end{bmatrix}}_{\mathbf{y}_0^{(k+1)}} = \underbrace{\begin{bmatrix} y_{00}^{(k)} \\ y_{10}^{(k)} \end{bmatrix}}_{\mathbf{y}_0^{(k)}} - \underbrace{\begin{bmatrix} \dfrac{\partial y_0}{\partial y_{00}} - 1 & \dfrac{\partial y_0}{\partial y_{10}} \\ \dfrac{\partial y_1}{\partial y_{00}} & \dfrac{\partial y_1}{\partial y_{10}} - 1 \end{bmatrix}_\tau^{-1}}_{\mathbf{J}^{-1}[\mathbf{y}_0^{(k)}]} \underbrace{\begin{bmatrix} y_0(\tau) - y_{00}^{(k)} \\ y_1(\tau) - y_{10}^{(k)} \end{bmatrix}}_{\mathbf{g}[\mathbf{y}_0^{(k)}]}
$$

4. Assign $\mathbf{y}_0^{(k+1)} \to \mathbf{y}_0^{(k)}$, increment k by one, and go to Step 2 on the previous page until convergence when either the absolute relative change in $\mathbf{y}_0^{(k)}$, or the magnitude of $\mathbf{g}[\mathbf{y}_0^{(k)}]$ (or the maximum absolute value among its elements) becomes less than a specified, small positive number (i.e., accuracy).

Figure 7.18 on the next page shows the species concentrations obtained from the above algorithm for the parameters of Table 7.7 on the previous page. With $y_0^{(0)} = y_1^{(0)} = 1$, and the specified accuracy of (i) machine epsilon, of $O(10^{-16})$, for the maximum, absolute relative change in an element of $\mathbf{y}_0^{(k)}$, and (ii) 10^{-8} for the maximum absolute value among the elements of $\mathbf{g}[\mathbf{y}_0^{(k)}]$, the algorithm converged in one iteration.

❏

Simplified Approaches

The shooting Newton–Raphson method may be simplified by using the secant method to approximate the Jacobian in Equation (7.49) on p. 268. Another approach is to use the method of *successive substitution*. In this method, the equations are integrated using guessed initial conditions. The final values of the dependent variables are substituted for the initial conditions. The integration and substitution are repeated until the initial conditions are sufficiently close to the final values of dependent variables.

7.6 Programming of Derivatives

Derivatives have a significant role in (i) the solution of non-linear algebraic equations, and differential equations that are stiff, or have periodic or split boundary conditions, (ii) non-linear regression, and (iii) optimization. With regards to accuracy and computational effort, it is considerably more efficient to use analytical expressions for derivatives than their approximations as done, for example, in the secant method. There is a caveat though.

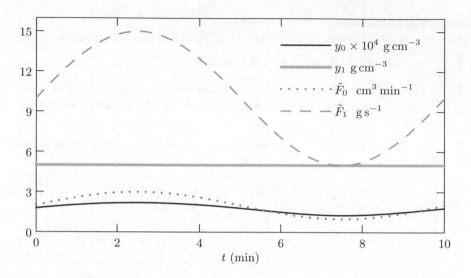

Figure 7.18 The reactant and catalyst concentrations (respectively, y_0 and y_1) in the periodic CSTR at the convergence of the shooting Newton–Raphson algorithm in Example 7.5.1 on p. 270

Even small errors in derivative expressions may lead to erroneous results. Thus, it is of utmost importance in computer programs to have derivatives without errors.

Analytical expressions for the derivatives of functions can be flawlessly programmed with the help of finite difference approximation. This can be accomplished using a robust method, which involves systematic evaluation of a derivative followed by its scrutiny. We explain this method with the help of the following function:

$$f(T) \;=\; \frac{ak}{\rho}$$

where

$$k \;=\; k_0 e^{-b/T} \qquad \text{and} \qquad \rho \;=\; \frac{\rho_0}{1 + b(T - T_0)}$$

with a, k_0, b and ρ_0 as specified constants.

The goal is to accurately program the derivative $\mathrm{d}f/\mathrm{d}T$, which is given by

$$\frac{\mathrm{d}f}{\mathrm{d}T} \;=\; a\left[\frac{1}{\rho}\frac{\mathrm{d}k}{\mathrm{d}T} + k\frac{\mathrm{d}}{\mathrm{d}T}\left(\frac{1}{\rho}\right)\right] \;=\; a\left(\frac{1}{\rho}\frac{\mathrm{d}k}{\mathrm{d}T} - \frac{k}{\rho^2}\frac{\mathrm{d}\rho}{\mathrm{d}T}\right)$$

where

$$\frac{\mathrm{d}k}{\mathrm{d}T} \;=\; k_0 e^{-b/T}\frac{b}{T^2} \;=\; \frac{bk}{T^2}$$

and

$$\frac{\mathrm{d}\rho}{\mathrm{d}T} \;=\; -\frac{\rho_0 b}{[1 + b(T - T_0)]^2} \;=\; -\frac{b\rho^2}{\rho_0}$$

Evaluation of the Derivative

When evaluating complicated expressions, it is recommended to use dummy variables to preclude errors due to substitutions unless the latter are very simple. Thus, with this approach, the expression for the derivative df/dT in a computer program is evaluated using the following sequence of calculations:

$$d_1 = -b\rho^2/\rho_0 \qquad \text{(i.e., } d\rho/dT)$$

$$d_2 = \frac{bk}{T^2} \qquad \text{(i.e., } dk/dT)$$

$$f_T = a\left(\frac{d_1}{\rho} - \frac{kd_2}{\rho^2}\right) \qquad \text{(i.e., } df/dT)$$

where d_1 and d_2 are dummy variables.

Scrutiny of the Derivative

The above expression for the derivative (f_T) is checked using the following steps:

Step 1 For the specified values of constants, evaluate the function values $f_1 = f(T_1)$ and $f_2 = f(T_2)$ where $T_2 = T_1(1 + \alpha)$, and α is a small and adjustable positive fraction such as 10^{-4}.

Step 2 Evaluate the finite difference approximation of f_T at T_1 from

$$\tilde{f}_T = \frac{f_2 - f_1}{\alpha T_1}$$

Step 3 Ensure that the error between f_T and $\tilde{f}_T$, i.e.,

$$\epsilon = \begin{cases} \left|1 - \tilde{f}_T/f_T\right| & \text{if } |f_T| > 0 \\[2ex] \left|f_T - \tilde{f}_T\right| & \text{if } f_T = 0 \end{cases}$$

is less than a small positive number, say, 10^{-4}.

Step 4 Repeat calculations with smaller values of α, and ensure that ϵ reduces with α.

The above steps catch any errors right away in the expression of the analytical derivative. If a function is made of additive terms then the derivative of each term may be checked individually to pinpoint and correct any error. For the purpose of checking, the values of involved constants may be suitably changed to avoid any subtractive cancellation [see p. 277]. Analytical derivative expressions that are scrutinized in this manner are error free. With this method, we can be completely assured of the accuracy of derivative expressions in computer programs.

7.7   Miscellanea

In this section, we present Simpson's rules for the integration of discrete data, and methods to find roots of quadratic and cubic equations.

7.7.1   Integration of Discrete Data

Integration is the calculation of an integral, i.e., the area under a curve. For example, the integral

$$I = \int_a^b f(x)\,\mathrm{d}x$$

is the area between the integrand, i.e., the function $f(x)$, and the x-axis in the interval $[a, b]$. If $f(x) = \mathrm{d}y/\mathrm{d}x$ then $I = y(b) - y(a)$. Hence, solving the differential equation,

$$\frac{\mathrm{d}y}{\mathrm{d}x} = f(x)$$

with the initial condition $y(a) = 0$ to find $y(b)$ is equivalent to calculating the integral.

Numerical integration methods use Taylor expansions to approximate the integrand $f(x)$. These methods are employed when either the integrand is known at discrete values of the independent variable, or analytical integration is hard or impossible to do. The simplest method is the trapezoidal rule, which uses the first order Taylor expansion, and yields the integral as the area of the resultant trapezium. A more accurate method is Simpson's 1/3 rule.

Simpson's 1/3 Rule

This method uses the third order Taylor expansion [Equation (8.12) on p. 322 for $n = 3$] of the integrand for the data set

$$\left\{ [x_0, f(x_0)] \quad [x_1, f(x_1)] \quad [x_2, f(x_2)] \right\}$$

at $x = x_1$ with equispaced x_is. Thus, the integral is given by

$$I = \int_{x_0}^{x_2} \left[f(x_1) + hf^{(1)}(x_1) + \frac{h^2}{2}f^{(2)}(x_1) + \frac{h^3}{3!}f^{(3)}(x_1) + \frac{h^4}{4!}f^{(4)}(\zeta) \right]\mathrm{d}x$$

where h is the constant step size given by

$$h = x_1 - x_0 = x_2 - x_1 = (x_2 - x_0)/2$$

and $x_1 < \zeta < x_1 + h$. The above integral with the help of the finite difference formula,

$$f^{(2)}(x_1) = \frac{f(x_2) - 2f(x_1) + f(x_0)}{h^2} + O(h^2)$$

simplifies to Simpson's 1/3 rule, i.e.,

$$I = \underbrace{\frac{h}{3}\left[f(x_0) + 4f(x_1) + f(x_2)\right]}_{\text{Simpson's 1/3 rule}} + \underbrace{O(h^5)}_{\text{truncation error}}$$

The accuracy of numerical integration can be increased by increasing the order of the involved Taylor expansion, which would require more data points. However, when more data points are present, a preferred approach to increase the accuracy of calculation is that of composite application of integration formulas based on low-order Taylor expansions. In this approach, the formulas are applied sequentially over the sub-intervals. The accuracy is higher because of smaller step sizes, and smaller truncation errors as a consequence.

Composite Application

The integral over the interval $[x_0, x_1, \ldots, x_n]$ can be expressed as

$$I = \int_{x_0}^{x_n} f(x)\,dx = \int_{x_0}^{x_2} f(x)\,dx + \int_{x_2}^{x_4} f(x)\,dx + \cdots + \int_{x_{n-2}}^{x_n} f(x)\,dx$$

For an even n, the application of Simpson's 1/3 rule to each integral on the right-hand side of the above equation yields the **composite Simpson's 1/3 rule**, i.e.,

$$I = \frac{h}{3}\left[f(x_0) + 4\sum_{i=1,3,5,\ldots}^{n-1} f(x_i) + 2\sum_{i=2,4,6,\ldots}^{n-2} f(x_i) + f(x_n)\right]$$

If the number of data points $(n+1)$ is even, and greater than five then Simpson's 1/3 rule can be applied, excluding the last three intervals. The latter are covered by Simpson's 3/8 rule, which has the same order of accuracy.

Simpson's 3/8 rule

Proceeding similar to the derivation of the 1/3 rule but with an additional data point $[x_3, f(x_3)]$, and $f(x)$ approximated by the fourth order Taylor expansion at $(x_3 - x_0)/2$, the integral is given by

$$I = \underbrace{\frac{3h}{8}\left[f(x_0) + 3f(x_1) + 3f(x_2) + f(x_3)\right]}_{\text{Simpson's 3/8 rule}} + \underbrace{O(h^5)}_{\text{truncation error}}$$

Example 7.7.1

Gas flow rate in a pipeline is measured every hour. The data over a 9 h period are:

t (h)	0	1	2	3	4	5	6	7	8	9
$F \times 10^{-6}$ (m³ h⁻¹)	0	0.12	0.24	0.55	0.98	1.22	1.34	1.31	1.28	1.25

Calculate the volume of gas transported by the pipeline during this time period.

Solution

The volume of gas transported is given by

$$I = \int_0^9 F(t)\,dt$$

where $F(t)$ is given at nine, equispaced values of t with $h = 1$. Since the number of data points is even, and greater than five, we apply Simpson's 3/8 rule to the last three x-intervals. For the remaining intervals, we apply composite Simpson's 1/3 rule. Thus,

$$I = \underbrace{\frac{1}{3}\big\{ F(0) + 4[F(1) + F(3) + F(5)] + 2F[(2) + F(4)] + F(6) \big\}}_{\int_0^6 F(t)\,dt}$$

$$\underbrace{+ \frac{3}{8}\big[F(6) + 3F(7) + 3F(8) + F(9) \big]}_{\int_6^9 F(t)\,dt} = 7.6650 \times 10^6 \text{ m}^3$$

❏

7.7.2   *Roots of a Single Algebraic Equation*

Process simulation often requires finding the roots of a single non-linear equation of the form $f(x) = 0$. Roots could be real, or complex numbers.

Quadratic Equations

The roots of a quadratic equation

$$ax^2 + bx + c = 0$$

are given by the standard formulas,

$$x_1 = \frac{-b + \sqrt{b^2 - 4ac}}{2a} \quad \text{and} \quad x_2 = \frac{-b - \sqrt{b^2 - 4ac}}{2a} \tag{7.54}$$

If $4ac$ is very small in comparison to b^2 and $b > 0$ then

$$x_1 = \frac{-b + \overbrace{\sqrt{b^2 - 4ac}}^{\equiv\, b_1}}{2a}$$

In this case, the calculation of x_1 would involve subtracting b from a very close, positive number b_1 as denoted above. On calculators and computers, which use floating-point arithmetic, subtracting nearly equal numbers increases the relative error, and causes loss in significance. This phenomenon is called **subtractive cancellation**.

For the calculation of x_1 in this case, subtractive cancellation can be avoided by multiplying the numerator and denominator of the right-hand side of the equation for x_1 by the factor, $(-b - \sqrt{b^2 - 4ac})$. After simplification, we get

$$x_1 = \frac{-2c}{b + \sqrt{b^2 - 4ac}} \tag{7.55}$$

which does not involve the subtraction between b and $\sqrt{b^2 - 4ac}$, both of which are positive.

If $4ac \ll b^2$ as before but $b < 0$ then Equation (7.54) on the previous page for x_2 would risk subtractive cancellation. Similar to the previous case, the use of the multiplicative factor $(-b + \sqrt{b^2 - 4ac})$ results in

$$x_2 = \frac{-2c}{b - \sqrt{b^2 - 4ac}} \tag{7.56}$$

which is free of subtractive cancellation.

Example 7.7.2

Find the roots of the quadratic equation,

$$0.5x^2 + x + 1.0000005 \times 10^{-6} = 0.$$

Solution

In this equation, $a = 0.5$, $b = 1$ and $c = 1.0000005 \times 10^{-6}$. Note that $b \approx \sqrt{b^2 - 4ac}$, so the calculation of x_1 using Equation (7.54) on the previous page involves subtractive cancellation. Table 7.8 below shows the roots obtained with single and double precisions. While x_1 and x_2 are obtained from Equation (7.54), $\tilde{x}_1$ is obtained from Equation (7.55).

Note that the numerical accuracy of the calculated roots increases with precision. Equation (7.55), which precludes subtractive cancellation, gives a more accurate result for the same precision.

Table 7.8 Roots calculated with single and double precisions (correct to 15 significant figures) along with their exact values

root	single precision	double precision
x_1	$9.536743164062500 \times 10^{-7}$	$9.999999999177334 \times 10^{-7}$
$\tilde{x}_1$	$9.999999974752427 \times 10^{-7}$	$1.000000000000000 \times 10^{-6}$
	(exact value of x_1: 10^{-6})	
x_2	-2.000000953674316	-2.000001000000000
	(exact value of x_2: -2.000002000001)	

□

7.7.3 Cubic Equations

A number of thermodynamic equations of state are cubic equations. The general form of a cubic equation is

$$ax^3 + bx^2 + cx + d = 0$$

The procedure to solve the above equation defines

$$p \equiv \frac{3ac - b^2}{3a^2}, \qquad q \equiv \frac{2b^3 + 27a^2d - 9abc}{27a^3} \quad \text{and} \quad r \equiv \frac{4p^3 + 27q^2}{108} \qquad (7.57)$$

and involves the following three cases based on the sign of r:

Case 1

If $r < 0$ then there are three real unequal roots. They are given by

$$x_j = -\frac{2q}{|q|}\sqrt{-\frac{p}{3}}\cos\left[\frac{\phi}{3} + \frac{2\pi}{3}(j-1)\right] - \frac{b}{3a}, \qquad j = 1,2,3 \qquad (7.58)$$

where $\quad \phi = \cos^{-1}\sqrt{-\frac{27q^2}{4p^3}} \qquad \text{(rad)}$ $\qquad\qquad\qquad (7.59)$

If $r \not< 0$ then the roots are given by

$$x_1 = u + v - \frac{b}{3a} \qquad (7.60)$$

$$x_{2,3} = -\frac{u+v}{2} - \frac{b}{3a} \pm \frac{i\sqrt{3}}{2}(u-v) \qquad (7.61)$$

where $\quad u = \left(-\frac{q}{2} + \sqrt{r}\right)^{1/3}, \qquad v = \left(-\frac{q}{2} - \sqrt{r}\right)^{1/3}$ $\qquad (7.62)$

and $i = \sqrt{-1}$.

Case 2

If $r = 0$ then $u = v$, and Equations (7.60) and (7.61) above yield three real roots, of which at least two are equal.

Case 3

Finally, if $r > 0$, and a, b, c and d are all real then Equations (7.60)–(7.62) yield one real root (x_1), and two complex conjugate roots (x_2 and x_3).

Example 7.7.3

Find the roots of the following cubic equation:

$$ax^3 + bx^2 + cx + d = 0$$

for the following parameter sets:

1. $a = -2$, $b = 3$, $c = 2$, and $d = -2$
2. $a = 2$, $b = 1$, $c = 4$, and $d = -2$

Solution

For the first set of parameters, Equation (7.57) on the previous page provides

$$p = -1.7500, \qquad q = 0.2500 \qquad \text{and} \qquad r = -0.1829$$

Since $r < 0$, we use Equation (7.59) on the previous page, and obtain $\phi = 1.2864$. Equation (7.58) then yields

$$x_1 = -0.8892, \qquad x_2 = 1.7446 \text{ and } x_3 = 0.6446$$

For the second set of parameters, Equation (7.57) provides

$$p = 1.9167, \qquad q = -1.3241 \qquad \text{and} \qquad r = 0.6991$$

Since $r > 0$, we use Equation (7.62) to obtain $u = 1.1442$ and $v = -0.5584$. Equation (7.60) and Equation (7.61) then yield

$$x_1 = 0.4192, \qquad x_2 = -0.4596 + 1.4745i \qquad \text{and}$$

$$x_3 = -0.4596 - 1.4745i$$

where $i = \sqrt{-1}$.

❏

General Equations

One method to find the real and any imaginary roots of a non-linear algebraic equation is **Muller's method**. This method carries out the following algorithm:

1. Select three initial values of x as the roots of a given function.

2. Fit a quadratic to the function at the three x values.

3. Find the roots of the quadratic. Select one of the roots as the new x value, which is expected to be closer to the root of the function.

4. Go to Step 2 above, and repeat the calculations with the new x value, and two other x values until convergence when the absolute value of the function becomes less than a specified, small positive number (or accuracy).

Fitting of a Quadratic

This is the second step of the above algorithm in which the quadratic

$$f(x) = a(x - x_2)^2 + b(x - x_2) + c$$

is expressed at x_0, x_1 and x_2 to yield

$$f(x_0) = a(x_0 - x_2)^2 + b(x_0 - x_2) + c$$
$$f(x_1) = a(x_1 - x_2)^2 + b(x_1 - x_2) + c$$
$$f(x_2) = c$$

Then in terms of

$$h_0 \equiv x_1 - x_0, \qquad h_1 \equiv x_2 - x_1,$$

$$d_0 \equiv \frac{f(x_1) - f(x_0)}{h_0} \quad \text{and} \quad d_1 \equiv \frac{f(x_2) - f(x_1)}{h_1}$$

the coefficients of the quadratic are

$$a = \frac{d_1 - d_0}{h_1 + h_0}, \qquad b = ah_1 + d_1 \quad \text{and} \quad c = f(x_2)$$

From Equations (7.55) and (7.56) on p. 277, the root of the quadratic $f(x_3)$ is

$$x_3 - x_2 = \frac{-2c}{b \pm \sqrt{b^2 - 4ac}} \quad \Rightarrow \quad x_3 = x_2 - \frac{2c}{b \pm \sqrt{b^2 - 4ac}} \tag{7.63}$$

For the next iteration, the value of x_3 that is closest to x_2 is selected. That value is calculated from the above equation by having the sign of the square root term the same as that of b. Then, x_1 is assigned to x_0, x_2 is assigned to x_1, and x_3 is assigned to x_2. If all x_is are real then the value of x (among x_0, x_1 and x_2) that is furthest from x_3 is discarded. In this manner, the iterations are continued. When they converge, the root (either a real root, or a complex conjugate pair of imaginary roots) is given by x_3.

Example 7.7.4

Apply Muller's method to obtain the roots of the function

$$f(x) = x^4 + 14x^3 + 159x^2 + 554x + 1378$$

Solution

We use Equation (7.63) above to obtain the roots from the quadratic fitted on three x-values, namely, x_0, x_1 and x_2. With their initial values as, respectively, 0, $1 + i$, and $2 + 2i$, we obtain

$$
\begin{aligned}
h_0 &= 1 + i & a &= 201 + 56i \\
h_1 &= 1 + i & b &= 1146 + 960i \\
d_0 &= 711 + 189i & c &= 2198 + 2604i \\
d_1 &= 1001 + 703i & x_3 &= -2.4912 + 1.2178i
\end{aligned}
$$

Calculations are repeated after the assignments: $x_1 \rightarrow x_0$, $x_2 \rightarrow x_1$, and $x_3 \rightarrow x_2$. With the accuracy of 10^{-8} in the absolute value of the function, the method converges in seven iterations to the pair of imaginary roots: $x_{r1} = (-2 + 3i)$ and $x_{r2} = (-2 - 3i)$.

To find the remaining roots, this method needs to be applied repeatedly to the given function after dividing it each time by all the known factors, i.e., the $(x - x_{ri})$s.

❑

7.A   Partial Pivoting for Matrix Inverse

Partial pivoting causes row switches in the lower-upper decomposition of a matrix. To account for this effect when finding the matrix inverse, we make use of the following equivalence:

Switching the rows of an $n \times n$ matrix (or an n vector) is equivalent to its pre-multiplication by the $n \times n$ identity matrix $\mathbf{I}$ with the same rows switched. For example, given a 3×3 matrix with the second and third rows switched,

$$
\begin{bmatrix} a_{00} & a_{01} & a_{02} \\ a_{20} & a_{21} & a_{22} \\ a_{10} & a_{11} & a_{12} \end{bmatrix} = \begin{bmatrix} 1 & 0 & 0 \\ 0 & 0 & 1 \\ 0 & 1 & 0 \end{bmatrix} \begin{bmatrix} a_{00} & a_{01} & a_{02} \\ a_{10} & a_{11} & a_{12} \\ a_{20} & a_{21} & a_{22} \end{bmatrix}
$$

Now during the lower-upper decomposition of a square matrix $\mathbf{A}$, suppose that each time when we switch rows during partial pivoting, we switch the identical rows in $\mathbf{I}$. Thus, along with $\mathbf{L}$ and $\mathbf{U}$, we obtain $\mathbf{I}'$, which is $\mathbf{I}$ with all row switches that happened during the lower-upper decomposition. Note that the product of $\mathbf{L}$ and $\mathbf{U}$ is $\mathbf{A}'$ (see p. 230), which is $\mathbf{A}$ with the same row switches. Based on the aforementioned equivalence, $\mathbf{I}'\mathbf{A} = \mathbf{A}'$, which means

$$
\mathbf{A} = (\mathbf{I}')^{-1}\mathbf{A}'
$$

Moreover, from Equation (7.5) on p. 234, the ith column of $\mathbf{A}^{-1}$ is $\mathbf{x}_i$. This vector is given by the solution of

$$
\mathbf{A}\mathbf{x}_i = \mathbf{b}_i
$$

where $\mathbf{b}_i$ is a vector with all elements zero, except the ith one, which is unity.

From the last two equations,

$$
(\mathbf{I}')^{-1}\mathbf{A}'\mathbf{x}_i = \mathbf{b}_i
$$

Multiplying the above equation by $\mathbf{I}'$, and considering that $\mathbf{I}'(\mathbf{I}')^{-1} = \mathbf{I}$, we obtain

$$
\mathbf{A}'\mathbf{x}_i = \mathbf{I}'\mathbf{b}_i = \mathbf{b}_k
$$

where k is the index of a row in $\mathbf{I}'$ (or $\mathbf{A}'$, or $\mathbf{b}'$) to which the ith row of $\mathbf{I}$ (or $\mathbf{A}$, or $\mathbf{b}$) eventually got switched during the lower-upper decomposition of $\mathbf{A}$.

The above result simply means that if the ith row of $\mathbf{b}$ eventually becomes the kth row of $\mathbf{b}'$ due to partial pivoting in the lower-upper decomposition then $\mathbf{b}_k$ is to be used to get the ith column of $\mathbf{A}$ from $\mathbf{A}'\mathbf{x}_i = \mathbf{b}_k$.

7.B   Derivation of Newton–Raphson Method

Given a set of non-linear equations in the form $\mathbf{f}(\mathbf{x}) = \mathbf{0}$, the first order Taylor expansion of $\mathbf{f}$ at $\mathbf{x}^{(k+1)}$ near a reference point $\mathbf{x}^{(k)}$ is given by [see Section 8.8.2, p. 323]

$$
\mathbf{f}[\mathbf{x}^{(k+1)}] = \mathbf{f}[\mathbf{x}^{(k)}] + \mathbf{J}[\mathbf{x}^{(k)}]\left[\mathbf{x}^{(k+1)} - \mathbf{x}^{(k)}\right]
$$

where $J[x^{(k)}]$ is the matrix of partial derivatives of f, or Jacobian, evaluated at $x^{(k)}$, i.e.,

$$J[x^{(k)}] = \begin{bmatrix} \frac{\partial f_0}{\partial x_0}\Big|_{x^{(k)}} & \frac{\partial f_0}{\partial x_1}\Big|_{x^{(k)}} & \cdots & \frac{\partial f_0}{\partial x_{n-1}}\Big|_{x^{(k)}} \\ \frac{\partial f_1}{\partial x_0}\Big|_{x^{(k)}} & \frac{\partial f_1}{\partial x_1}\Big|_{x^{(k)}} & \cdots & \frac{\partial f_1}{\partial x_{n-1}}\Big|_{x^{(k)}} \\ \vdots & \vdots & \cdots & \vdots \\ \frac{\partial f_{n-1}}{\partial x_0}\Big|_{x^{(k)}} & \frac{\partial f_{n-1}}{\partial x_1}\Big|_{x^{(k)}} & \cdots & \frac{\partial f_{n-1}}{\partial x_{n-1}}\Big|_{x^{(k)}} \end{bmatrix}$$

When $x^{(k+1)}$ is sufficiently close to the root, say, x_r, the corresponding $f[x^{(k+1)}]$ is close to 0, and the Taylor expansion above can be approximated by

$$0 = f[x^{(k)}] + J[x^{(k)}]\left[x^{(k+1)} - x^{(k)}\right] \tag{7.64}$$

where $x^{(k)}$ is obviously farther from x_r than $x^{(k+1)}$. The above equation, upon rearrangement, yields

$$x^{(k+1)} = x^{(k)} - J^{-1}[x^{(k)}]\,f[x^{(k)}]$$

which, as assumed, is the new value of x that is closer to the root than $x^{(k)}$. In other words, $x^{(k+1)}$ is the improvement on $x^{(k)}$. Beginning with a guessed $x^{(0)}$, the iterative application of the above equation to improve x with k is known as the Newton–Raphson method.

Abnormal Behavior

Depending on the non-linearity, and the location of function evaluation, the Newton–Raphson method may behave abnormally. Figure 7.19 on the next page shows the method stuck in an infinite loop because the improved value of x leads to its old value.

For another function [see Figure 7.20, p. 284] the improved value of x leads the method away from the root. Sometimes the Jacobian may become singular. When solving equations with the computer program of this method, these abnormalities are dealt with by specifying appropriate convergence criteria, maximum number of iterations, and different initial guesses.

7.B.1   Quadratic Convergence

Sufficiently close to the root x_r, we have from the second order Taylor expansion [see Equation (8.13), p. 323],

$$f(x_r) = 0 = f[x^{(k)}] + J[x^{(k)}]\underbrace{\left[x_r - x^{(k)}\right]}_{e} + a \tag{7.65}$$

where $\quad a = \dfrac{1}{2}\left[\underbrace{D^2[f_1(\zeta)]}_{=\,e^\top H_1 e} \quad \underbrace{D^2[f_2(\zeta)]}_{=\,e^\top H_2 e} \quad \cdots \quad \underbrace{D^2[f_n(\zeta)]}_{=\,e^\top H_n e}\right]^\top,$

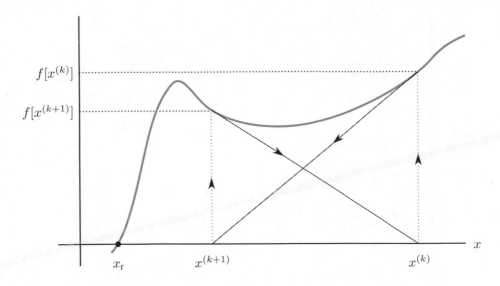

Figure 7.19 Example of the abnormal behavior of the Newton–Raphson method – infinite number of iterations

$\mathbf{e}$ is the vector of current errors, and $\mathbf{H}_i(\zeta)$ is the Hessian of the ith function f_i evaluated at ζ in the interval $(\mathbf{x}_0, \mathbf{x}_0 + \mathbf{h})$.

Subtracting Equation (7.64) on the previous page from Equation (7.65), we obtain

$$\underbrace{\mathbf{x}_r - \mathbf{x}^{(k+1)}}_{\hat{\mathbf{e}}} = -\mathbf{J}^{-1}[\mathbf{x}^{(k)}]\,\mathbf{a}$$

where $\hat{\mathbf{e}}$ is the vector of new errors, and $\mathbf{J}[\mathbf{x}^{(k)}]$ is assumed to be invertible. Taking the norm on both sides of the above equation, we get

$$\|\hat{\mathbf{e}}\| = \left\| -\mathbf{J}^{-1}[\mathbf{x}^{(k)}]\,\mathbf{a} \right\| \leq \underbrace{\left\| -\mathbf{J}^{-1}[\mathbf{x}^{(k)}] \right\|}_{\equiv p} \|\mathbf{a}\|$$

where we have utilized the operator inequality [see Section 8.5.3, p. 305] in the last step.

Now, let n be the number of elements of $\mathbf{x}$, and q be the maximum absolute value among the elements of all $\mathbf{H}_i$s. Noting that $a_i = \mathbf{e}^\top \mathbf{H}_i \mathbf{e}$,

$$\|\mathbf{a}\| \leq q\sqrt{n}\left(\sum_{i=1}^{n} e_i \right)^2 \leq qn^{3/2}\|\mathbf{e}\|^2$$

where in the last step we have used Cauchy–Schwarz inequality [Section 8.4.2, p. 302] with $\mathbf{e}$ as the first vector, and all elements of the second vector equal to one. Combining the above inequalities involving $\|\hat{\mathbf{e}}\|$ and $\|\mathbf{a}\|$, we obtain

$$\|\hat{\mathbf{e}}\| \leq pqn^{3/2}\|\mathbf{e}\|^2$$

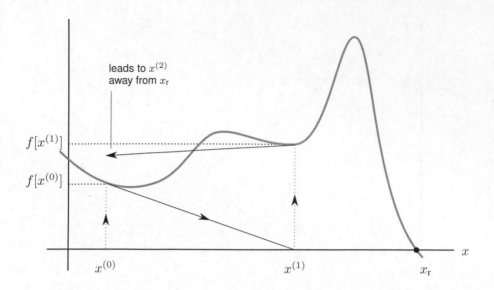

Figure 7.20   Example of the abnormal behavior of the Newton–Raphson method – divergence

This result shows that in the Newton–Raphson method, when $\mathbf{x}^{(k)}$ is sufficiently close to $\mathbf{x}_r$, the magnitude of the new error is proportional to the square of that of the previous error. This phenomenon is called quadratic convergence. For example, close enough to the root, if the magnitude of error is 10^{-2} then the magnitude of the next error would be of the order 10^{-4}.

7.C   General Derivation of Finite Difference Formulas

In this section, we present a general formula to obtain finite difference formulas of specified accuracy, and number of equispaced grid points ahead of, and behind a reference grid point.

Consider l and r equispaced grid points, respectively, to the left and right of the reference grid point denoted by i, as shown in Figure 7.21 below.

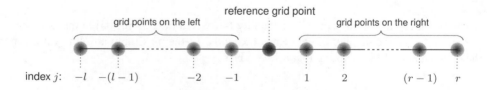

Figure 7.21   Grid points for a general finite difference formula

Thus, we have

$$x_{i+j} - x_i = jh; \qquad j = -l, -(l-1), \ldots, -2, -1, 1, 2, \ldots, (r-1), r$$

where h is the distance between two consecutive grid points. Using Taylor expansion [Equation (8.12), p. 322], the function value at the jth grid point is given by

$$f_{i+j} = f_i + (jh)f_i^{(1)} + \frac{(jh)^2}{2!}f_i^{(2)} + \frac{(jh)^3}{3!}f_i^{(3)} + \cdots + \frac{(jh)^n}{n!}f_i^{(n)} + O(h^{n+1})$$

$$j = -l, -(l-1), \ldots, -2, -1, 1, 2, \ldots, (r-1), r$$

where $f_k \equiv f(x_k)$ and $f_i^{(k)} \equiv f^{(k)}(x_i)$.

Multiplying each jth equation of the above expansion by a constant p_j, and summing the resulting equations yields

$$\underbrace{\sum_{\substack{j=-l \\ j\neq 0}}^{r} p_j(f_{i+j} - f_i)}_{\equiv Q} = \underbrace{\left(\sum_{\substack{j=-l \\ j\neq 0}}^{r} p_j j\right)}_{\equiv q_1} hf_i^{(1)} + \underbrace{\left(\sum_{\substack{j=-l \\ j\neq 0}}^{r} p_j j^2\right)}_{\equiv q_2} \frac{h^2}{2!}f_i^{(2)}$$

$$+ \underbrace{\left(\sum_{\substack{j=-l \\ j\neq 0}}^{r} p_j j^3\right)}_{\equiv q_3} \frac{h^3}{3!}f_i^{(3)} + \cdots + \underbrace{\left(\sum_{\substack{j=-l \\ j\neq 0}}^{r} p_j j^n\right)}_{\equiv q_n} \frac{h^n}{n!}f_i^{(n)}$$

$$+ O(h^{n+1})$$

In terms of Q and q_is as denoted above,

$$Q = q_1 hf_i^{(1)} + q_2 \frac{h^2}{2!}f_i^{(2)} + q_3 \frac{h^3}{3!}f_i^{(3)} + \cdots + q_n \frac{h^n}{n!}f_i^{(n)} + O(h^{n+1}) \qquad (7.66)$$

The above equation generates all finite difference formulas with appropriate choices of q_is. We show this with the help of a few examples.

7.C.1 *First Derivative, Centered Second Order Formula*

To obtain this formula, we express Equation (7.66) above with the remainder term of order $O(h^3)$, i.e.,

$$Q = q_1 hf_i^{(1)} + q_2 \frac{h^2}{2!}f_i^{(2)} + O(h^3)$$

Note that the above equation will lead to the $O(h^2)$-accurate formula as

$$f_i^{(1)} = \frac{1}{q_1 h}\left(Q - q_2 \frac{h^2}{2!}f_i^{(2)}\right) + \frac{\overbrace{O(h^3)}^{O(h^2)}}{-q_1 h}$$

In the above equation, specifying

$$q_1 = 1 \qquad\qquad\qquad (7.67)$$

$$q_2 = 0 \qquad\qquad\qquad (7.68)$$

yields the formula

$$f_i^{(1)} = \frac{Q}{h} + O(h^2) \tag{7.69}$$

To satisfy the two conditions – Equations (7.67) and (7.68) on the previous page – we need two grid points $(x_j s)$. For the centered formula, they are x_{i-1} and x_{i+1} with the corresponding unknown coefficients p_{-1} and p_1, respectively. Thus, $j = \{-1, 1\}$, and the two equations expand, respectively, to

$$\underbrace{p_{-1}(-1) + p_1(1)}_{q_1} = 1 \quad \text{and} \quad \underbrace{p_{-1}(-1)^2 + p_1(1)^2}_{q_2} = 0$$

whose solution is $p_{-1} = -1/2$ and $p_1 = 1/2$. Thus,

$$Q = p_{-1}(f_{i-1} - f_i) + p_1(f_{i+1} - f_i) = \frac{f_{i+1} - f_{i-1}}{2}$$

so that, from Equation (7.69) above, we get the desired finite difference formula,

$$f_i^{(1)} = \frac{f_{i+1} - f_{i-1}}{2h} + O(h^2)$$

7.C.2   Second Derivative, Forward Second Order Formula

In this case, $O(h^3)$-accuracy requires that we express Equation (7.66) on the previous page with the remainder term of order $O(h^4)$ so that

$$Q = q_1 h f_i^{(1)} + q_2 \frac{h^2}{2!} f_i^{(2)} + q_3 \frac{h^3}{3!} f_i^3 + O(h^4)$$

Note that the above equation will lead to the $O(h^2)$-accurate formula as

$$f_i^{(2)} = \frac{2}{q_2 h^2} \left(Q - q_1 h f_i^{(1)} - q_3 \frac{h^3}{3!} f_i^{(3)} \right) + \overbrace{\frac{O(h^4)}{-q_2 h^2/2}}^{O(h^2)}$$

In the above equation, specifying the following three conditions:

$$q_2 = 1, \quad q_1 = 0, \quad \text{and} \quad q_3 = 0$$

yields the formula

$$f_i^{(2)} = \frac{2Q}{h^2} + O(h^2) \tag{7.70}$$

The three conditions above require three grid points, which for the forward scheme are x_{i+1},

x_{i+2} and x_{i+3}. Thus, $j = \{1, 2, 3\}$, and the three conditions expand as

$$q_2 = p_1(1)^2 + p_2(2)^2 + p_3(3)^2 = 1$$
$$q_1 = p_1(1) + p_2(2) + p_3(3) = 0$$
$$q_3 = p_1(1)^3 + p_2(2)^3 + p_3(3)^3 = 0$$

Solving the above equations we get

$$p_1 = -5/2 \qquad p_2 = 2, \qquad \text{and} \qquad p_3 = -1/2$$

Thus, we have

$$Q = p_1(f_{i+1} - f_i) + p_2(f_{i+2} - f_i) + p_3(f_{i+3} - f_i)$$
$$= \frac{-f_{i+3} + 4f_{i+2} - 5f_{i+1} + 2f_i}{2}$$

so that from Equation (7.70) on the previous page, we get the desired finite difference formula,

$$f_i^{(2)} = \frac{-f_{i+3} + 4f_{i+2} - 5f_{i+1} + 2f_i}{h^2} + O(h^2)$$

7.C.3   Third Derivative, Mixed Fourth Order Formula

In this case, $O(h^4)$-accuracy for the third derivative requires Equation (7.66) on p. 285 with the remainder term of order $O(h^7)$ so that

$$Q = q_1 h f_i^{(1)} + q_2 \frac{h^2}{2!} f_i^{(2)} + q_3 \frac{h^3}{3!} f_i^3 + q_4 \frac{h^4}{4!} f_i^4 + q_5 \frac{h^5}{5!} f_i^5 + q_6 \frac{h^6}{6!} f_i^6 + O(h^7)$$

leads to

$$f_i^{(3)} = \frac{6}{q_3 h^3} \left(Q - q_1 h f_i^{(1)} - q_2 \frac{h^2}{2!} f_i^{(2)} - q_4 \frac{h^4}{4!} f_i^4 - q_5 \frac{h^5}{5!} f_i^5 - q_6 \frac{h^6}{6!} f_i^6 \right) + \overbrace{\frac{O(h^7)}{-q_3 h^3/6}}^{O(h^4)}$$

In the above equation, specifying the six conditions

$$q_3 = 1 \qquad \text{and} \qquad q_i = 0; \qquad i = 1, 2, 4, 5, 6$$

yields the formula

$$f_i^{(3)} = \frac{6Q}{h^3} + O(h^4) \tag{7.71}$$

To satisfy the six conditions above, we require six grid points. Let us choose them to be x_{i-4}, x_{i-3}, x_{i-2}, x_{i-1}, x_{i+1} and x_{i+2} in a mixed scheme of two forward, and four backward grid points, relative to the reference grid point x_i. Then the six conditions expand to

$$q_3 = \sum_{\substack{j=-4 \\ j \neq 0}}^{2} p_j(j)^3 = 1 \quad \text{and} \quad q_i = \sum_{\substack{j=-4 \\ j \neq 0}}^{2} p_j(j)^i = 0; \qquad i = 1, 2, 4, 5, 6$$

The above equations in the matrix form are

$$\begin{bmatrix} (-4)^3 & (-3)^3 & (-2)^3 & (-1)^3 & 1^3 & 2^3 \\ -4 & -3 & -2 & -1 & 1 & 2 \\ (-4)^2 & (-3)^2 & (-2)^2 & (-1)^2 & 1^2 & 2^2 \\ (-4)^4 & (-3)^4 & (-2)^4 & (-1)^4 & 1^4 & 2^4 \\ (-4)^5 & (-3)^5 & (-2)^5 & (-1)^5 & 1^5 & 2^5 \\ (-4)^6 & (-3)^6 & (-2)^6 & (-1)^6 & 1^6 & 2^6 \end{bmatrix} \begin{bmatrix} p_{-4} \\ p_{-3} \\ p_{-2} \\ p_{-1} \\ p_1 \\ p_2 \end{bmatrix} = \begin{bmatrix} 1 \\ 0 \\ 0 \\ 0 \\ 0 \\ 0 \end{bmatrix}$$

whose solution is

$$\begin{bmatrix} p_{-4} \\ p_{-3} \\ p_{-2} \\ p_{-1} \\ p_1 \\ p_2 \end{bmatrix} = \begin{bmatrix} -1/48 \\ 1/6 \\ -29/48 \\ 1 \\ 1/6 \\ 1/48 \end{bmatrix}$$

Thus,

$$Q = \sum_{\substack{j=-4 \\ j \neq 0}}^{2} p_j(f_{i+j} - f_i) = \frac{f_{i+2} + 8f_{i+1} - 35f_i + 48f_{i-1} - 29f_{i-2} + 8f_{i-3} - f_{i-4}}{48}$$

Substituting the above in Equation (7.71) on the previous page, we get

$$f_i^{(3)} = \frac{f_{i+2} + 8f_{i+1} - 35f_i + 48f_{i-1} - 29f_{i-2} + 8f_{i-3} - f_{i-4}}{8h^3} + O(h^4)$$

which is the desired finite difference formula.

7.C.4   Common Finite Difference Formulas

In this section, we list $O(h)$- and $O(h^2)$-accurate finite difference formulas for the first to fourth order derivatives. The number of grid points required by a finite difference formula is proportional to the order of the derivative as well as the order of accuracy.

In any given problem, the finite difference formulas that are used should of the same order of accuracy. Generally, the $O(h^2)$-accurate finite difference formulas are preferred. The backward and forward finite difference formulas find use at the terminal values of the independent variables.

Formulas of $O(h)$

First Derivative

$$f^{(1)}(x_i) = \frac{f(x_{i+1}) - f(x_i)}{h}$$

$$f^{(1)}(x_i) = \frac{f(x_i) - f(x_{i-1})}{h}$$

Second Derivative

$$f^{(2)}(x_i) = \frac{f(x_{i+2}) - 2f(x_{i+1}) + f(x_i)}{h^2}$$

$$f^{(2)}(x_i) = \frac{f(x_i) - 2f(x_{i-1}) + f(x_{i-2})}{h^2}$$

Third Derivative

$$f^{(3)}(x_i) = \frac{1}{h^3}\left[f(x_{i+3}) - 3f(x_{i+2}) + 3f(x_{i+1}) - f(x_i)\right]$$

$$f^{(3)}(x_i) = \frac{1}{h^3}\left[f(x_i) - 3f(x_{i-1}) + 3f(x_{i-2}) - f(x_{i-3})\right]$$

Fourth Derivative

$$f^{(4)}(x_i) = \frac{1}{h^4}\left[f(x_{i+4}) - 4f(x_{i+3}) + 6f(x_{i+2}) - 4f(x_{i+1}) + f(x_i)\right]$$

$$f^{(4)}(x_i) = \frac{1}{h^4}\left[f(x_i) - 4f(x_{i-1}) + 6f(x_{i-2}) - 4f(x_{i-3}) + 4f(x_{i-4})\right]$$

Formulas of $O(h^2)$

First Derivative

$$f^{(1)}(x_i) = \frac{-f(x_{i+2}) + 4f(x_{i+1}) - 3f(x_i)}{2h}$$

$$f^{(1)}(x_i) = \frac{3f(x_i) - 4f(x_{i-1}) + f(x_{i-2})}{2h}$$

$$f^{(1)}(x_i) = \frac{f(x_{i+1}) - f(x_{i-1})}{2h}$$

Second Derivative

$$f^{(2)}(x_i) = \frac{1}{h^2}\left[-f(x_{i+3}) + 4f(x_{i+2}) - 5f(x_{i+1}) + 2f(x_i)\right]$$

$$f^{(2)}(x_i) = \frac{1}{h^2}\left[2f(x_i) - 5f(x_{i-1}) + 4f(x_{i-2}) - f(x_{i-3})\right]$$

$$f^{(2)}(x_i) = \frac{f(x_{i+1}) - 2f(x_i) + f(x_{i-1})}{h^2}$$

Third Derivative

$$f^{(3)}(x_i) = \frac{1}{2h^3}\left[-3f(x_{i+4}) + 14f(x_{i+3}) - 24f(x_{i+2}) + 18f(x_{i+1}) - 5f(x_i)\right]$$

$$f^{(3)}(x_i) = \frac{1}{2h^3}\left[5f(x_i) - 18f(x_{i-1}) + 24f(x_{i-2}) - 14f(x_{i-3}) + 3f(x_{i-4})\right]$$

$$f^{(3)}(x_i) = \frac{1}{2h^3}\left[f(x_{i+2}) - 2f(x_{i+1}) + 2f(x_{i-1}) - f(x_{i-2})\right]$$

Fourth Derivative

$$f^{(4)}(x_i) = \frac{1}{h^4}\left[-2f(x_{i+5}) + 11f(x_{i+4}) - 24f(x_{i+3}) + 26f(x_{i+2}) - 14f(x_{i+1})\right.$$
$$\left. + 3f(x_i)\right]$$

$$f^{(4)}(x_i) = \frac{1}{h^4}\left[3f(x_i) - 14f(x_{i-1}) + 26f(x_{i-2}) - 24f(x_{i-3}) + 11f(x_{i-4})\right.$$
$$\left. - 2f(x_{i-5})\right]$$

$$f^{(4)}(x_i) = \frac{1}{h^4}\left[f(x_{i+2}) - 4f(x_{i+1}) + 6f(x_i) - 4f(x_{i-1}) + f(x_{i-2})\right]$$

References

[1] B.S. Sundaram, S.R. Upreti, and A. Lohi. "Optimal Control of Batch MMA Polymerization with Specified Time, Monomer Conversion, and Average Polymer Molecular Weights". In: *Macromolecular Theory and Simulations* 14 (2005), pp. 374–386.

[2] W.H. Press et al. *Numerical Recipes in C++ The Art of Scientific Computing.* 2nd edition. p. 748. Cambridge, U.K.: Cambridge University Press, 2002.

Bibliography

[1] S.C. Chapra and R.P. Canale. *Numerical Methods for Engineers.* 6th edition. New York: McGraw-Hill, 2010.

[2] S.K. Gupta. *Numerical Methods for Engineers.* New Delhi: New Age International Pvt. Ltd. Publishers, 2015.

[3] O.T. Hanna and O.C. Sandall. *Computational Methods in Chemical Engineering.* New Jersey: Prentice-Hall, Inc., 1995.

[4] W.E. Schiesser. *The Numerical Method of Lines: Integration of Partial Differential Equations.* San Diego: Academic Press Inc., 1991.

[5] E. Hairer, S.P. Nørsett, and G. Wanner. *Solving Differential Equations I: Nonstiff Problems.* 2nd edition. Berlin: Springer-Verlag, 2000.

[6] E. Hairer, S.P. Nørsett, and G. Wanner. *Solving Differential Equations II: Stiff and Differential-Algebraic Problems.* 2nd edition. Berlin: Springer-Verlag, 2010.

[7] W.H. Press et al. *Numerical Recipes The Art of Scientific Computing.* 3rd edition. New York: Cambridge University Press, 2007.

[8] R.G. Rice and D.D. Do. *Applied Mathematics and Modeling for Chemical Engineers.* 2nd edition. New Jersey: John Wiley & Sons, 2012.

Exercises

7.1 Repeat Example 7.1.3 on p. 238 with Peng–Robinson equation of state:

$$z^3 + (B-1)z^2 + (A - 3B^2 - 2B)z - (AB - B^2 - B^3) = 0$$

where

$$A = 0.45724\frac{P_r}{T_r}F, \qquad B = 0.0778\frac{P_r}{T_r}, \quad \text{and}$$

$$F = \frac{1}{T_r}\left[1 + (0.37464 + 1.5422\omega - 0.26992\omega^2)(1 - \sqrt{T_r})\right]^2.$$

7.2 Write the algorithm to find the determinant of a square matrix using lower-upper decomposition.

7.3 The solution of the tridiagonal system of equations

$$
\begin{bmatrix}
q_1 & r_1 & 0 & 0 & \cdots & \cdots & 0 \\
p_2 & q_2 & r_2 & 0 & \cdots & \cdots & 0 \\
0 & p_3 & q_3 & r_3 & 0 & \cdots & \vdots \\
0 & 0 & \ddots & \ddots & \ddots & \vdots & \vdots \\
\vdots & \vdots & 0 & p_{n-2} & q_{n-2} & r_{n-2} & 0 \\
\vdots & \vdots & \vdots & 0 & p_{n-1} & q_{n-1} & r_{n-1} \\
0 & 0 & 0 & \cdots & 0 & p_n & q_n
\end{bmatrix}
\begin{bmatrix}
x_1 \\
x_2 \\
\vdots \\
\vdots \\
\vdots \\
x_{n-1} \\
x_n
\end{bmatrix}
=
\begin{bmatrix}
b_1 \\
b_2 \\
\vdots \\
\vdots \\
\vdots \\
b_{n-1} \\
b_n
\end{bmatrix}
$$

can be efficiently obtained using **Thomas algorithm** as

$$x_n = \tilde{b}_n \quad \text{and}$$

$$x_i = \tilde{b}_i - \tilde{r}_i x_{i+1}; \quad i = n-1, n-2, \ldots, 1$$

where $\tilde{b}_i$s and $\tilde{r}_i$s are calculated as follows:

$$\tilde{b}_1 = \frac{b_1}{q_1} \qquad \tilde{b}_i = \frac{b_i - p_i \tilde{b}_{i-1}}{q_i - p_i \tilde{r}_{i-1}}; \qquad i = 2, 3, \ldots, n$$

$$\tilde{r}_1 = \frac{r_1}{q_1} \qquad \tilde{r}_i = \frac{r_i}{q_i - p_i \tilde{r}_{i-1}}; \qquad i = 2, 3, \ldots, n-1$$

Implement the above algorithm to fit cubic splines using Equation (6.35) on p. 217, and Equation (6.36) on p. 218. Test them using the data of Example 6.2.5 on p. 216.

7.4 Apply finite differences to the following process model:

$$\frac{1}{\text{Pe}} \frac{d^2 y}{dx^2} - \frac{dy}{dx} - \text{Da}\, y^2 = 0$$

with the following boundary conditions:

$$\frac{dy}{dx} = \text{Pe}(y - 1) \quad \text{at} \quad x = 0$$

$$\frac{dy}{dx} = 0 \qquad\qquad \text{at} \quad x = 1$$

7.5 Solve the finite-differenced equations obtained in Exercise 7.4 above for $\text{Pe} = 10^4$, and $\text{Da} = 1.1$. Obtain y versus x for different number of grid points starting with 11, i.e., 10 x-intervals. Find the effect of increasing the number of grid points.

7.6 The concentration of a species during reaction inside a porous spherical catalyst is described by the following model:

$$\frac{\partial c}{\partial t} = \frac{D}{r^2} \frac{\partial}{\partial r}\left(r^2 \frac{\partial c}{\partial r}\right) - kc$$

The above partial differential equation has the following initial and boundary conditions:

$$c(r, 0) = 0$$

$$\frac{\partial c}{\partial r} = 0 \qquad \text{at} \quad r = 0$$

$$-D\frac{\partial c}{\partial r} = k_g(c - c_a) \qquad \text{at} \quad r = R$$

Apply finite differences to the above model, and obtain a set of ordinary differential equations for concentrations on the grid points along the radial direction.

7.7 Derive the expressions for the accumulated truncation errors involved in solution of Equation (7.19) on p. 248 obtained from the explicit and implicit Euler's methods.

7.8 A composite function is given by

$$f(T) = ak/\rho \qquad \text{where} \qquad k = k_0 e^{-b/T}, \qquad \rho = \rho_0 c^{-(1-T/d)^e},$$

and a, k_0, b, ρ_0, c, d and e are some constants. Obtain the analytical expression for the derivative of the function with respect to T. Write a computer program to check the expression.

7.9 Simulate the original and simplified plug flow reactor models in Section 6.1.1 on p. 190. Compare the results.

7.10 Solve the ordinary differential equations obtained in Exercise 7.6 on the previous page for different number (N) of radial grid points, and parameters given in Table 7.9 below. Comment on the results.

Table 7.9   Values of parameters for Exercise 7.10

parameter	value	parameter	value
D	10^{-8} m^2 s^{-1}	k	2 s^{-1}
k_g	2×10^{-3} m s^{-1}	c_a	0.3 mol m^{-3}
R	5×10^{-3} m	N	$11, 21, 41, \ldots, 321$

7.11 The reactant concentration c, and the temperature T in a tubular reactor under steady state, and laminar flow are given by the following model:

$$\frac{\partial c}{\partial z} = -\frac{k_0 c^2 R^2}{2\bar{v}(R^2 - r^2)} \exp\left(-\frac{a}{T}\right)$$

$$\frac{\partial T}{\partial z} = \frac{R^2}{2\bar{v}(R^2 - r^2)\rho\hat{C}_P}\left[k\left(\frac{\partial^2 T}{\partial r^2} + \frac{1}{r}\frac{\partial T}{\partial r}\right) + (-\Delta H_r)k_0 \exp\left(-\frac{a}{T}\right)c^2\right]$$

where z is axial coordinate, k_0 is Arrhenius rate constant, r is radial distance, $\bar{v}$ is average axial velocity, R is the inside radius of the reactor, a is the ratio of activation energy of the reaction to universal gas constant, and $(-\Delta H_r)$ is the heat of reaction. The density, thermal conductivity, and specific heat capacity of the reaction mixture are ρ, k and $\hat{C}_P$, respectively. The model equations are subject to the following conditions:

$$c = c_f \quad \text{at} \quad z = 0, \qquad \frac{\partial T}{\partial r} = 0 \quad \text{at} \quad r = 0,$$

$$T = T_f \quad \text{at} \quad z = 0, \quad \text{and} \quad T = T_w \quad \text{at} \quad r = R$$

where c_f and T_f are, respectively, the values of c and T at the reactor inlet, and T_w is the wall temperature of the reactor.

Apply the method of lines on the above reactor model, and obtain a set of ordinary differential equations.

7.12 Solve the ordinary differential equations obtained in Exercise 7.11 above for different number (N) of radial grid points, and parameters given in Table 7.10 below. Comment on the results.

Table 7.10   Values of parameters for Exercise 7.12

parameter	value	parameter	value
c_f	1 mol m^{-3}	T_f	293 K
k_0	$1.6 \text{ m}^3 \text{ mol}^{-1} \text{ s}^{-1}$	$\bar{v}$	0.02 m s^{-1}
a	10^3 K	T_w	333 K
R	0.03 m	ρ	10^3 kg m^{-3}
$\hat{C}_P$	$12 \text{ J kg}^{-1} \text{ K}^{-1}$	$-\Delta H_r$	$-1.2 \times 10^8 \text{ J mol}^{-1}$
k	$10 \text{ W m}^{-1} \text{ K}^{-1}$	N	$11, 21, 41$

7.13 Using finite differences, convert the model in Exercise 7.6 on p. 292 to a set of algebraic equations. Solve these equations for the parameters given in Table 7.9 on the previous page. Compare the solution with that obtained in Exercise 7.10.

7.14 Express the model in Exercise 7.6 at steady state, and solve it using the shooting Newton–Raphson method.

7.15 Write a computer program to implement composite Simpson's 1/3 rule. The program should utilize Simpson's 3/8 rule if the number of data points is (i) four, or (ii) even and greater than five.

7.16 Find the remaining roots in Example 7.7.4 on p. 280.

8

Mathematical Review

This chapter reviews the important mathematical concepts underpinning the topics covered in this book.

8.1   Order of Magnitude

The order of magnitude of a real number is the power of ten included in the number. Thus, the order of magnitude of y is the integer part of the $\log_{10} y$.

The n orders-of-magnitude difference between two real numbers means that the ratio of the bigger number to the other is $m \times 10^n$ where $1 \leq m < 10$.

8.2   Big-O Notation

Consider a function $f(x)$ in the limit of x tending to some value x_0. If the upper bound of $|f(x)|$ is the product of a positive constant c and $|x|$ when x is sufficiently close to x_0 then

$$\lim_{x \to x_0} |f(x)| \leq c|x|, \quad |x - x_0| < \delta > 0$$

A brief form of the above expression is

$$\lim_{x \to x_0} f(x) = O(x)$$

where $O(x)$ represents the upper bound, or the order of the function f with respect to the variable x.

8.3   Analytical Function

An analytical function is an infinitely differentiable function, which can be represented as a convergent Taylor series at any point in the domain of the function. Examples of analytical functions are polynomials, and exponential, logarithmic and trigonometric functions.

The sums and products of analytical functions are analytical functions. So are the functions of analytical functions as well as the inverses of analytical functions whose first derivatives are not zero.

Process Modeling and Simulation for Chemical Engineers: Theory and Practice, First Edition. Simant Ranjan Upreti.
© 2017 John Wiley & Sons Ltd. Published 2017 by John Wiley & Sons Ltd.
Companion website: www.wiley.com/go/upreti/pms_for_chemical_engineers

Analytical Methods

These methods solve equations, and provide answers in terms of analytical functions. For example, an analytical method would solve

$$\frac{dy}{dt} = f(y), \qquad y(0) = \bar{y}$$

providing answer as an analytical function $y = y(t, \bar{y})$. In this function, we could plug values for t and $\bar{y}$ to obtain the associated value of y. However, there are limited types of equations that could be solved by analytical methods.

Numerical Methods

These methods approximate equations, or their parts using Taylor expansions with finite terms. These methods frequently require iterative calculations, and provide answers as direct numerical values.

While answers from analytical methods are exact, those from numerical methods are accurate to the degree of the involved approximation. But the latter can be applied to solve any equations including those that cannot be solved by analytical methods.

8.4   Vectors

A vector in general is an ordered list of finite number of elements. A vector is usually denoted by a lowercase boldface letter, e.g., $\mathbf{a}$. Thus, in the vector notation

$$\mathbf{a} = \begin{bmatrix} a_1 \\ a_2 \\ \vdots \\ a_n \end{bmatrix}$$

where a_is are the elements or components of $\mathbf{a}$. By default, a vector is considered as a column vector. The transpose of a column vector is a row vector, and vice versa. Thus, the transpose of $\mathbf{a}$ is

$$\mathbf{a}^\top = \begin{bmatrix} a_1 & a_2 & \dots & a_n \end{bmatrix}$$

In a Euclidean space such as the three-dimensional Cartesian coordinate system, a vector is an entity characterized by a magnitude, and a direction. As shown in Figure 8.1 on the next page, a vector $\mathbf{a}$ is represented by an arrow. While the length of the arrow represents the magnitude or norm of the vector, the arrow itself points to the direction of the vector. The tail of the arrow is implicitly assumed to lie on the origin. The endpoint of the arrow is the set of coordinates for the vector components. In terms of the unit vectors ($\hat{\mathbf{x}}_i$s) along the coordinate axes, the vector is expressed as

$$\mathbf{a} = a_1\hat{\mathbf{x}}_1 + a_2\hat{\mathbf{x}}_2 + a_3\hat{\mathbf{x}}_3$$

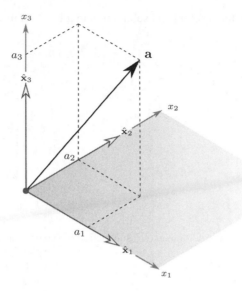

Figure 8.1 A vector **a** in the three-dimensional Cartesian coordinate system

where a_is are the vector components. As shown in Figure 8.1 above, these components are the projections of the vector on the coordinate axes. Thus, the length of **a**, which is also the magnitude or norm, is given by

$$\|\mathbf{a}\| \;=\; \sqrt{a_1^2 + a_2^2 + a_3^2}$$

In general, the *Euclidean norm* of a vector $\mathbf{v} = \begin{bmatrix} v_1 & v_2 & \cdots & v_n \end{bmatrix}^\top$ is given by

$$\|\mathbf{v}\| \;=\; \sqrt{\sum_{i=1}^{n} v_i^2}$$

8.4.1 Vector Operations

The operations defined for vectors are as follows.

Addition and Subtraction

The sum of two vectors of the same size is the vector with each element equal to the sum of corresponding elements of the added vectors. The order of addition is not important. Thus,

$$\mathbf{a} + \mathbf{b} \;=\; \mathbf{c} \;=\; \mathbf{b} + \mathbf{a} \qquad \text{where} \qquad c_i \;=\; a_i + b_i; \qquad i = 1, 2, \ldots, n$$

where **a**, **b** and **c** are vectors of the same size n, and have elements a_i, b_i and c_i, respectively. Similarly for subtraction,

$$\mathbf{a} - \mathbf{b} \;=\; \mathbf{c} \;=\; -\mathbf{b} + \mathbf{a} \qquad \text{where} \qquad c_i \;=\; a_i - b_i; \qquad i = 1, 2, \ldots, n$$

Figure 8.2 below shows vector addition and subtraction in a Euclidean space.

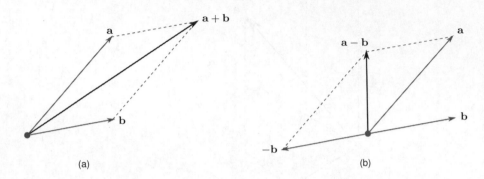

(a)   (b)

Figure 8.2   (a) Addition and (b) subtraction of vectors **a** and **b**

Multiplication by a Scalar

Multiplication of a vector by scalar results in each vector element multiplied by the scalar.

Differentiation

The derivative of a vector

$$\mathbf{y} = \begin{bmatrix} y_1 & y_2 & \cdots & y_m \end{bmatrix}^{\top}$$

with respect to a scalar variable x is written as the column vector

$$\frac{\mathrm{d}\mathbf{y}}{\mathrm{d}x} = \begin{bmatrix} \dfrac{\mathrm{d}y_1}{\mathrm{d}x} & \dfrac{\mathrm{d}y_2}{\mathrm{d}x} & \cdots & \dfrac{\mathrm{d}y_m}{\mathrm{d}x} \end{bmatrix}^{\top}$$

The derivative of a scalar y with respect to a vector

$$\mathbf{x} = \begin{bmatrix} x_1 & x_2 & \cdots & x_n \end{bmatrix}^{\top}$$

is written as the row vector

$$\frac{\partial y}{\partial \mathbf{x}} = \begin{bmatrix} \dfrac{\partial y}{\partial x_1} & \dfrac{\partial y}{\partial x_2} & \cdots & \dfrac{\partial y}{\partial x_n} \end{bmatrix}$$

For $n = 3$,

$$\frac{\partial}{\partial \mathbf{x}} \equiv \nabla = \begin{bmatrix} \dfrac{\partial}{\partial x_1} & \dfrac{\partial}{\partial x_2} & \dfrac{\partial}{\partial x_3} \end{bmatrix}$$

is called the **gradient operator** in the coordinates of Cartesian coordinate system (x_1, x_2, x_3).
The derivative of the vector $\mathbf{y}$ with respect to the vector $\mathbf{x}$ is written as the matrix

$$\nabla \mathbf{y} \equiv \frac{\partial \mathbf{y}}{\partial \mathbf{x}} = \begin{bmatrix} \dfrac{\partial y_1}{\partial x_1} & \dfrac{\partial y_1}{\partial x_2} & \cdots & \dfrac{\partial y_1}{\partial x_n} \\[2ex] \dfrac{\partial y_2}{\partial x_1} & \dfrac{\partial y_2}{\partial x_2} & \cdots & \dfrac{\partial y_2}{\partial x_n} \\[2ex] \vdots & \vdots & \ddots & \vdots \\[2ex] \dfrac{\partial y_m}{\partial x_1} & \dfrac{\partial y_m}{\partial x_2} & \cdots & \dfrac{\partial y_m}{\partial x_n} \end{bmatrix} \qquad (8.1)$$

The above matrix follows the *numerator layout convention*. According to this convention,
the row index is that of the numerator entity, i.e., ∂y_i, and the column index is that of the
denominator entity, i.e., ∂x_j.

Dot or Scalar or Inner Product

This product for two vectors $\mathbf{a}$ and $\mathbf{b}$ is defined as

$$\mathbf{a} \cdot \mathbf{b} = \|\mathbf{a}\| \|\mathbf{b}\| \cos \theta$$

where θ is the angle between the two vectors. Figure 8.3 below shows the geometrical
interpretation of the dot product. Note that $\mathbf{a} \cdot \mathbf{a} = \|\mathbf{a}\|^2$.

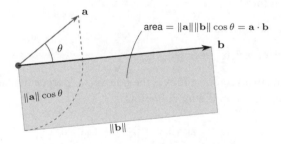

Figure 8.3   Dot product of vectors $\mathbf{a}$ and $\mathbf{b}$ shown as the shaded area

The dot product has the following properties:

Commutativity	$\mathbf{a} \cdot \mathbf{b} = \mathbf{b} \cdot \mathbf{a}$
Distributivity	$\mathbf{c} \cdot (\mathbf{a} + \mathbf{b}) = \mathbf{c} \cdot \mathbf{a} + \mathbf{c} \cdot \mathbf{b}$
Non-Associativity	$(\mathbf{a} \cdot \mathbf{b})\mathbf{c} \neq \mathbf{a}(\mathbf{b} \cdot \mathbf{c})$
Scaling Homogeneity	$(\alpha \mathbf{a}) \cdot \mathbf{b} = \mathbf{a} \cdot (\alpha \mathbf{b}) = \alpha(\mathbf{a} \cdot \mathbf{b})$
Non-Cancellation	$\mathbf{a} \cdot \mathbf{b} = \mathbf{a} \cdot \mathbf{c}$ (where $\mathbf{a} \neq 0$) $\not\Rightarrow \mathbf{b} = \mathbf{c}$

In an orthogonal coordinate system (y_1, y_2, y_3) having unit vectors $\hat{\mathbf{y}}_1$, $\hat{\mathbf{y}}_2$ and $\hat{\mathbf{y}}_3$, using the distributive property of the dot product,

$$
\begin{aligned}
\mathbf{a} \cdot \mathbf{b} \; &= \; (a_1\hat{\mathbf{y}}_1 + a_2\hat{\mathbf{y}}_2 + a_3\hat{\mathbf{y}}_3) \cdot (b_1\hat{\mathbf{y}}_1 + b_2\hat{\mathbf{y}}_2 + b_3\hat{\mathbf{y}}_3) \\[4pt]
&= \; a_1b_1\hat{\mathbf{y}}_1 \cdot \hat{\mathbf{y}}_1 + a_1b_2\hat{\mathbf{y}}_1 \cdot \hat{\mathbf{y}}_2 + a_1b_3\hat{\mathbf{y}}_1 \cdot \hat{\mathbf{y}}_3 \\[4pt]
&\quad + \; a_2b_1\hat{\mathbf{y}}_2 \cdot \hat{\mathbf{y}}_1 + a_2b_2\hat{\mathbf{y}}_2 \cdot \hat{\mathbf{y}}_2 + a_2b_3\hat{\mathbf{y}}_2 \cdot \hat{\mathbf{y}}_3 \\[4pt]
&\quad + \; a_3b_1\hat{\mathbf{y}}_3 \cdot \hat{\mathbf{y}}_1 + a_3b_2\hat{\mathbf{y}}_3 \cdot \hat{\mathbf{y}}_2 + a_3b_3\hat{\mathbf{y}}_3 \cdot \hat{\mathbf{y}}_3
\end{aligned}
$$

Note that from the definition of the dot product,

$$\hat{\mathbf{y}}_1 \cdot \hat{\mathbf{y}}_1 \; = \; \hat{\mathbf{y}}_2 \cdot \hat{\mathbf{y}}_2 \; = \; \hat{\mathbf{y}}_3 \cdot \hat{\mathbf{y}}_3 \; = \; 1$$

$$\hat{\mathbf{y}}_1 \cdot \hat{\mathbf{y}}_2 \; = \; \hat{\mathbf{y}}_2 \cdot \hat{\mathbf{y}}_1 \; = \; \hat{\mathbf{y}}_2 \cdot \hat{\mathbf{y}}_3 \; = \; \hat{\mathbf{y}}_3 \cdot \hat{\mathbf{y}}_2 \; = \; \hat{\mathbf{y}}_3 \cdot \hat{\mathbf{y}}_1 \; = \; \hat{\mathbf{y}}_1 \cdot \hat{\mathbf{y}}_3 \; = \; 0$$

Therefore,

$$\mathbf{a} \cdot \mathbf{b} \; = \; a_1b_1 + a_2b_2 + a_3b_3$$

Differentiation

The differentiation of the dot product follows the product rule of differentiation, i.e.,

$$\frac{\mathrm{d}}{\mathrm{d}x}(\mathbf{a} \cdot \mathbf{b}) \; = \; \frac{\mathrm{d}\mathbf{a}}{\mathrm{d}x} \cdot \mathbf{b} + \mathbf{a} \cdot \frac{\mathrm{d}\mathbf{b}}{\mathrm{d}x}$$

Cross or Vector or Outer Product

This product for two vectors $\mathbf{a}$ and $\mathbf{b}$ is defined as

$$\mathbf{a} \times \mathbf{b} \; = \; (\|\mathbf{a}\|\|\mathbf{b}\| \sin\theta)\hat{\mathbf{n}} \; \equiv \; \mathbf{c}$$

where θ is the angle between the two vectors, and $\hat{\mathbf{n}}$ is the unit vector perpendicular to $\mathbf{a}$ and $\mathbf{b}$ in the right-handed system, as shown in Figure 8.4 below.

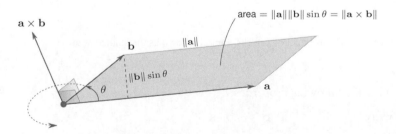

Figure 8.4 Cross product of vectors $\mathbf{a}$ and $\mathbf{b}$

The result of the cross product is the vector $\mathbf{c}$ such that looking opposite to it, θ is the angle obtained by turning counter-clockwise from $\mathbf{a}$ to $\mathbf{b}$. Thus, $\mathbf{c}$ is along the upward pointing right-hand thumb with the remaining fingers curled counter-clockwise in the direction of θ increasing from $\mathbf{a}$ to $\mathbf{b}$.

From the definition of cross product, $\mathbf{a} \times \mathbf{a} = \mathbf{0}$. The cross product has the following properties:

Anti-Commutativity $\qquad \mathbf{a} \times \mathbf{b} = -(\mathbf{b} \times \mathbf{a})$

Distributivity $\qquad \mathbf{c} \times (\mathbf{a} + \mathbf{b}) = \mathbf{c} \times \mathbf{a} + \mathbf{c} \times \mathbf{b}$

Non-Associativity $\qquad (\mathbf{a} \times \mathbf{b}) \times \mathbf{c} \neq \mathbf{a} \times (\mathbf{b} \times \mathbf{c})$

Scaling Homogeneity $\qquad (\alpha \mathbf{a}) \times \mathbf{b} = \mathbf{a} \times (\alpha \mathbf{b}) = \alpha(\mathbf{a} \times \mathbf{b})$

Non Cancellation $\qquad \mathbf{a} \times \mathbf{b} = \mathbf{a} \times \mathbf{c} \ \text{(where a} \neq 0) \quad \nRightarrow \mathbf{b} = \mathbf{c}$

In an orthogonal coordinate system (y_1, y_2, y_3) having unit vectors $\hat{\mathbf{y}}_1$, $\hat{\mathbf{y}}_2$ and $\hat{\mathbf{y}}_3$, using the distributive property of the cross product,

$$
\begin{aligned}
\mathbf{a} \times \mathbf{b} &= (a_1\hat{\mathbf{y}}_1 + a_2\hat{\mathbf{y}}_2 + a_3\hat{\mathbf{y}}_3) \times (b_1\hat{\mathbf{y}}_1 + b_2\hat{\mathbf{y}}_2 + b_3\hat{\mathbf{y}}_3) \\
&= a_1b_1\hat{\mathbf{y}}_1 \times \hat{\mathbf{y}}_1 + a_1b_2\hat{\mathbf{y}}_1 \times \hat{\mathbf{y}}_2 + a_1b_3\hat{\mathbf{y}}_1 \times \hat{\mathbf{y}}_3 \\
&\quad + a_2b_1\hat{\mathbf{y}}_2 \times \hat{\mathbf{y}}_1 + a_2b_2\hat{\mathbf{y}}_2 \times \hat{\mathbf{y}}_2 + a_2b_3\hat{\mathbf{y}}_2 \times \hat{\mathbf{y}}_3 \\
&\quad + a_3b_1\hat{\mathbf{y}}_3 \times \hat{\mathbf{y}}_1 + a_3b_2\hat{\mathbf{y}}_3 \times \hat{\mathbf{y}}_2 + a_3b_3\hat{\mathbf{y}}_3 \times \hat{\mathbf{y}}_3
\end{aligned}
$$

Note that from the definition of the cross product,

$$
\begin{aligned}
\hat{\mathbf{y}}_1 \times \hat{\mathbf{y}}_1 &= \hat{\mathbf{y}}_2 \times \hat{\mathbf{y}}_2 = \hat{\mathbf{y}}_3 \times \hat{\mathbf{y}}_3 = \mathbf{0} \\
\hat{\mathbf{y}}_1 \times \hat{\mathbf{y}}_2 &= -\hat{\mathbf{y}}_2 \times \hat{\mathbf{y}}_1 = \hat{\mathbf{y}}_3 \\
\hat{\mathbf{y}}_2 \times \hat{\mathbf{y}}_3 &= -\hat{\mathbf{y}}_3 \times \hat{\mathbf{y}}_2 = \hat{\mathbf{y}}_1 \\
\hat{\mathbf{y}}_3 \times \hat{\mathbf{y}}_1 &= -\hat{\mathbf{y}}_1 \times \hat{\mathbf{y}}_3 = \hat{\mathbf{y}}_2
\end{aligned}
$$

Therefore,

$$
\begin{aligned}
\mathbf{a} \times \mathbf{b} &= (a_2b_3 - a_3b_2)\hat{\mathbf{y}}_1 - (a_1b_3 - a_3b_1)\hat{\mathbf{y}}_2 + (a_1b_2 - a_2b_1)\hat{\mathbf{y}}_3 \\[2mm]
&= \hat{\mathbf{y}}_1 \begin{vmatrix} a_2 & a_3 \\ b_2 & b_3 \end{vmatrix} - \hat{\mathbf{y}}_2 \begin{vmatrix} a_1 & a_3 \\ b_1 & b_3 \end{vmatrix} + \hat{\mathbf{y}}_3 \begin{vmatrix} a_1 & a_2 \\ b_1 & b_2 \end{vmatrix} \\[2mm]
&= \begin{vmatrix} \hat{\mathbf{y}}_1 & \hat{\mathbf{y}}_2 & \hat{\mathbf{y}}_3 \\ a_1 & a_2 & a_3 \\ b_1 & b_2 & b_3 \end{vmatrix}
\end{aligned} \qquad (8.2)
$$

Differentiation

The differentiation of the cross product follows the product rule of differentiation, i.e.,

$$\frac{d}{dx}(\mathbf{a} \times \mathbf{b}) = \frac{d\mathbf{a}}{dx} \times \mathbf{b} + \mathbf{a} \times \frac{d\mathbf{b}}{dx}$$

Triple Cross Product

The triple cross product is the cross product of a vector by another vector resulting from the cross product of two vectors, i.e., $\mathbf{a} \times (\mathbf{b} \times \mathbf{c})$. This product satisfies the identity

$$\mathbf{a} \times (\mathbf{b} \times \mathbf{c}) = \mathbf{b}(\mathbf{a} \cdot \mathbf{c}) - \mathbf{c}(\mathbf{a} \cdot \mathbf{b})$$

which can be verified by expanding both of its sides using the definitions of the cross and dot vector products. By setting

$$\mathbf{a} = \mathbf{b} = \nabla$$

where ∇ is the gradient operator (a vector), the above identity yields

$$\nabla \times (\nabla \times \mathbf{c}) = \nabla(\nabla \cdot \mathbf{c}) - \mathbf{c}\underbrace{(\nabla \cdot \nabla)}_{=\nabla^2}$$

Upon rearrangement, the above result provides an important formula for the **Laplacian** of a vector $\mathbf{c}$, i.e.,

$$\nabla^2 \mathbf{c} = \nabla(\nabla \cdot \mathbf{c}) - \nabla \times (\nabla \times \mathbf{c}) \tag{8.3}$$

8.4.2   Cauchy–Schwarz Inequality

According to this inequality,

$$(\mathbf{x} \cdot \mathbf{y})^2 \leq \|\mathbf{x}\|^2 \|\mathbf{y}\|^2$$

where $\mathbf{x}$ and $\mathbf{y}$ are two vectors having the same number of elements, n.
   For $n = 2$, let

$$\mathbf{x} \equiv \begin{bmatrix} x_1 & x_2 \end{bmatrix}^\top \quad \text{and} \quad \mathbf{y} \equiv \begin{bmatrix} y_1 & y_2 \end{bmatrix}^\top$$

In this case,

$$\underbrace{(x_1^2 + x_2^2)}_{\|\mathbf{x}\|^2} \underbrace{(y_1^2 + y_2^2)}_{\|\mathbf{y}\|^2} - \underbrace{(x_1 y_1 + x_2 y_2)^2}_{(\mathbf{x} \cdot \mathbf{y})^2} = \underbrace{(x_1 y_2 - x_2 y_1)^2}_{\geq 0}$$

Thus, $(\mathbf{x} \cdot \mathbf{y})^2 \leq \|\mathbf{x}\|^2 \|\mathbf{y}\|^2$, which can be easily generalized for higher dimensions.

8.5   Matrices

Matrices are two-dimensional collections of elements. They are arranged in rows and columns, and impart brevity in dealing with sets of linear equations.

8.5.1   Terminology

A *square matrix* has equal numbers of rows and columns. A *diagonal matrix* has a principal diagonal, and all remaining elements 0, e.g.,

$$
\begin{bmatrix}
A_{11} & 0 & 0 \\
0 & A_{22} & 0 \\
0 & 0 & A_{33}
\end{bmatrix}
$$

The *trace* of a matrix is the sum of its diagonal elements, $\sum_{i=1}^{n} A_{ii}$. An *identity matrix* is the diagonal matrix $\mathbf{I}$ with elements $I_{ii} = 1$ for all values of i. A *symmetric matrix* has $A_{ij} = A_{ji}$ for all i and j, e. g.,

$$
\begin{bmatrix}
1 & 4 & 5 \\
4 & 2 & 6 \\
5 & 6 & 3
\end{bmatrix}
$$

An *upper triangular matrix* has all elements 0 in the triangular part *below* the diagonal, e.g.,

$$
\begin{bmatrix}
A_{11} & A_{12} & A_{13} \\
0 & A_{22} & A_{23} \\
0 & 0 & A_{33}
\end{bmatrix}
$$

On the other hand, a *lower triangular matrix* has all elements 0 in the triangular part *above* the diagonal, e.g.,

$$
\begin{bmatrix}
A_{11} & 0 & 0 \\
A_{21} & A_{22} & 0 \\
A_{31} & A_{32} & A_{33}
\end{bmatrix}
$$

The inverse $\mathbf{A}^{-1}$ of a square matrix $\mathbf{A}$ is another square matrix of the same size, which satisfies

$$
\mathbf{A}^{-1} \times \mathbf{A} = \mathbf{A} \times \mathbf{A}^{-1} = \mathbf{I}
$$

where $\mathbf{I}$ is the identity matrix of the same size.

The *transpose* of a matrix $\mathbf{A}$ is the matrix denoted by $\mathbf{A}^{\top}$, which is obtained by turning the rows of $\mathbf{A}$ into its columns.

Lastly, similar to the Euclidean norm of a vector, the *Frobenius norm* of a matrix $\mathbf{A}$ of n rows and m columns is defined as

$$
\|\mathbf{A}\| = \sum_{i=1}^{n} \sum_{j=1}^{m} A_{ij}^{2}
$$

where A_{ij} is the element in the ith row and j^{th} column of the matrix.

8.5.2   Matrix Operations

The operations defined for matrices are as follows.

Addition and Subtraction

The sum of two matrices of the same size is the matrix with each element equal to the sum of corresponding elements in the matrices being added. The order of addition is not important. Thus, for matrices with n rows and m columns,

$$\mathbf{A} + \mathbf{B} \;=\; \mathbf{C} \;=\; \mathbf{B} + \mathbf{A} \qquad \text{where} \qquad C_{ij} \;=\; A_{ij} + B_{ij}$$

$$i = 1, 2, \ldots, n; \quad j = 1, 2, \ldots, m$$

where C_{ij}, A_{ij} and B_{ij} are, respectively, the elements in the ith row and j^{th} column of $\mathbf{C}$, $\mathbf{A}$ and $\mathbf{B}$. Similarly for subtraction,

$$\mathbf{A} - \mathbf{B} \;=\; \mathbf{C} \;=\; -\mathbf{B} + \mathbf{A} \qquad \text{where} \qquad C_{ij} \;=\; A_{ij} - B_{ij}$$

$$i = 1, 2, \ldots, n; \quad j = 1, 2, \ldots, m$$

Multiplication by a Scalar

Multiplication of a matrix by a scalar results in the multiplication of each matrix element by the scalar.

Multiplication of Matrices

Multiplication of two matrices is defined if the number of columns of the first matrix is equal to the number of rows of the second matrix. The result is a matrix that has number of rows of the first matrix, and the number of columns of the second matrix. Thus, if $\mathbf{A}$ has n rows and m columns, and $\mathbf{B}$ has m rows and l columns then

$$\mathbf{A}_{n \times m} \times \mathbf{B}_{m \times l} \;=\; \mathbf{C}_{n \times l}$$

The elements of the product $\mathbf{C}$ are given by

$$C_{ij} \;=\; \sum_{k=1}^{m} A_{ik} B_{kj}; \qquad i = 1, 2, \ldots, n; \qquad j = 1, 2, \ldots, m$$

The product of two matrices is associative, i.e., $(\mathbf{AB})\mathbf{C} = \mathbf{A}(\mathbf{BC})$ as well as distributive, i.e., $(\mathbf{A} + \mathbf{B})\mathbf{C} = \mathbf{AC} + \mathbf{BC}$. Note that $\mathbf{A} \times \mathbf{B} \neq \mathbf{B} \times \mathbf{A}$.

Differentiation

The derivative of a matrix $\mathbf{A}$ (of n rows and m columns) with respect to a scalar x is written as

$$
\frac{d\mathbf{A}}{dx} = \begin{bmatrix} \dfrac{dA_{11}}{dx} & \dfrac{dA_{12}}{dx} & \cdots & \dfrac{dA_{1m}}{dx} \\[2ex] \dfrac{dA_{21}}{dx} & \dfrac{dA_{22}}{dx} & \cdots & \dfrac{dA_{2m}}{dx} \\[2ex] \vdots & \vdots & \ddots & \vdots \\[2ex] \dfrac{dA_{n1}}{dx} & \dfrac{dA_{n2}}{dx} & \cdots & \dfrac{dA_{nm}}{dx} \end{bmatrix}
$$

8.5.3 *Operator Inequality*

According to this inequality

$$
\|\mathbf{A}\mathbf{b}\| \leq \|\mathbf{A}\|\|\mathbf{b}\|
$$

where $\mathbf{A}$ is an $n \times m$ operator (matrix), and $\mathbf{b}$ is an m-dimensional vector.

For $n = m = 2$, let

$$
\mathbf{A} \equiv \begin{bmatrix} A_{11} & A_{12} \\ A_{21} & A_{22} \end{bmatrix} \quad \text{and} \quad \mathbf{b} \equiv \begin{bmatrix} b_1 \\ b_2 \end{bmatrix}
$$

Then

$$
\mathbf{A}\mathbf{b} = \begin{bmatrix} A_{11}b_1 + A_{12}b_2 \\ A_{21}b_1 + A_{22}b_2 \end{bmatrix}
$$

and

$$
\|\mathbf{A}\mathbf{b}\|^2 = \underbrace{(A_{11}b_1 + A_{12}b_2)^2}_{p} + \underbrace{(A_{21}b_1 + A_{22}b_2)^2}_{q}
$$

Because

$$
(A_{11}^2 + A_{12}^2)(b_1^2 + b_2^2) - p = (A_{11}b_2 - A_{12}b_1)^2 \geq 0
$$

we obtain

$$
p \leq (A_{11}^2 + A_{12}^2)(b_1^2 + b_2^2)
$$

Similarly we get,

$$
q \leq (A_{21}^2 + A_{22}^2)(b_1^2 + b_2^2)
$$

Combining the last two inequalities yields

$$\underbrace{p+q}_{\|\mathbf{Ab}\|^2} \leq \underbrace{(A_{11}^2 + A_{12}^2 + A_{21}^2 + A_{22}^2)}_{\|\mathbf{A}\|^2} \underbrace{(b_1^2 + b_2^2)}_{\|\mathbf{b}\|^2}$$

The above result in terms of $\mathbf{A}$ and $\mathbf{b}$ is

$$\|\mathbf{Ab}\| \leq \|\mathbf{A}\|\|\mathbf{b}\|$$

which easily generalizes to higher dimensions.

8.6   Tensors

A tensor is a multilinear and coordinate-independent relation between scalars, vectors, or higher-dimensional objects. Examples are

1. the relation s that yields the distance d between two position vectors ($\mathbf{a}$ and $\mathbf{b}$), i.e.,

$$d = s(\mathbf{a}, \mathbf{b})$$

2. the relation f between the force $\mathbf{F}$ acting on a particle moving with an acceleration $\mathbf{a}$, i.e.,

$$\mathbf{F} = f(\mathbf{a}) = m\mathbf{a}$$

3. the relation T between the differential force $d\mathbf{F}$ acting on a differential area element $d\mathbf{A}$, i.e.,

$$d\mathbf{F} = T(d\mathbf{A}) \tag{8.4}$$

Given a coordinate system, a tensor is represented as collection of numerical values in one or more dimensions. The number of dimensions needed for this purpose is called the order of the tensor.

8.6.1   *Multilinearity*

Multilinearity of a tensor means that it is linear with respect to all of its arguments. Thus, for example,

$$s(\alpha_1 \mathbf{a} + \alpha_2 \mathbf{a}, \mathbf{b}) = \alpha_1 s(\mathbf{a}, \mathbf{b}) + \alpha_2 s(\mathbf{a}, \mathbf{b})$$

$$s(\mathbf{a}, \beta_1 \mathbf{b} + \beta_2 \mathbf{b}) = \beta_1 s(\mathbf{a}, \mathbf{b}) + \beta_2 s(\mathbf{a}, \mathbf{b})$$

where α_is and β_is are any scalars.

8.6.2   *Coordinate-Independence*

Coordinate-independence of tensor means that its output is preserved under coordinate transformation. For example, the shortest distance between two position vectors stays the same regardless of the coordinate system. However, note that the representation of a tensor *does* change according to the coordinate system. Thus, the shortest distance d between the position vectors $\mathbf{a}$ and $\mathbf{b}$ in Cartesian coordinate system is given by the tensor

$$s(\mathbf{a}, \mathbf{b}) = \sqrt{(a_1 - b_1)^2 + (a_2 - b_2)^2 + (a_3 - b_3)^2}$$

In cylindrical coordinate system (r, θ, z),

$$\mathbf{a} = \begin{bmatrix} r_1 = \sqrt{a_1^2 + a_2^2} \\ \theta_1 = \tan^{-1}\left(\dfrac{a_2}{a_1}\right) \\ z_1 = a_3 \end{bmatrix}, \qquad \mathbf{b} = \begin{bmatrix} r_2 = \sqrt{b_1^2 + b_2^2} \\ \theta_2 = \tan^{-1}\left(\dfrac{b_2}{b_1}\right) \\ z_2 = b_3 \end{bmatrix}$$

and the shortest distance d between the two vectors is given by the tensor

$$s(\mathbf{a}, \mathbf{b}) = \sqrt{r_1^2 + r_2^2 - 2r_1 r_2 \cos(\theta_2 - \theta_1) + (z_1 - z_2)^2}$$

Note that the above representation of s is different from that in Cartesian coordinate system. But the output d (readers can easily verify) for any given pair of $\mathbf{a}$ and $\mathbf{b}$ stays the same in both coordinate systems.

8.6.3   *Representation of Second Order Tensor*

We are interested in finding the representation of the tensor T of Equation (8.4) on the previous page in an orthogonal coordinate system (y_1, y_2, y_3). We will find out shortly that T is a second order tensor.

Equation (8.4) on the previous page can be written as

$$\underbrace{dF_1\hat{\mathbf{y}}_1 + dF_2\hat{\mathbf{y}}_2 + dF_3\hat{\mathbf{y}}_3}_{d\mathbf{F}} = T(\underbrace{dA_1\hat{\mathbf{y}}_1 + dA_2\hat{\mathbf{y}}_2 + dA_3\hat{\mathbf{y}}_3}_{d\mathbf{A}})$$

where dF_is and dA_is are, respectively, the components of $d\mathbf{F}$ and $d\mathbf{A}$ along the unit vector $\hat{\mathbf{y}}_i$ of the coordinate system. Using the linearity property of T, the above equation becomes

$$dF_1\hat{\mathbf{y}}_1 + dF_2\hat{\mathbf{y}}_2 + dF_3\hat{\mathbf{y}}_3 = dA_1 T(\hat{\mathbf{y}}_1) + dA_2 T(\hat{\mathbf{y}}_2) + dA_3 T(\hat{\mathbf{y}}_3) \qquad (8.5)$$

Consistency requires that $T(\hat{\mathbf{y}}_i)$s should be vectors. Like any vector, $T(\hat{\mathbf{y}}_i)$s can be expressed in terms of unit vectors as

$$T(\hat{\mathbf{y}}_i) = T_{1i}\hat{\mathbf{y}}_1 + T_{2i}\hat{\mathbf{y}}_2 + T_{3i}\hat{\mathbf{y}}_3; \qquad i = 1, 2, 3$$

with T_{1i}, T_{2i} and T_{3i} as the components of $T(\hat{\mathbf{y}}_i)$, respectively, along the y_1-, y_2- and y_3-directions. Substituting the above expressions in Equation (8.5), we obtain after some rearrangement,

$$\begin{aligned} dF_1\hat{\mathbf{y}}_1 + dF_2\hat{\mathbf{y}}_2 + dF_3\hat{\mathbf{y}}_3 = & (T_{11}dA_1 + T_{12}dA_2 + T_{13}dA_3)\hat{\mathbf{y}}_1 \\ & + (T_{21}dA_1 + T_{22}dA_2 + T_{23}dA_3)\hat{\mathbf{y}}_2 \\ & + (T_{31}dA_1 + T_{32}dA_2 + T_{33}dA_3)\hat{\mathbf{y}}_3 \end{aligned}$$

Upon comparing the coefficients of $\hat{\mathbf{y}}_i$s on both sides of the above equation, we obtain

$$\begin{aligned} dF_1 &= T_{11}dA_1 + T_{12}dA_2 + T_{13}dA_3 \\ dF_2 &= T_{21}dA_1 + T_{22}dA_2 + T_{23}dA_3 \\ dF_3 &= T_{31}dA_1 + T_{32}dA_2 + T_{33}dA_3 \end{aligned}$$

The above relations can be written succinctly as the matrix equation

$$
\underbrace{\begin{bmatrix} dF_1 \\ dF_2 \\ dF_3 \end{bmatrix}}_{d\mathbf{F}} = \underbrace{\begin{bmatrix} T_{11} & T_{12} & T_{13} \\ T_{21} & T_{22} & T_{23} \\ T_{31} & T_{32} & T_{33} \end{bmatrix}}_{\mathbf{T}} \underbrace{\begin{bmatrix} dA_1 \\ dA_2 \\ dA_3 \end{bmatrix}}_{d\mathbf{A}}
\tag{8.6}
$$

where the matrix $\mathbf{T}$ is called the representation of the tensor T, or just the tensor.

Order of a Tensor

Since the number of dimensions involved in the above representation is two, the tensor is said to be of the second order. By the same token, a tensor represented by a vector is of the first order. The tensor having only one component is a scalar, and is of the zeroth order.

Alternative Representation

Using the identities

$$(\mathbf{ab}) \cdot \mathbf{c} = \mathbf{a}(\mathbf{b} \cdot \mathbf{c})$$

$$\hat{\mathbf{y}}_i \cdot \hat{\mathbf{y}}_j = \begin{cases} 1 & \text{if } i = j \\ 0 & \text{if } i \neq j \end{cases}$$

it is straightforward to verify that Equation (8.6) above can also be expressed as

$$d\mathbf{F} = \mathbf{T} \cdot d\mathbf{A}$$

with $\mathbf{T}$ represented alternatively by

$$
\mathbf{T} = T_{11}\hat{\mathbf{y}}_1\hat{\mathbf{y}}_1 + T_{12}\hat{\mathbf{y}}_1\hat{\mathbf{y}}_2 + T_{13}\hat{\mathbf{y}}_1\hat{\mathbf{y}}_3 + T_{21}\hat{\mathbf{y}}_2\hat{\mathbf{y}}_1 + T_{22}\hat{\mathbf{y}}_2\hat{\mathbf{y}}_2 + T_{23}\hat{\mathbf{y}}_2\hat{\mathbf{y}}_3
$$
$$
+ T_{31}\hat{\mathbf{y}}_3\hat{\mathbf{y}}_1 + T_{32}\hat{\mathbf{y}}_3\hat{\mathbf{y}}_2 + T_{33}\hat{\mathbf{y}}_3\hat{\mathbf{y}}_3
$$

To summarize, a second order tensor in an orthogonal coordinate system is represented as

$$
\mathbf{T} = \begin{bmatrix} T_{11} & T_{12} & T_{13} \\ T_{21} & T_{22} & T_{23} \\ T_{31} & T_{32} & T_{33} \end{bmatrix} = \sum_{i=1}^{3}\sum_{j=1}^{3} T_{ij}\hat{\mathbf{y}}_i\hat{\mathbf{y}}_j
\tag{8.7}
$$

The transpose and trace of $\mathbf{T}$ are similar to those for matrices.

8.6.4   Einstein or Index Notation

This notation obviates the use of summation symbol to imply summation. In this notation, two instances of an index in a term mean its summation over all index values. Hence,

$$a_i b_i \equiv \sum_{i=1}^{n} a_i b_i = a_1 b_1 + a_2 b_2 + \cdots + a_n b_n$$

where the index i has values $1, 2, \ldots, n$.

In the index notation, multiple indices can repeat in a term, but not more than once. For example, consider the term $A_{ij}B_{jk}c_k$, which has indices i, j and k where the last two repeat once. Let these indices take values from the set $\{1, 2, 3\}$. Then,

$$A_{ij}B_{jk}c_k = \left(\underbrace{\sum_{j=1}^{3} A_{ij}B_{jk}}_{\equiv d_k} \right) c_k = \sum_{k=1}^{3} d_k c_k = \sum_{k=1}^{3} \left(\sum_{j=1}^{3} A_{ij}B_{jk} \right) c_k$$

$$= (A_{i1}B_{11} + A_{i2}B_{21} + A_{i3}B_{31})c_1 + (A_{i1}B_{12} + A_{i2}B_{22} + A_{i3}B_{32})c_2$$
$$+ (A_{i1}B_{13} + A_{i2}B_{23} + A_{i3}B_{33})c_3$$

where we first summed over j and then k. The order of summation does not matter. Thus, summing first over k and then j gives the same result, i.e.,

$$A_{ij}B_{jk}c_k = A_{ij}\left(\sum_{k=1}^{3} B_{jk}c_k \right) = \sum_{j=1}^{3} A_{ij}\left(\sum_{k=1}^{3} B_{jk}c_k \right)$$

$$= A_{i1}(B_{11}c_1 + B_{12}c_2 + B_{13}c_3) + A_{i2}(B_{21}c_1 + B_{22}c_2 + B_{23}c_3)$$
$$+ A_{i3}(B_{31}c_1 + B_{32}c_2 + B_{33}c_3)$$

$$= (A_{i1}B_{11} + A_{i2}B_{21} + A_{i3}B_{31})c_1 + (A_{i1}B_{12} + A_{i2}B_{22} + A_{i3}B_{32})c_2$$
$$+ (A_{i1}B_{13} + A_{i2}B_{23} + A_{i3}B_{33})c_3$$

Dummy and Free Indices

In index notation, renaming the repeating indices j and k to, respectively, l and m changes the term $A_{ij}B_{jk}c_k$ to $A_{il}B_{lm}c_m$ but not the expansion. A once-repeating index such as j is therefore called a *dummy index*.

The dummy indices take up all values from the specified set. In contrast, any non-repeating index such as i in $A_{ij}B_{jk}c_k$ is free to take any value, and is called a *free index*.

Range of Indices

All indices in index notation take values from the same set unless specified otherwise. Thus, with the free index i taking values from the same set $\{1, 2, 3\}$, the term $A_{ij}B_{jk}c_k$ stands for the set $\{A_{1j}B_{jk}c_k, A_{2j}B_{jk}c_k, A_{3j}B_{jk}c_k\}$.

Putting it all together, the term $A_{ij}B_{jk}c_k$ represents

$$A_{1j}B_{jk}c_k = (A_{11}B_{11} + A_{12}B_{21} + A_{13}B_{31})c_1 + (A_{11}B_{12} + A_{12}B_{22} + A_{13}B_{32})c_2$$
$$+ (A_{11}B_{13} + A_{12}B_{23} + A_{13}B_{33})c_3$$

$$A_{2j}B_{jk}c_k = (A_{21}B_{11} + A_{22}B_{21} + A_{23}B_{31})c_1 + (A_{21}B_{12} + A_{22}B_{22} + A_{23}B_{32})c_2$$
$$+ (A_{21}B_{13} + A_{22}B_{23} + A_{23}B_{33})c_3$$

$$A_{3j}B_{jk}c_k = (A_{31}B_{11} + A_{32}B_{21} + A_{33}B_{31})c_1 + (A_{31}B_{12} + A_{32}B_{22} + A_{33}B_{32})c_2$$
$$+ (A_{31}B_{13} + A_{32}B_{23} + A_{33}B_{33})c_3$$

Note how index notation enables brevity. It is necessary for efficient handling of expressions involving vector and tensors.

Substitution of Expressions

In index notation, when substituting one term into another, the dummy indices of one of the terms should be renamed suitably to avoid any duplication between the terms. Consider for example the substitution of the term $D_{ik}e_i$ for c_k in the main term $A_{ij}B_{jk}c_k$. The two terms have the index i in common, which is a dummy index in the first term, i.e., $D_{ik}e_i$. Before doing the substitution, this index should be renamed to, say, l, which does not appear in the main term. Thus,

$$A_{ij}B_{jk}c_k = A_{ij}B_{jk}D_{lk}e_l$$

Representation of Vectors and Tensors

In index notation, a vector $\mathbf{a}$ of an orthogonal coordinate system is represented by $a_i\hat{\mathbf{y}}_i$ where a_i is the component of $\mathbf{a}$ along the unit vector $\hat{\mathbf{y}}_i$ associated with the y_i-coordinate. A second order tensor $\mathbf{T}$ in the same system is represented by $T_{ij}\hat{\mathbf{y}}_i\hat{\mathbf{y}}_j$ where T_{ij} is the tensor component associated with the y_i- and y_j-coordinates.

8.6.5   *Kronecker Delta*

This term is defined as

$$\delta_{ij} = \begin{cases} 1 & \text{for } i = j \\ 0 & \text{for } i \neq j \end{cases}$$

Hence, when present in a term, the Kronecker delta knocks out terms with unequal indices. For example, when the index values are 1, 2 and 3,

$$a_i b_j \delta_{ij} = a_1 b_j \delta_{1j} + a_2 b_j \delta_{2j} + a_3 b_j \delta_{3j}$$
$$= a_1 b_1 \delta_{11} + a_1 b_2 \delta_{12} + a_1 b_3 \delta_{13} + a_2 b_1 \delta_{21} + a_2 b_2 \delta_{22} + a_2 b_3 \delta_{23}$$
$$+ a_3 b_1 \delta_{31} + a_3 b_2 \delta_{32} + a_3 b_3 \delta_{33}$$
$$= a_1 b_1 + a_2 b_2 + a_3 b_3 = a_i b_i$$

8.6.6   *Operations Involving Vectors and Second Order Tensors*

These operations in an orthogonal coordinate system are best understood with the help of tensors called dyads, and the unit dyadic.

Dyads

A dyad is a tensor resulting from the multiplication of two vectors as follows:

$$\mathbf{ab} \equiv \mathbf{ab}^\top = \begin{bmatrix} a_1 \\ a_2 \\ a_3 \end{bmatrix} \begin{bmatrix} b_1 & b_2 & b_3 \end{bmatrix} = \begin{bmatrix} a_1 b_1 & a_1 b_2 & a_1 b_3 \\ a_2 b_1 & a_2 b_2 & a_2 b_3 \\ a_3 b_1 & a_3 b_2 & a_3 b_3 \end{bmatrix}$$

If $\mathbf{a}$ and $\mathbf{b}$ are unit vectors then their product yields a matrix with one unit element, and the rest of the elements zero. For example, in an orthogonal coordinate system,

$$\hat{\mathbf{y}}_1 \hat{\mathbf{y}}_1 = \begin{bmatrix} 1 & 0 & 0 \\ 0 & 0 & 0 \\ 0 & 0 & 0 \end{bmatrix}, \qquad \hat{\mathbf{y}}_1 \hat{\mathbf{y}}_2 = \begin{bmatrix} 0 & 1 & 0 \\ 0 & 0 & 0 \\ 0 & 0 & 0 \end{bmatrix},$$

$$\hat{\mathbf{y}}_1 \hat{\mathbf{y}}_3 = \begin{bmatrix} 0 & 0 & 1 \\ 0 & 0 & 0 \\ 0 & 0 & 0 \end{bmatrix}, \qquad \dots, \qquad \hat{\mathbf{y}}_3 \hat{\mathbf{y}}_3 = \begin{bmatrix} 0 & 0 & 0 \\ 0 & 0 & 0 \\ 0 & 0 & 1 \end{bmatrix}$$

where $\hat{\mathbf{y}}_i \hat{\mathbf{y}}_j$ is a **unit dyad**. It is a 3×3 matrix with all elements zero except one, which is unity and lies on the ith row and j^{th} column.

A tensor can be expressed as a linear combination of unit dyads as follows:

$$\mathbf{T} = T_{11} \begin{bmatrix} 1 & 0 & 0 \\ 0 & 0 & 0 \\ 0 & 0 & 0 \end{bmatrix} + T_{12} \begin{bmatrix} 0 & 1 & 0 \\ 0 & 0 & 0 \\ 0 & 0 & 0 \end{bmatrix}$$

$$+ T_{13} \begin{bmatrix} 0 & 0 & 1 \\ 0 & 0 & 0 \\ 0 & 0 & 0 \end{bmatrix} + \cdots + T_{33} \begin{bmatrix} 0 & 0 & 0 \\ 0 & 0 & 0 \\ 0 & 0 & 1 \end{bmatrix}$$

$$= \begin{bmatrix} T_{11} & T_{12} & T_{13} \\ T_{21} & T_{22} & T_{23} \\ T_{31} & T_{32} & T_{33} \end{bmatrix}$$

where T_{ij}s are scalar coefficients. Or, in index notation,

$$\mathbf{T} = T_{ij} \hat{\mathbf{y}}_i \hat{\mathbf{y}}_j$$

Dyadic

A dyadic is a linear combination of dyads with scalar coefficients, and is a second order tensor. A unit dyadic (or unit tensor) is a tensor denoted by δ such that for any second order tensor $\mathbf{T}$,

$$\delta \cdot \mathbf{T} = \mathbf{T}$$

In an orthogonal coordinate system,

$$\delta = \hat{\mathbf{y}}_1\hat{\mathbf{y}}_1 + \hat{\mathbf{y}}_2\hat{\mathbf{y}}_2 + \hat{\mathbf{y}}_3\hat{\mathbf{y}}_3 = \begin{bmatrix} 1 & 0 & 0 \\ 0 & 1 & 0 \\ 0 & 0 & 1 \end{bmatrix} \tag{8.8}$$

Operations

The following operations apply to the second order tensors in an orthogonal coordinate system:

Addition and Subtraction

These two operations are similar to that for matrices.

Product of a Tensor with a Vector

The product of a tensor $\mathbf{A}$ with a vector $\mathbf{b}$ is a vector

$$\mathbf{a} = \mathbf{A} \cdot \mathbf{b}$$

In matrix representation, the product is $\mathbf{Ab}$, i.e.,

$$\underbrace{\begin{bmatrix} a_1 \\ a_2 \\ a_3 \end{bmatrix}}_{\mathbf{a}} = \underbrace{\begin{bmatrix} A_{11} & A_{12} & A_{13} \\ A_{21} & A_{22} & A_{23} \\ A_{31} & A_{32} & A_{33} \end{bmatrix}}_{\mathbf{A}} \underbrace{\begin{bmatrix} b_1 \\ b_2 \\ b_3 \end{bmatrix}}_{\mathbf{b}} = \begin{bmatrix} A_{11}a_1 + A_{12}a_2 + A_{13}a_3 \\ A_{21}a_1 + A_{22}a_2 + A_{23}a_3 \\ A_{31}a_1 + A_{32}a_2 + A_{33}a_3 \end{bmatrix}$$

In index notation, this product is denoted by

$$a_i = A_{ij}a_j$$

Product of a Vector with a Tensor

The product of a vector $\mathbf{b}$ with a tensor $\mathbf{A}$ is a vector

$$\mathbf{a} = \mathbf{b} \cdot \mathbf{A}$$

In matrix representation, the product is $\mathbf{b}^\top \mathbf{A}$, i.e.,

$$\begin{bmatrix} a_1 \\ a_2 \\ a_3 \end{bmatrix} = \begin{bmatrix} b_1 \\ b_2 \\ b_3 \end{bmatrix}^\top \begin{bmatrix} A_{11} & A_{12} & A_{13} \\ A_{21} & A_{22} & A_{23} \\ A_{31} & A_{32} & A_{33} \end{bmatrix} = \begin{bmatrix} b_1 A_{11} + b_2 A_{21} + b_3 A_{31} \\ b_1 A_{12} + b_2 A_{22} + b_3 A_{32} \\ b_1 A_{13} + b_2 A_{23} + b_3 A_{33} \end{bmatrix}$$

In index notation, this product is denoted by

$$a_i = b_j A_{ji}$$

Thus, $\mathbf{A} \cdot \mathbf{b} \neq \mathbf{b} \cdot \mathbf{A}$ unless $\mathbf{A}$ is symmetric, i.e., $A_{ij} = A_{ji}$.

Product of Two Tensors

The product of two tensors $\mathbf{A}$ and $\mathbf{B}$ is written as $\mathbf{A} \cdot \mathbf{B}$. It is tensor $\mathbf{C}$ that is obtained by multiplying $\mathbf{A}$ and $\mathbf{B}$ as matrices. In index notation, the product is denoted by

$$C_{ij} = A_{ik} B_{kj}$$

Similar to the product of two matrices, $\mathbf{A} \cdot \mathbf{B} \neq \mathbf{B} \cdot \mathbf{A}$.

Scalar or Double Dot Product

The scalar product of two tensors results in a scalar, and is based on the following definition of the product of two dyads:

$$\hat{\mathbf{y}}_i \hat{\mathbf{y}}_j : \hat{\mathbf{y}}_k \hat{\mathbf{y}}_l = (\hat{\mathbf{y}}_i \cdot \hat{\mathbf{y}}_l)(\hat{\mathbf{y}}_j \cdot \hat{\mathbf{y}}_k) = \delta_{il}\delta_{jk} \quad \text{(scalar)}$$

Using the above definition, the product is given by

$$\begin{aligned}
\mathbf{A} : \mathbf{B} &= A_{ij}\hat{\mathbf{y}}_i\hat{\mathbf{y}}_j : B_{kl}\hat{\mathbf{y}}_k\hat{\mathbf{y}}_l = A_{ij}B_{kl}\hat{\mathbf{y}}_i\hat{\mathbf{y}}_j : \hat{\mathbf{y}}_k\hat{\mathbf{y}}_l = A_{ij}B_{kl}\delta_{il}\delta_{jk} \\
&= A_{ij}B_{ji} \quad \text{(scalar)} \\
&= A_{i1}B_{1i} + A_{i2}B_{2i} + A_{i3}B_{3i} \\
&= A_{11}B_{11} + A_{21}B_{12} + A_{31}B_{13} + A_{12}B_{21} + A_{22}B_{22} + A_{32}B_{23} \\
&\quad + A_{13}B_{31} + A_{23}B_{32} + A_{33}B_{33}
\end{aligned}$$

Tensor Product

The tensor product of two tensors results in a tensor, and is based on the following definition of the dot product of two dyads:

$$\begin{aligned}
\hat{\mathbf{y}}_i\hat{\mathbf{y}}_j \cdot \hat{\mathbf{y}}_k\hat{\mathbf{y}}_l &= \hat{\mathbf{y}}_i(\hat{\mathbf{y}}_j \cdot \hat{\mathbf{y}}_k)\hat{\mathbf{y}}_l = \hat{\mathbf{y}}_i(\delta_{jk})\hat{\mathbf{y}}_l = \hat{\mathbf{y}}_i\hat{\mathbf{y}}_l \quad \text{(dyad, i.e., tensor)} \\
&= \hat{\mathbf{y}}_i\hat{\mathbf{y}}_1 + \hat{\mathbf{y}}_i\hat{\mathbf{y}}_2 + \hat{\mathbf{y}}_i\hat{\mathbf{y}}_3 \\
&= \hat{\mathbf{y}}_1\hat{\mathbf{y}}_1 + \hat{\mathbf{y}}_2\hat{\mathbf{y}}_1 + \hat{\mathbf{y}}_3\hat{\mathbf{y}}_1 + \hat{\mathbf{y}}_1\hat{\mathbf{y}}_2 + \hat{\mathbf{y}}_2\hat{\mathbf{y}}_2 + \hat{\mathbf{y}}_3\hat{\mathbf{y}}_2 \\
&\quad + \hat{\mathbf{y}}_1\hat{\mathbf{y}}_3 + \hat{\mathbf{y}}_2\hat{\mathbf{y}}_3 + \hat{\mathbf{y}}_3\hat{\mathbf{y}}_3
\end{aligned}$$

Thus,

$$
\begin{aligned}
\mathbf{A} \cdot \mathbf{B} &= A_{ij}\hat{\mathbf{y}}_i\hat{\mathbf{y}}_j \cdot B_{kl}\hat{\mathbf{y}}_k\hat{\mathbf{y}}_l = A_{ij}B_{kl}\hat{\mathbf{y}}_i(\hat{\mathbf{y}}_j \cdot \hat{\mathbf{y}}_k)\hat{\mathbf{y}}_l = A_{ij}B_{kl}\hat{\mathbf{y}}_i(\delta_{jk})\hat{\mathbf{y}}_l \\
&= A_{ij}B_{jl}\hat{\mathbf{y}}_i\hat{\mathbf{y}}_l = \underbrace{\left(A_{i1}B_{1l} + A_{i2}B_{2l} + A_{i3}B_{3l}\right)}_{\equiv C_{il}}\hat{\mathbf{y}}_i\hat{\mathbf{y}}_l \\
&= C_{il}\hat{\mathbf{y}}_i\hat{\mathbf{y}}_l \quad (\text{tensor}) \\
&= C_{i1}\hat{\mathbf{y}}_i\hat{\mathbf{y}}_1 + C_{i2}\hat{\mathbf{y}}_i\hat{\mathbf{y}}_2 + C_{i3}\hat{\mathbf{y}}_i\hat{\mathbf{y}}_3 \\
&= C_{11}\hat{\mathbf{y}}_1\hat{\mathbf{y}}_1 + C_{21}\hat{\mathbf{y}}_2\hat{\mathbf{y}}_1 + C_{31}\hat{\mathbf{y}}_3\hat{\mathbf{y}}_1 \;+\; C_{12}\hat{\mathbf{y}}_1\hat{\mathbf{y}}_2 + C_{22}\hat{\mathbf{y}}_2\hat{\mathbf{y}}_2 + C_{32}\hat{\mathbf{y}}_3\hat{\mathbf{y}}_2 \\
&\quad + C_{13}\hat{\mathbf{y}}_1\hat{\mathbf{y}}_3 + C_{23}\hat{\mathbf{y}}_2\hat{\mathbf{y}}_3 + C_{33}\hat{\mathbf{y}}_3\hat{\mathbf{y}}_3
\end{aligned}
$$

Dot Product of a Tensor and a Vector

This product results in a vector, and is based on the following definition of the dot product of a dyad, and a unit vector:

$$
\hat{\mathbf{y}}_i\hat{\mathbf{y}}_j \cdot \hat{\mathbf{y}}_k = \hat{\mathbf{y}}_i(\hat{\mathbf{y}}_j \cdot \hat{\mathbf{y}}_k) = \hat{\mathbf{y}}_i(\delta_{jk}) = \hat{\mathbf{y}}_i \quad (\text{unit vector})
$$

Thus,

$$
\begin{aligned}
\mathbf{A} \cdot \mathbf{b} &= A_{ij}\hat{\mathbf{y}}_i\hat{\mathbf{y}}_j \cdot b_k\hat{\mathbf{y}}_k = A_{ij}b_k\hat{\mathbf{y}}_i(\hat{\mathbf{y}}_j \cdot \hat{\mathbf{y}}_k) = A_{ij}b_k\hat{\mathbf{y}}_i(\delta_{jk}) \\
&= A_{ij}b_j\hat{\mathbf{y}}_i \quad (\text{vector}) \\
&= A_{i1}b_1\hat{\mathbf{y}}_i + A_{i2}b_2\hat{\mathbf{y}}_i + A_{i3}b_3\hat{\mathbf{y}}_i \\
&= A_{11}b_1\hat{\mathbf{y}}_1 + A_{21}b_1\hat{\mathbf{y}}_2 + A_{31}b_1\hat{\mathbf{y}}_3 \;+\; A_{12}b_2\hat{\mathbf{y}}_1 + A_{22}b_2\hat{\mathbf{y}}_2 + A_{32}b_2\hat{\mathbf{y}}_3 \\
&\quad + A_{13}b_3\hat{\mathbf{y}}_1 + A_{23}b_3\hat{\mathbf{y}}_2 + A_{33}b_3\hat{\mathbf{y}}_3 \\
&= (A_{11}b_1 + A_{12}b_2 + A_{13}b_3)\hat{\mathbf{y}}_1 \;+\; (A_{21}b_1 + A_{22}b_2 + A_{23}b_3)\hat{\mathbf{y}}_2 \\
&\quad + (A_{31}b_1 + A_{32}b_2 + A_{33}b_3)\hat{\mathbf{y}}_3
\end{aligned}
$$

Similarly,

$$
\begin{aligned}
\mathbf{b} \cdot \mathbf{A} &= (A_{11}b_1 + A_{21}b_2 + A_{31}b_3)\hat{\mathbf{y}}_1 \;+\; (A_{12}b_1 + A_{22}b_2 + A_{32}b_3)\hat{\mathbf{y}}_2 \\
&\quad + (A_{13}b_1 + A_{23}b_2 + A_{33}b_3)\hat{\mathbf{y}}_3
\end{aligned}
$$

Thus, unless $\mathbf{A}$ is symmetric (i.e., $A_{ij} = A_{ji}$), $\mathbf{A} \cdot \mathbf{b} \neq \mathbf{b} \cdot \mathbf{A}$.

The operations described next involve the gradient operator ∇, which is a vector. In index notation, this operator is written as

$$
\nabla = \hat{\mathbf{y}}_i \frac{\partial}{\partial y_i}
$$

Gradient of a Vector

The gradient of a vector is a tensor given by Equation (8.1) on p. 299. In index notation,

$$
\nabla \mathbf{a} = \left(\underbrace{\hat{\mathbf{y}}_i \frac{\partial}{\partial y_i}}_{\nabla} \right) \underbrace{(a_j \hat{\mathbf{y}}_j)}_{\mathbf{a}} = \hat{\mathbf{y}}_i \frac{\partial a_j}{\partial y_i} \hat{\mathbf{y}}_j = \frac{\partial a_j}{\partial y_i} \hat{\mathbf{y}}_i \hat{\mathbf{y}}_j
$$

$$
= \left(\frac{\partial a_j}{\partial y_1} \hat{\mathbf{y}}_1 + \frac{\partial a_j}{\partial y_2} \hat{\mathbf{y}}_2 + \frac{\partial a_j}{\partial y_3} \hat{\mathbf{y}}_3 \right) \hat{\mathbf{y}}_j = \frac{\partial a_j}{\partial y_1} \hat{\mathbf{y}}_1 \hat{\mathbf{y}}_j + \frac{\partial a_j}{\partial y_2} \hat{\mathbf{y}}_2 \hat{\mathbf{y}}_j + \frac{\partial a_j}{\partial y_3} \hat{\mathbf{y}}_3 \hat{\mathbf{y}}_j
$$

$$
= \frac{\partial a_1}{\partial y_1} \hat{\mathbf{y}}_1 \hat{\mathbf{y}}_1 + \frac{\partial a_2}{\partial y_1} \hat{\mathbf{y}}_1 \hat{\mathbf{y}}_2 + \frac{\partial a_3}{\partial y_1} \hat{\mathbf{y}}_1 \hat{\mathbf{y}}_3 + \frac{\partial a_1}{\partial y_2} \hat{\mathbf{y}}_2 \hat{\mathbf{y}}_1 + \frac{\partial a_2}{\partial y_2} \hat{\mathbf{y}}_2 \hat{\mathbf{y}}_2 + \frac{\partial a_3}{\partial y_2} \hat{\mathbf{y}}_2 \hat{\mathbf{y}}_3
$$

$$
+ \frac{\partial a_1}{\partial y_3} \hat{\mathbf{y}}_3 \hat{\mathbf{y}}_1 + \frac{\partial a_2}{\partial y_3} \hat{\mathbf{y}}_3 \hat{\mathbf{y}}_2 + \frac{\partial a_3}{\partial y_3} \hat{\mathbf{y}}_3 \hat{\mathbf{y}}_3
$$

Divergence of a Tensor

The divergence of a tensor $\mathbf{A}$ is given by

$$
\nabla \cdot \mathbf{A} = \underbrace{\hat{\mathbf{y}}_k \frac{\partial}{\partial y_k}}_{\nabla} \cdot \underbrace{A_{ij} \hat{\mathbf{y}}_i \hat{\mathbf{y}}_j}_{\mathbf{A}} = \hat{\mathbf{y}}_k \cdot \frac{\partial A_{ij}}{\partial y_k} \hat{\mathbf{y}}_i \hat{\mathbf{y}}_j = \frac{\partial A_{ij}}{\partial y_k} (\hat{\mathbf{y}}_k \cdot \hat{\mathbf{y}}_i) \hat{\mathbf{y}}_j
$$

$$
= \frac{\partial A_{ij}}{\partial y_k} (\delta_{ki}) \hat{\mathbf{y}}_j = \frac{\partial A_{ij}}{\partial y_i} \hat{\mathbf{y}}_j
$$

$$
= \left(\frac{\partial A_{1j}}{\partial y_1} + \frac{\partial A_{2j}}{\partial y_2} + \frac{\partial A_{3j}}{\partial y_3} \right) \hat{\mathbf{y}}_j = \frac{\partial A_{1j}}{\partial y_1} \hat{\mathbf{y}}_j + \frac{\partial A_{2j}}{\partial y_2} \hat{\mathbf{y}}_j + \frac{\partial A_{3j}}{\partial y_3} \hat{\mathbf{y}}_j
$$

$$
= \left(\frac{\partial A_{11}}{\partial y_1} \hat{\mathbf{y}}_1 + \frac{\partial A_{12}}{\partial y_1} \hat{\mathbf{y}}_2 + \frac{\partial A_{13}}{\partial y_1} \hat{\mathbf{y}}_3 \right) + \left(\frac{\partial A_{21}}{\partial y_2} \hat{\mathbf{y}}_1 + \frac{\partial A_{22}}{\partial y_2} \hat{\mathbf{y}}_2 + \frac{\partial A_{23}}{\partial y_2} \hat{\mathbf{y}}_3 \right)
$$

$$
+ \left(\frac{\partial A_{31}}{\partial y_3} \hat{\mathbf{y}}_1 + \frac{\partial A_{32}}{\partial y_3} \hat{\mathbf{y}}_2 + \frac{\partial A_{33}}{\partial y_3} \hat{\mathbf{y}}_3 \right)
$$

$$
= \left(\frac{\partial A_{11}}{\partial y_1} + \frac{\partial A_{21}}{\partial y_2} + \frac{\partial A_{31}}{\partial y_3} \right) \hat{\mathbf{y}}_1 + \left(\frac{\partial A_{12}}{\partial y_1} + \frac{\partial A_{22}}{\partial y_2} + \frac{\partial A_{32}}{\partial y_3} \right) \hat{\mathbf{y}}_2
$$

$$
+ \left(\frac{\partial A_{13}}{\partial y_1} + \frac{\partial A_{23}}{\partial y_2} + \frac{\partial A_{33}}{\partial y_3} \right) \hat{\mathbf{y}}_3
$$

Divergence of a Dyad

The divergence of a dyad $\mathbf{ab}$ is given by the identity

$$
\nabla \cdot \mathbf{ab} = \mathbf{a} \cdot \nabla \mathbf{b} + \mathbf{b}(\nabla \cdot \mathbf{a})
$$

$$
\tag{8.9}
$$

We prove the validity of the Equation (8.9) on the previous page by expanding its terms. The first right-hand side term of this equation is given by

$$
\mathbf{a} \cdot \nabla \mathbf{b} = \sum_{i=1}^{3} a_i \hat{\mathbf{y}}_i \cdot \sum_{j=1}^{3} \hat{\mathbf{y}}_j \frac{\partial}{\partial y_j} \sum_{k=1}^{3} b_k \hat{\mathbf{y}}_k = \sum_{i=1}^{3} \hat{\mathbf{y}}_i a_i \cdot \sum_{j=1}^{3} \sum_{k=1}^{3} \hat{\mathbf{y}}_j \frac{\partial (b_k \hat{\mathbf{y}}_k)}{\partial y_j}
$$

$$
= \sum_{i=1}^{3} \sum_{j=1}^{3} \sum_{k=1}^{3} a_i \underbrace{\hat{\mathbf{y}}_i \cdot \hat{\mathbf{y}}_j}_{\delta_{ij}} \frac{\partial (b_k \hat{\mathbf{y}}_k)}{\partial y_j} = \sum_{i=1}^{3} \sum_{k=1}^{3} a_i \frac{\partial (b_k \hat{\mathbf{y}}_k)}{\partial y_i}
$$

$$
= \sum_{i=1}^{3} \sum_{k=1}^{3} \left(a_i b_k \underbrace{\frac{\partial \hat{\mathbf{y}}_k}{\partial y_i}}_{=0} + a_i \frac{\partial b_k}{\partial y_i} \hat{\mathbf{y}}_k \right) = \sum_{i=1}^{3} \sum_{k=1}^{3} a_i \frac{\partial b_k}{\partial y_i} \hat{\mathbf{y}}_k
$$

The second right-hand term of Equation (8.9) is given by

$$
\mathbf{b} (\nabla \cdot \mathbf{a}) = \sum_{k=1}^{3} \hat{\mathbf{y}}_k b_k \left(\sum_{j=1}^{3} \hat{\mathbf{y}}_j \frac{\partial}{\partial y_j} \cdot \sum_{i=1}^{3} \hat{\mathbf{y}}_i a_i \right) = \sum_{k=1}^{3} \hat{\mathbf{y}}_k b_k \sum_{j=1}^{3} \sum_{i=1}^{3} \hat{\mathbf{y}}_j \cdot \frac{\partial (\hat{\mathbf{y}}_i a_i)}{\partial y_j}
$$

$$
= \sum_{k=1}^{3} \hat{\mathbf{y}}_k b_k \sum_{j=1}^{3} \sum_{i=1}^{3} \hat{\mathbf{y}}_j \cdot \left(\underbrace{\frac{\partial \hat{\mathbf{y}}_i}{\partial y_j} a_i + \hat{\mathbf{y}}_i \frac{\partial a_i}{\partial y_j}}_{=0} \right)
$$

$$
= \sum_{k=1}^{3} \hat{\mathbf{y}}_k b_k \sum_{j=1}^{3} \sum_{i=1}^{3} \underbrace{\hat{\mathbf{y}}_j \cdot \hat{\mathbf{y}}_i}_{\delta_{ji}} \frac{\partial a_i}{\partial y_j}
$$

$$
= \sum_{k=1}^{3} \hat{\mathbf{y}}_k b_k \sum_{i=1}^{3} \frac{\partial a_i}{\partial y_i} = \sum_{k=1}^{3} \sum_{i=1}^{3} \frac{\partial a_i}{\partial y_i} b_k \hat{\mathbf{y}}_k
$$

Finally, the left-hand side of Equation (8.9) is given by

$$
\nabla \cdot \mathbf{a} \mathbf{b} = \sum_{i=1}^{3} \hat{\mathbf{y}}_i \frac{\partial}{\partial y_i} \cdot \sum_{j=1}^{3} \sum_{k=1}^{3} \hat{\mathbf{y}}_j \hat{\mathbf{y}}_k a_j b_k = \sum_{i=1}^{3} \sum_{j=1}^{3} \sum_{k=1}^{3} \hat{\mathbf{y}}_i \cdot \frac{\partial}{\partial y_i} (\hat{\mathbf{y}}_j \hat{\mathbf{y}}_k a_j b_k)
$$

$$
= \sum_{i=1}^{3} \sum_{j=1}^{3} \sum_{k=1}^{3} \left[\hat{\mathbf{y}}_i \cdot \underbrace{\frac{\partial (\hat{\mathbf{y}}_j \hat{\mathbf{y}}_k)}{\partial y_i}}_{=0} a_j b_k + \underbrace{\hat{\mathbf{y}}_i \cdot \hat{\mathbf{y}}_j}_{\delta_{ij}} \hat{\mathbf{y}}_k \frac{\partial}{\partial y_i} (a_j b_k) \right]
$$

$$
= \sum_{i=1}^{3} \sum_{k=1}^{3} \hat{\mathbf{y}}_k \frac{\partial}{\partial y_i} (a_i b_k) = \underbrace{\sum_{i=1}^{3} \sum_{k=1}^{3} \frac{\partial a_i}{\partial y_i} b_k \hat{\mathbf{y}}_k}_{= \mathbf{b} (\nabla \cdot \mathbf{a})} + \underbrace{\sum_{i=1}^{3} \sum_{k=1}^{3} a_i \frac{\partial b_k}{\partial y_i} \hat{\mathbf{y}}_k}_{= \mathbf{a} \cdot \nabla \mathbf{b}}
$$

which is the sum of the two right-hand side terms derived above.

Dot Product of a Vector and Gradient of a Vector

This product for vectors **a** and **b** is given by

$$\mathbf{a} \cdot \nabla \mathbf{b} = a_k \hat{\mathbf{y}}_k \cdot \frac{\partial b_j}{\partial y_i} \hat{\mathbf{y}}_i \hat{\mathbf{y}}_j = a_k \frac{\partial b_j}{\partial y_i} (\hat{\mathbf{y}}_k \cdot \hat{\mathbf{y}}_i) \hat{\mathbf{y}}_j = a_k \frac{\partial b_j}{\partial y_i} (\delta_{ki}) \hat{\mathbf{y}}_j$$

$$= a_i \frac{\partial b_j}{\partial y_i} \hat{\mathbf{y}}_j \quad \text{(vector)}$$

$$= \left(a_1 \frac{\partial b_j}{\partial y_1} + a_2 \frac{\partial b_j}{\partial y_2} + a_3 \frac{\partial b_j}{\partial y_3} \right) \hat{\mathbf{y}}_j$$

$$= \left(a_1 \frac{\partial b_1}{\partial y_1} + a_2 \frac{\partial b_1}{\partial y_2} + a_3 \frac{\partial b_1}{\partial y_3} \right) \hat{\mathbf{y}}_1 + \left(a_1 \frac{\partial b_2}{\partial y_1} + a_2 \frac{\partial b_2}{\partial y_2} + a_3 \frac{\partial b_2}{\partial y_3} \right) \hat{\mathbf{y}}_2$$

$$+ \left(a_1 \frac{\partial b_3}{\partial y_1} + a_2 \frac{\partial b_3}{\partial y_2} + a_3 \frac{\partial b_3}{\partial y_3} \right) \hat{\mathbf{y}}_3$$

Double Dot Product of a Tensor and the Gradient of a Vector

This product for a tensor **A**, and a vector **b** is given by

$$\mathbf{A} : \nabla \mathbf{b} = A_{kl} \hat{\mathbf{y}}_k \hat{\mathbf{y}}_l : \frac{\partial b_j}{\partial y_i} \hat{\mathbf{y}}_i \hat{\mathbf{y}}_j = A_{kl} \frac{\partial b_j}{\partial y_i} \hat{\mathbf{y}}_k \hat{\mathbf{y}}_l : \hat{\mathbf{y}}_i \hat{\mathbf{y}}_j = A_{kl} \frac{\partial b_j}{\partial y_i} (\hat{\mathbf{y}}_k \cdot \hat{\mathbf{y}}_j)(\hat{\mathbf{y}}_l \cdot \hat{\mathbf{y}}_i)$$

$$= A_{kl} \frac{\partial b_j}{\partial y_i} \delta_{kj} \delta_{li} = A_{ji} \frac{\partial b_j}{\partial y_i} \quad \text{(scalar)}$$

$$= A_{1i} \frac{\partial b_1}{\partial y_i} + A_{2i} \frac{\partial b_2}{\partial y_i} + A_{3i} \frac{\partial b_3}{\partial y_i}$$

$$= A_{11} \frac{\partial b_1}{\partial y_1} + A_{12} \frac{\partial b_1}{\partial y_2} + A_{13} \frac{\partial b_1}{\partial y_3} + A_{21} \frac{\partial b_2}{\partial y_1} + A_{22} \frac{\partial b_2}{\partial y_2} + A_{23} \frac{\partial b_2}{\partial y_3}$$

$$+ A_{31} \frac{\partial b_3}{\partial y_1} + A_{32} \frac{\partial b_3}{\partial y_2} + A_{33} \frac{\partial b_3}{\partial y_3}$$

It may be noted that the above product is the trace of the product of the two matrices, namely, **A** and $\nabla \mathbf{b}$.

Divergence of the Product of a Scalar and a Vector

Using the product rule of differentiation,

$$\nabla \cdot a\mathbf{b} = \sum_{j=1}^{3} \frac{\partial (ab_j)}{\partial x_j} = a \sum_{j=1}^{3} \frac{\partial b_j}{\partial x_j} + \sum_{j=1}^{3} b_j \frac{\partial a}{\partial x_j} = a \nabla \cdot \mathbf{b} + \mathbf{b} \cdot \nabla a \qquad (8.10)$$

8.7 Differential

Consider a function $f(x)$. Then the differential of f at $x = x_0$ is defined as the change $\mathrm{d}f$ in f associated with the change h in x_0 that satisfies the following requirements:

1. Compared to h, the change $\mathrm{d}f$ is closer to the actual function change

$$f(x_0 + h) - f(x)$$

2. The change $\mathrm{d}f$ is a linear and continuous function of h at x_0.

Remarks

The motivation for the above requirements is to obtain a simple function $\mathrm{d}f$ relating function change to the variable change with little or no error. Utilizing this function, we can calculate the new function value $f(x_0 + h)$ simply from[*]

$$f(x_0 + h) = f(x_0) + \mathrm{d}f(x_0; h)$$

for sufficiently small changes h around x_0. Note that $\mathrm{d}f$ is evaluated at x_0, and is a linear function of h. The error involved in the above calculation is the remainder term of the first order Taylor expansion. As per the first requirement, this error must reduce faster than h. Thus, if h is reduced progressively then at some h_0 there would be no error left, and $\mathrm{d}f$ would accurately represent the function changes in the interval $[x_0, x_0 + h_0]$.

8.7.1 Derivative

The definition of the derivative stems from that of the differential. Consider the last equation in the limit of h tending to zero when there is no error involved. Then multiplying and dividing $\mathrm{d}f(x_0; h)$ by h in the equation yields

$$f(x_0 + h) = f(x_0) + \underbrace{\frac{\mathrm{d}f(x_0; h)}{h}}_{\text{derivative}} h; \qquad \lim h \to 0$$

where the derivative as shown above is the coefficient of h. More precisely, the derivative of $f(x)$ at x_0 is defined as

$$\left.\frac{\mathrm{d}f}{\mathrm{d}x}\right|_{x_0} \equiv \lim_{h \to 0} \frac{\mathrm{d}f(x_0; h)}{h} = \lim_{h \to 0} \frac{f(x_0 + h) - f(x_0)}{h}$$

In terms of the derivative, the differential of $f(x)$ for a change $\mathrm{d}x$ in x is given by

$$\mathrm{d}f = \frac{\mathrm{d}f}{\mathrm{d}x}\mathrm{d}x$$

8.7.2 Partial Derivative and Differential

For a multivariable function $f(\mathbf{x})$, where $\mathbf{x} \begin{bmatrix} x_1 & x_2 & \cdots & x_n \end{bmatrix}^\top$, the partial derivative of f with respect to the x_i is the derivative

$$\left.\frac{\mathrm{d}f}{\mathrm{d}x_i}\right|_{\tilde{\mathbf{x}}} \equiv \frac{\partial f}{\partial x_i} = \lim_{h \to 0} \frac{f(x_i + h, \tilde{\mathbf{x}}) - f(x_i, \tilde{\mathbf{x}})}{h}$$

[*]Instead of evaluating $f(x_0 + h)$ all over again from $f(x)$.

that is obtained keeping all variables except x_i (i.e., the vector $\tilde{\mathbf{x}}$) constant. Thus, the partial differential with respect to x_i is

$$\frac{\partial f}{\partial x_i} \mathrm{d}x_i$$

The sum of all partial differentials of $f(x)$, i.e.,

$$\sum_{i=1}^{n} \frac{\partial f}{\partial x_i} \mathrm{d}x_i$$

is called the total differential.

8.7.3   *Chain Rule of Differentiation*

This rule provides the derivative of a function with respect to an indirectly related variable. The derivative is obtained by combining and simplifying the involved differentials.

For example, consider $f = f(y)$ where $y = y(x)$. Here f is indirectly related to x. To obtain $\mathrm{d}f/\mathrm{d}x$, we first write down the differentials of the involved functions. The differential of f is given by

$$\mathrm{d}f = \frac{\mathrm{d}f}{\mathrm{d}y}\mathrm{d}y$$

Similarly, the differential of y is given by

$$\mathrm{d}y = \frac{\mathrm{d}y}{\mathrm{d}x}\mathrm{d}x$$

Combining the last two equations, and dividing the result by $\mathrm{d}x$, we obtain

$$\frac{\mathrm{d}f}{\mathrm{d}x} = \frac{\mathrm{d}f}{\mathrm{d}y}\frac{\mathrm{d}y}{\mathrm{d}x}$$

Multivariable Function

Let $f = f(x, y)$ with $x = x(t)$ and $y = y(t)$. Thus,

$$f = f[x(t), y(t)]$$

Here, f is indirectly related to t. To obtain the derivative of f with respect to t, we first write down the differentials of the involved functions. The differential of f is given by

$$\mathrm{d}f = \frac{\partial f}{\partial x}\mathrm{d}x + \frac{\partial f}{\partial y}\mathrm{d}y$$

The differentials of x and y in turn are, respectively, given by

$$\mathrm{d}x = \frac{\mathrm{d}x}{\mathrm{d}t}\mathrm{d}t \quad \text{and} \quad \mathrm{d}y = \frac{\mathrm{d}y}{\mathrm{d}t}\mathrm{d}t$$

Combining the last three equations, and dividing the result by dt we obtain

$$\frac{df}{dt} = \frac{\partial f}{\partial x}\frac{dx}{dt} + \frac{\partial f}{\partial y}\frac{dy}{dt}$$

If f additionally depends explicitly on t, i.e.,

$$f = f[x(t), y(t), t]$$

then

$$df = \frac{\partial f}{\partial x}dx + \frac{\partial f}{\partial y}dy + \frac{\partial f}{\partial t}dt$$

so that

$$\frac{df}{dt} = \frac{\partial f}{\partial x}\frac{dx}{dt} + \frac{\partial f}{\partial y}\frac{dy}{dt} + \frac{\partial f}{\partial t}$$

Function of Several Multivariable Functions

Let $f = f(x, y)$ with $x = x(s, t)$ and $y = (s, t)$. Thus,

$$f = f[x(s, t), y(s, t)]$$

Here, f is indirectly related to t and s, and the last two variables are independent of each other. To obtain the derivative of f with respect to s, we first write down the differentials of the involved functions. The differentials are

$$df = \frac{\partial f}{\partial x}dx + \frac{\partial f}{\partial y}dy, \qquad dx = \frac{\partial x}{\partial s}ds + \frac{\partial x}{\partial t}dt, \qquad \text{and} \qquad dy = \frac{\partial y}{\partial s}ds + \frac{\partial y}{\partial t}dt.$$

Combining the last three equations, and dividing the result by ds, we obtain

$$\frac{df}{ds} = \frac{\partial f}{\partial x}\left(\frac{\partial x}{\partial s} + \frac{\partial x}{\partial t}\underbrace{\frac{dt}{ds}}_{=0}\right) + \frac{\partial f}{\partial y}\left(\frac{\partial y}{\partial s} + \frac{\partial y}{\partial t}\underbrace{\frac{dt}{ds}}_{=0}\right) = \frac{\partial f}{\partial x}\frac{\partial x}{\partial s} + \frac{\partial f}{\partial y}\frac{\partial y}{\partial s}$$

since t does not depend on s.

In the same manner, the derivative of f with respect to t is given by

$$\frac{df}{dt} = \frac{\partial f}{\partial x}\frac{\partial x}{\partial t} + \frac{\partial f}{\partial y}\frac{\partial y}{\partial t}$$

If f additionally depends explicitly on t, i.e.,

$$f = f[x(s, t), y(s, t), t]$$

then

$$\frac{df}{dt} = \frac{\partial f}{\partial x}\frac{\partial x}{\partial t} + \frac{\partial f}{\partial y}\frac{\partial y}{\partial t} + \frac{\partial f}{\partial t}$$

8.7.4 Material and Total Derivatives

Consider a particle moving in space, as shown in Figure 8.5 below. At time t, the position of the particle is given by

$$x_i(t) \;=\; \underbrace{x_i(0)}_{\bar{x}_i} + \int_0^t \frac{\mathrm{d}x_i}{\mathrm{d}t}\,\mathrm{d}t; \qquad i = 1, 2, 3$$

or, in vector notation,

$$\mathbf{x} \;=\; \bar{\mathbf{x}} + \int_0^t \frac{\mathrm{d}\mathbf{x}}{\mathrm{d}t}\,\mathrm{d}t$$

where $\mathbf{x} = \begin{bmatrix} x_1 & x_2 & x_3 \end{bmatrix}^{\mathsf{T}}$ is the spatial position vector, and $\bar{\mathbf{x}} = \begin{bmatrix} \bar{x}_1 & \bar{x}_2 & \bar{x}_3 \end{bmatrix}^{\mathsf{T}}$ is the material position vector. The latter identifies a particle that occupies a position at a reference time, $t = 0$. This identification is unique because only one particle can occupy a position at a given time.

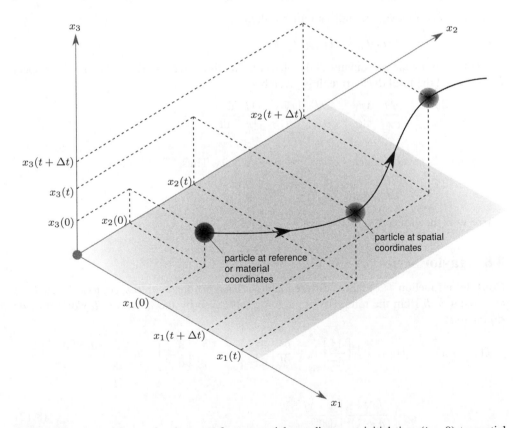

Figure 8.5 A particle moving in space from material coordinates at initial time ($t = 0$) to spatial coordinates at time t

We express the particle position symbolically as

$$\mathbf{x} = \mathbf{x}(\bar{\mathbf{x}}, t)$$

and consider a property, say temperature T, which varies with $\mathbf{x}$ as well as t. Then

$$T = T(\mathbf{x}, t) = T[\mathbf{x}(\bar{\mathbf{x}}, t), t] = T[x_1(\bar{\mathbf{x}}, t), x_2(\bar{\mathbf{x}}, t), x_3(\bar{\mathbf{x}}, t), t]$$

For a given particle, $\bar{\mathbf{x}}$ is constant so that

$$T = T[x_1(t), x_2(t), x_3(t), t]$$

and the derivative of T with respect to t is called the **material** or **substantial derivative**. Using the chain rule of differentiation, the substantial derivative is given by

$$\frac{dT}{dt} \equiv \frac{DT}{Dt} = \frac{\partial T}{\partial x_1}\frac{dx_1}{dt} + \frac{\partial T}{\partial x_2}\frac{dx_2}{dt} + \frac{\partial T}{\partial x_3}\frac{dx_3}{dt} + \frac{\partial T}{\partial t}$$

Note that dx_i/dt is the ith component of the particle velocity, i.e., v_i. Thus,

$$\frac{DT}{Dt} = \frac{\partial T}{\partial x_1}v_1 + \frac{\partial T}{\partial x_2}v_2 + \frac{\partial T}{\partial x_3}v_2 + \frac{\partial T}{\partial t} = \nabla T \cdot \mathbf{v} + \frac{\partial T}{\partial t} \qquad (8.11)$$

If an observer moves around, sampling T then

$$T = T[x_1(t), x_2(t), x_3(t), t]$$

where the x_is are the coordinates of the observer. In this case, the derivative of T with respect to t is called the **total derivative**. It is given by

$$\frac{dT}{dt} = \frac{\partial T}{\partial x_1}\underbrace{\frac{dx_1}{dt}}_{w_1} + \frac{\partial T}{\partial x_2}\underbrace{\frac{dx_2}{dt}}_{w_2} + \frac{\partial T}{\partial x_3}\underbrace{\frac{dx_3}{dt}}_{w_3} + \frac{\partial T}{\partial t}$$

where the w_is are components of the observer velocity. Thus,

$$\frac{dT}{dt} = \nabla T \cdot \mathbf{w} + \frac{\partial T}{\partial t}$$

where $\mathbf{w}$ is the observer velocity.

8.8 Taylor Series

Consider a function $f(x)$ that is differentiable at least n times with respect to x. Then $f(x)$ at a distance h from the reference point $x = x_0$ is given by the nth *order Taylor series or expansion*

$$f(x_0 + h) = f(x_0) + h\left[\frac{df}{dx}\right]_{x_0} + \frac{h^2}{2!}\left[\frac{d^2 f}{dx^2}\right]_{x_0} + \frac{h^3}{3!}\left[\frac{d^3 f}{dx^3}\right]_{x_0} + \cdots + \frac{h^n}{n!}\left[\frac{d^n f}{dx^n}\right]_{x_0}$$

$$+ \underbrace{\frac{h^{n+1}}{(n+1)!}\left[\frac{d^{n+1} f}{dx^{n+1}}\right]_{\zeta}}_{\text{remainder of } O(h^{n+1})} \qquad (8.12)$$

where ζ lies in the interval $(x_0, x_0 + h)$. The last term of Equation (8.12) on the previous page is called the remainder. It is the sum of the remaining infinite number of terms, i.e.,

$$\frac{h^{n+1}}{(n+1)!}\left[\frac{d^{n+1}f}{dx^{n+1}}\right]_{x_0} + \frac{h^{n+2}}{(n+2)!}\left[\frac{d^{n+2}f}{dx^{n+2}}\right]_{x_0} + \cdots$$

Note that when h is a fraction, the remainder decreases with n. For sufficiently large n, or small h, the remainder becomes negligible, and is therefore discarded. Equation (8.12) without the remainder is called the nth order Taylor approximation of $f(x)$ at x_0.

8.8.1 Multivariable Taylor Series

Consider a function $f(\mathbf{x})$ where $\mathbf{x}$ is a vector of m independent variables: $x_1, x_2, \ldots, x_m$. The function is differentiable at least n times with respect to $\mathbf{x}$. Let $\mathbf{x}_0$ be the vector of reference values of $\mathbf{x}$, and $\mathbf{x}_1 = (\mathbf{x}_0 + \mathbf{h})$ be a point in the neighborhood of $\mathbf{x}_0$. Then the nth-order, *multivariable Taylor series or expansion* of $f(\mathbf{x}_1)$ is given by

$$f(\mathbf{x}_0 + \mathbf{h}) = f(\mathbf{x}_0) + \mathcal{D}\big[f(\mathbf{x}_0)\big] + \frac{1}{2!}\mathcal{D}^2\big[f(\mathbf{x}_0)\big] + \frac{1}{3!}\mathcal{D}^3\big[f(\mathbf{x}_0)\big] + \cdots$$

$$+ \frac{1}{n!}\mathcal{D}^n\big[f(\mathbf{x}_0)\big] + \underbrace{\frac{1}{(n+1)!}\mathcal{D}^{n+1}\big[f(\zeta)\big]}_{\text{remainder}} \qquad (8.13)$$

where ζ lies in the interval $(\mathbf{x}_0, \mathbf{x}_0 + \mathbf{h})$, and $\mathcal{D}^i\big[f(\mathbf{x}_0)\big]$ is the ith differential of f evaluated at $\mathbf{x}_0$, i.e.,

$$\mathcal{D}^i\big[f(\mathbf{x}_0)\big] = \sum_{k_1=1}^{m}\sum_{k_2=1}^{m}\cdots\sum_{k_i=1}^{m}\left\{ h_{k_1}h_{k_2}\cdots h_{k_i}\left[\frac{\partial^i f}{\partial x_{k_1}\partial x_{k_2}\cdots\partial x_{k_i}}\right]_{\mathbf{x}_0}\right\} \qquad (8.14)$$

For sufficiently large n, or small $\|\mathbf{h}\|$, the remainder term of Equation (8.13) above becomes negligible, and is therefore discarded. Equation (8.13) without the remainder is the nth-*order Taylor approximation* of $f(\mathbf{x})$ at $\mathbf{x}_0$.

8.8.2 First Order Taylor Expansion

A function $f(x)$ that is infinitely differentiable at a reference point $x = x_0$ can be expressed at $x = (x_0 + h)$ by the Taylor series, i.e.,

$$f(x_0 + h) = f(x_0) + \sum_{i=1}^{\infty}\left[\frac{d^i f}{dx^i}\right]_{x_0}(h)^i$$

In the above series, the smaller the h the less significant the terms containing its higher powers. In fact, for fractional h, the terms with powers of h two or more vanish faster than h. When those terms disappear with h tending to zero, we can write

$$f(x_0 + h) = f(x_0) + \left[\frac{df}{dx}\right]_{x_0}h$$

In general, for any reference point x, and h tending to zero

$$f(x+h) = f(x) + \frac{\mathrm{d}f}{\mathrm{d}x}h$$

The above equation is called the **first order Taylor expansion**. Given a function whose derivative exists at a reference point, this expansion provides the function value in the vicinity of the reference point.

Expansion of a Vector Function

For a vector function $\mathbf{f}(x) \equiv \begin{bmatrix} f_1(x) & f_2(x) & \cdots & f_n(x) \end{bmatrix}^\top$, the first order Taylor expansion at a distance h away from a given x is

$$\mathbf{f}(x+h) = \mathbf{f}(x) + \begin{bmatrix} \dfrac{\mathrm{d}f_1}{\mathrm{d}x} & \dfrac{\mathrm{d}f_2}{\mathrm{d}x} & \cdots & \dfrac{\mathrm{d}f_n}{\mathrm{d}x} \end{bmatrix}^\top h$$

If the function argument is a vector, say, $\mathbf{x} \equiv \begin{bmatrix} x_1 & x_2 & \cdots & x_m \end{bmatrix}^\top$, then

$$\mathbf{f}(\mathbf{x}+\mathbf{h}) = \mathbf{f}(\mathbf{x}) + \mathbf{J}\mathbf{h}$$

where

$$\mathbf{J} \equiv \begin{bmatrix} \dfrac{\partial f_1}{\partial x_1} & \dfrac{\partial f_1}{\partial x_2} & \cdots & \dfrac{\partial f_1}{\partial x_m} \\[2ex] \dfrac{\partial f_2}{\partial x_1} & \dfrac{\partial f_2}{\partial x_2} & \cdots & \dfrac{\partial f_2}{\partial x_m} \\[2ex] \vdots & \vdots & \ddots & \vdots \\[2ex] \dfrac{\partial f_n}{\partial x_1} & \dfrac{\partial f_n}{\partial x_2} & \cdots & \dfrac{\partial f_n}{\partial x_m} \end{bmatrix} \qquad \text{and} \qquad \mathbf{h} \equiv \begin{bmatrix} h_1 \\[2ex] h_2 \\[2ex] \vdots \\[2ex] h_m \end{bmatrix}$$

Example 8.8.1

Find the second order Taylor expansion of

$$f(x_1, x_2, x_3) = \frac{x_1^3}{x_2} + \sin x_3$$

in the vicinity of $\mathbf{x} = \mathbf{x}_0 = \begin{bmatrix} 1 & 1 & \dfrac{\pi}{2} \end{bmatrix}^\top$.

Solution

From Equation (8.13) on the previous page, the second order Taylor expansion is

$$f(\mathbf{x}_0 + \mathbf{h}) = f(\mathbf{x}_0) + \mathcal{D}\big[f(\mathbf{x}_0)\big] + \frac{1}{2!}\mathcal{D}^2\big[f(\mathbf{x}_0)\big]$$

In the last equation, $f(\mathbf{x}_0) = 2$. The differentials are obtained using Equation (8.14) on p. 323. Thus, the first differential is given by

$$
\mathcal{D}[f(\mathbf{x}_0)] = \sum_{k_1=1}^{3} h_{k_1} \left[\frac{\partial f}{\partial x_{k_1}} \right]_{\mathbf{x}_0} = \left[h_1 \frac{\partial f}{\partial x_1} + h_2 \frac{\partial f}{\partial x_2} + h_2 \frac{\partial f}{\partial x_2} \right]_{\mathbf{x}_0}
$$

$$
= \left[h_1 \left(\frac{3x_1^2}{x_2} \right) + h_2 \left(\frac{-x_1^3}{x_2^2} \right) + h_3 (\cos x_3) \right]_{\mathbf{x}_0}
$$

$$
= 3h_1 - h_2
$$

Similarly, the second differential is given by

$$
\mathcal{D}^2[f(\mathbf{x}_0)] = \sum_{k_1=1}^{3} \sum_{k_2=1}^{3} \left\{ h_{k_1} h_{k_2} \left[\frac{\partial^2 f}{\partial x_{k_1} \partial x_{k_2}} \right]_{\mathbf{x}_0} \right\}
$$

$$
= \left[h_1 \left(h_1 \frac{\partial^2 f}{\partial x_1^2} + h_2 \frac{\partial^2 f}{\partial x_1 \partial x_2} + h_3 \frac{\partial^2 f}{\partial x_1 \partial x_3} \right) \right.
$$

$$
+ h_2 \left(h_1 \frac{\partial^2 f}{\partial x_2 \partial x_1} + h_2 \frac{\partial^2 f}{\partial x_2^2} + h_3 \frac{\partial^2 f}{\partial x_2 \partial x_3} \right)
$$

$$
\left. + h_3 \left(h_1 \frac{\partial^2 f}{\partial x_3 \partial x_1} + h_2 \frac{\partial^2 f}{\partial x_3 \partial x_2} + h_3 \frac{\partial^2 f}{\partial x_3^2} \right) \right]_{\mathbf{x}_0}
$$

$$
= \left[h_1^2 \frac{\partial^2 f}{\partial x_1^2} + h_2^2 \frac{\partial^2 f}{\partial x_2^2} + h_3^2 \frac{\partial^2 f}{\partial x_3^2} + 2h_1 h_2 \frac{\partial^2 f}{\partial x_1 \partial x_2} + 2h_1 h_3 \frac{\partial^2 f}{\partial x_1 \partial x_3} \right.
$$

$$
\left. + 2h_2 h_3 \frac{\partial^2 f}{\partial x_2 \partial x_3} \right]_{\mathbf{x}_0} \qquad \left(\text{since} \quad \frac{\partial^2 f}{\partial x_i \partial x_j} = \frac{\partial^2 f}{\partial x_j \partial x_i} \right)
$$

$$
= \left[h_1^2 \left(\frac{6x_1}{x_2} \right) + h_2^2 \left(\frac{2x_1^3}{x_2^3} \right) + h_3^2 (-\sin x_3) + 2h_1 h_2 \left(\frac{-3x_1^2}{x_2^2} \right) + 2h_1 h_3 (0) \right.
$$

$$
\left. + 2h_2 h_3 (0) \right]_{\mathbf{x}_0}
$$

$$
= 6h_1^2 + 2h_2^2 - h_3^2
$$

With the help of the above function and differentials at $\mathbf{x}_0$, the second order Taylor expansion is given by

$$
f(\mathbf{x}_0 + \mathbf{h}) = 2 + 3h_1 - h_2 + 6h_1^2 + 2h_2^2 - h_3^2
$$

❑

8.9 L'Hôpital's Rule

As per this rule, the limit (if it exists) of a ratio of two differentiable functions, whose limits are either both zero or infinity, is the limit of the ratio of their derivatives. Thus,

$$\lim_{x \to a} \frac{f(x)}{g(x)} = \lim_{x \to a} \frac{df/dx}{dg/dx} = L$$

This rule helps circumvent "0/0" or "∞/∞" forms. For example,

$$\lim_{x \to 1} \frac{\ln x}{x - 1} = \frac{\ln 1}{1 - 1} = \frac{0}{0}$$

However, applying L'Hôpital's rule, we get

$$\lim_{x \to 1} \frac{\ln x}{x - 1} = \lim_{x \to 1} \frac{\dfrac{d}{dx}(\ln x)}{\dfrac{d}{dx}(x - 1)} = \lim_{x \to 1} \frac{1/x}{1} = 1$$

8.10 Leibniz's Rule

This rule provides the derivative of a definite integral with respect to its limit.
For the definite integral

$$I = \int_{l}^{u} f \, dx$$

the derivative of I with respect to the upper limit u is given by

$$\frac{dI}{du} = \lim_{\Delta u \to 0} \frac{I(u + \Delta u) - I(u)}{\Delta u} = \lim_{\Delta u \to 0} \frac{\displaystyle\int_{l}^{u+\Delta u} f \, dx - \int_{l}^{u} f \, dx}{\Delta u}$$

$$= \lim_{\Delta u \to 0} \frac{\displaystyle\int_{u}^{u+\Delta u} f \, dx}{\Delta u} = \lim_{\Delta u \to 0} \frac{f(u)\Delta u}{\Delta u} = f(u)$$

In the same manner, the derivative of I with respect to the lower limit l is given by

$$\frac{dI}{dl} = -f(l)$$

8.11   Integration by Parts

This is the integration of the product of two functions, and is given by the following formula:

$$\int_l^u yz \, \mathrm{d}x \;=\; \left[y \int z \, \mathrm{d}x \right]_l^u \;-\; \int_l^u \left[\frac{\mathrm{d}y}{\mathrm{d}x} \int z \, \mathrm{d}x \right] \mathrm{d}x$$

where y and z are continuous functions of x, and y is differentiable with respect to x. Note that the right-hand side of the above equation carries the integral of z. This feature enables simplification when z is a derivative of a function with respect to x.

8.12   Euler's Formulas

For a real number x, the Euler's formula is

$$e^{ix} \;=\; \cos x + i \sin x$$

where e is the base of natural logarithm, and i is $\sqrt{-1}$. Replacing x by $-x$, we get the second Euler's formula, i.e.,

$$e^{-ix} \;=\; \cos x - i \sin x$$

Solving the last two equations for $\sin x$ and $\cos x$ yields, respectively,

$$\sin x \;=\; \frac{e^{ix} - e^{-ix}}{2i} \qquad \text{and}$$

$$\cos x \;=\; \frac{e^{ix} + e^{-ix}}{2}$$

8.13   Solution of Linear Ordinary Differential Equations

The nth-order, linear ordinary differential equation is given by

$$f\!\left(\frac{\mathrm{d}^n y}{\mathrm{d}x^n}, \frac{\mathrm{d}^{n-1} y}{\mathrm{d}x^{n-1}}, \ldots, \frac{\mathrm{d}y}{\mathrm{d}x}, y, x \right) \;=\; 0$$

where f is a linear function of y, its derivatives with respect to x, and x. The above equation can be transformed into a set of first order equations by introducing a dependent variable for each derivative of order higher than one [see Section 5.4.1, p. 178].

8.13.1   Single First Order Equation

Consider the general first order differential equation expressed as

$$\frac{\mathrm{d}y}{\mathrm{d}x} + p(x)y \;=\; q(x) \tag{8.15}$$

Let $\mu(x)$ be a function that satisfies

$$\frac{\mathrm{d}\mu}{\mathrm{d}x} = \mu p$$

Suppose that $\mu > 0$, so that the integration of the above equation yields

$$\mu = e^{\int p \, \mathrm{d}x} \tag{8.16}$$

where the arbitrary constant of integration is taken to be zero. Note that $\mu > 0$ as supposed.

Multiplying both sides of Equation (8.15) on the previous page by μ, we get

$$\underbrace{\mu \frac{\mathrm{d}y}{\mathrm{d}x} + \overbrace{\mu p}^{\mathrm{d}\mu/\mathrm{d}x} y}_{\mathrm{d}(\mu y)/\mathrm{d}x} = \mu q \quad \Rightarrow \quad \frac{\mathrm{d}(\mu y)}{\mathrm{d}x} = \mu q$$

Integration of the above equation yields the following solution of Equation (8.15) on the previous page:

$$y = \frac{1}{\mu} \left(\int \mu q \, \mathrm{d}x + c \right)$$

In the above equation, c is the constant of integration, and μ is called the *integrating factor*, which is given by Equation (8.16) above.

8.13.2   Simultaneous First Order Equations

Consider the set of simultaneous first order linear equations

$$\frac{\mathrm{d}y_1}{\mathrm{d}x} = P_{11}(x)y_1 + P_{12}(x)y_2 + \cdots + P_{1n}(x)y_n + q_1(x)$$

$$\frac{\mathrm{d}y_2}{\mathrm{d}x} = P_{21}(x)y_1 + P_{22}(x)y_2 + \cdots + P_{2n}(x)y_n + q_2(x)$$

$$\vdots \quad \vdots \qquad\qquad\qquad\qquad \vdots$$

$$\frac{\mathrm{d}y_n}{\mathrm{d}x} = P_{n1}(x)y_1 + P_{n2}(x)y_2 + \cdots + P_{nn}(x)y_n + q_n(x)$$

where P_{ij}s and q_is are continuous functions of x. The above equations can be expressed as the matrix equation,

$$\frac{\mathrm{d}\mathbf{y}}{\mathrm{d}x} = \mathbf{P}(x)\mathbf{y} + \mathbf{q}(x) \tag{8.17}$$

The solution of the above equation stems from that of the corresponding homogenous equation, i.e., Equation (8.17) with $\mathbf{q} = \mathbf{0}$.

Solution of Homogenous Equation

Consider the homogenous equation

$$\frac{\mathrm{d}\mathbf{y}}{\mathrm{d}x} = \mathbf{P}(x)\mathbf{y} \tag{8.18}$$

Given a set of n independent solutions of the last equation,

$$\mathbf{y}_1(x), \quad \mathbf{y}_2(x), \quad \ldots, \quad \mathbf{y}_n(x)$$

which is called the **fundamental set of solutions**, their linear combination

$$\mathbf{y}(x) \;=\; c_1\mathbf{y}_1(x) + c_2\mathbf{y}_2(x) + \cdots + c_n\mathbf{y}_n(x)$$

is the **general solution** of Equation (8.18) on the previous page where c_is are some constants. The matrix of the fundamental set of solutions,

$$\Psi \;=\; \begin{bmatrix} \mathbf{y}_1(x) & \mathbf{y}_2(x) & \cdots & \mathbf{y}_n(x) \end{bmatrix} = \begin{bmatrix} y_{11}(x) & y_{12}(x) & \cdots & y_{1n}(x) \\ y_{21}(x) & y_{22}(x) & \cdots & y_{2n}(x) \\ \vdots & \vdots & \ddots & \vdots \\ y_{n1}(x) & y_{n2}(x) & \cdots & y_{nn}(x) \end{bmatrix}$$

is called **fundamental matrix**. It is non-singular since its columns are linearly independent.
   Suppose that the solution of Equation (8.18) is given by

$$\mathbf{y} \;=\; \zeta e^{rx}$$

where ζ is a vector of constants, and r is some exponent. Then

$$\frac{\mathrm{d}\mathbf{y}}{\mathrm{d}x} \;=\; r\zeta e^{rx}$$

Substituting the last two equations in Equation (8.18), we get

$$r\zeta e^{rx} \;=\; \mathbf{P}\zeta e^{rx} \qquad \Rightarrow \qquad \mathbf{P}\zeta \;=\; r\zeta$$

Thus,

$$(\mathbf{P} - r\mathbf{I})\zeta \;=\; 0 \tag{8.19}$$

where r and ζ are, respectively, the eigenvalue and eigenvector of $\mathbf{P}$, and $\mathbf{I}$ is the $n \times n$ identity matrix. For a non-zero ζ,

$$\mathbf{P} - r\mathbf{I} \;=\; 0$$

The above equation yields the values of r for which the associated ζ is obtained from Equation (8.19) above. For the system of n homogenous differential equations, the number of eigenvalues is n. When all eigenvalues $(r_1, r_2, \ldots, r_n)$ are distinct, the fundamental solutions are given by

$$\mathbf{y}_i \;=\; \zeta_i e^{r_i x}; \quad i = 1, 2, \ldots, n$$

Then the general solution of Equation (8.18) is given by

$$\mathbf{y} \;=\; c_1 \underbrace{\zeta_1 e^{r_1 x}}_{\mathbf{y}_1} + c_2 \underbrace{\zeta_2 e^{r_2 x}}_{\mathbf{y}_2} + \cdots + c_n \underbrace{\zeta_n e^{r_n x}}_{\mathbf{y}_n}$$

where the ζ_i is the eigenvector corresponding to the ith eigenvector, r_i.

When some eigenvalues are complex, they appear in conjugate pairs, and the solution is oscillatory. Sometimes, however, some eigenvalues may be identical, and need special treatment to determine the fundamental set of solutions. For example, consider the case when two eigenvalues have an identical value r, and only one eigenvector ζ. Then the corresponding solution is

$$\mathbf{y}_1 = \zeta e^{rx}$$

It can be easily verified that the solution given by

$$\mathbf{y}_2 = (\zeta x + \eta)e^{rx}$$

is linearly independent of the previous solution. In the above equation, η is called a **generalized eigenvector** corresponding to eigenvalue r. This eigenvector is determined from

$$(\mathbf{P} - r\mathbf{I})\eta = \zeta$$

Once η is known, the general solution is given by

$$\mathbf{y} = c_1 \underbrace{\zeta e^{rx}}_{\mathbf{y}_1} + c_2 \underbrace{(\zeta x + \eta)e^{rx}}_{\mathbf{y}_2}$$

Generalization

If m out of n eigenvalues have an identical value r with one eigenvector ζ, and the remaining eigenvalues are distinct then the general solution is given by

$$\mathbf{y} = \sum_{i=1}^{m} c_i \mathbf{y}_i + \underbrace{\sum_{i=m+1}^{n} c_i \mathbf{y}_i}_{\substack{\text{terms corresponding to} \\ \text{distinct eigenvalues}}}$$

where the m independent solutions corresponding to r are given by

$$\mathbf{y}_1 = \zeta e^{rx}$$

$$\mathbf{y}_2 = (\eta_1 + \zeta x)e^{rx}$$

$$\mathbf{y}_3 = \left[\eta_2 + \eta_1 x + \zeta \frac{x^2}{2!}\right]e^{rx}$$

$$\mathbf{y}_4 = \left[\eta_3 + \eta_2 x + \eta_1 \frac{x^2}{2!} + \zeta \frac{x^3}{3!}\right]e^{rx}$$

$$\vdots \quad \vdots \qquad\qquad \vdots$$

$$\mathbf{y}_m = \left[\eta_{m-1} + \eta_{m-2} x + \eta_{m-3}\frac{x^2}{2!} + \cdots + \eta_1 \frac{x^{m-2}}{(m-2)!} + \zeta \frac{x^{m-1}}{(m-1)!}\right]e^{rx}$$

For the last set of equations, ζ and the generalized eigenvectors η_is are obtained from

$$(\mathbf{P} - r\mathbf{I})\zeta = \mathbf{0}$$

$$(\mathbf{P} - r\mathbf{I})\eta_1 = \zeta$$

$$(\mathbf{P} - r\mathbf{I})\eta_2 = \eta_1$$

$$\vdots \qquad \vdots \quad \vdots$$

$$(\mathbf{P} - r\mathbf{I})\eta_{m-1} = \eta_{m-2}$$

For m eigenvalues that have the same value, several eigenvectors (say, $\zeta_1, \zeta_2, \ldots, \zeta_k$) up to m may be present. In that case, the above procedure is used with each ζ_i to find several η_is until m independent solutions are obtained.

Solution of Non-Homogenous Equation

We assume that the solution of the non-homogenous equation [Equation (8.17) on p. 328] is given by

$$\mathbf{y} = \boldsymbol{\Psi}(x)\mathbf{z}(x) \tag{8.20}$$

where $\boldsymbol{\Psi}$ is the fundamental matrix for the corresponding homogenous equation [Equation (8.18) on p. 328], and $\mathbf{z}(x)$ is the function that needs to be determined. Differentiating Equation (8.20) above with respect to x, we get

$$\frac{d\mathbf{y}}{dx} = \boldsymbol{\Psi}\frac{d\mathbf{z}}{dx} + \frac{d\boldsymbol{\Psi}}{dx}\mathbf{z}$$

Also, the substitution of Equation (8.20) into Equation (8.17) yields

$$\frac{d\mathbf{y}}{dx} = \mathbf{P}\boldsymbol{\Psi}\mathbf{z} + \mathbf{q}$$

Comparing the last two equations, and noting that $\boldsymbol{\Psi}$ is a fundamental matrix,

$$\frac{d\boldsymbol{\Psi}}{dx} = \mathbf{P}\boldsymbol{\Psi}$$

we obtain

$$\boldsymbol{\Psi}\frac{d\mathbf{z}}{dx} = \mathbf{q}$$

Since $\mathbf{P}(x)$ is continuous, $\boldsymbol{\Psi}$ is non-singular so that its inverse $\boldsymbol{\Psi}^{-1}$ exists. Therefore,

$$\frac{d\mathbf{z}}{dx} = \boldsymbol{\Psi}^{-1}\mathbf{q}$$

Integrating the above equation, we get

$$\mathbf{z} = \int \boldsymbol{\Psi}^{-1}\mathbf{q}\,dx + \mathbf{c}$$

where $\mathbf{c}$ is the vector of integration constants. The last equation upon substitution in Equation (8.20) on the previous page yields the solution of the non-homogenous equation [Equation (8.17), p. 328], i.e.,

$$y = \Psi \int \Psi^{-1} \mathbf{q}\, dx + \Psi \mathbf{c} \tag{8.21}$$

Initial Value Problem

In this problem,

$$\frac{d\mathbf{y}}{dx} = \mathbf{P}(x)\mathbf{y} + \mathbf{q}(x), \qquad \mathbf{y}(x = x_0) = \bar{\mathbf{y}}$$

where $\bar{\mathbf{y}}$ is $\mathbf{y}$ at the initial value of x, i.e., x_0. To obtain the solution of this problem, we select $\mathbf{c}$ in Equation (8.21) above such that the integral term becomes zero at x_0. This is done by expressing that equation as

$$y = \Psi \int_{x_0}^{x} \Psi^{-1} \mathbf{q}\, dx + \Psi \mathbf{c} \tag{8.22}$$

Thus, from the above equation at the initial x, when $x = x_0$,

$$\bar{\mathbf{y}} = \Psi(x_0)\mathbf{c} \quad \Rightarrow \quad \mathbf{c} = \Psi^{-1}(x_0)\bar{\mathbf{y}}$$

Substituting the above expression of $\mathbf{c}$ in Equation (8.22) above, we obtain

$$y = \Psi \int_{x_0}^{x} \Psi^{-1} \mathbf{q}\, dx + \Psi \Psi^{-1}(x_0)\bar{\mathbf{y}}$$

which is the desired solution.

Bibliography

[1]   R.C. Wrede. *Introduction to Vector and Tensor Analysis*. New York: John Wiley & Sons, Inc., 1963.

[2]   D. Lovelock and H. Rund. *Tensors, Differential Forms, and Variational Principles*. New York: Dover Publications Inc., 1989.

[3]   W.E. Boyce and R.C. DiPrima. *Elementary Differential Equations and Boundary Value Problems*. 10th edition. John Wiley & Sons, Inc., 2012.

[4]   A.E. Taylor and W.R. Mann. *Advanced Calculus*. 3rd edition. U.S.A.: John Wiley & Sons, Inc., 1983.

INDEX

In this index, a page number in bold face corresponds to the main information of the index entry. An underlined page number refers to an example.

Process Modeling and Simulation for Chemical Engineers: Theory and Practice, First Edition. Simant Ranjan Upreti.
© 2017 John Wiley & Sons Ltd. Published 2017 by John Wiley & Sons Ltd.
Companion website: www.wiley.com/go/upreti/pms_for_chemical_engineers